ENVIRONMENTAL SCIENCES AND APPLICATIONS

Series Editors: MARGARET R. BISWAS
ASIT K. BISWAS

Volume 11

NUCLEAR ENERGY AND THE ENVIRONMENT

ENVIRONMENTAL SCIENCES AND APPLICATIONS

Other titles in the series

Volume 1	UNITED NATIONS	Desertification: Its Causes and Consequences
Volume 2	FUKUSHIMA, Y.	Science for Better Environment
Volume 3	UNITED NATIONS INSTITUTE FOR TRAINING AND RESEARCH	Alternative Strategies for Desert Development and Management
Volume 4	BISWAS, A. K.	The Ozone Layer
Volume 5	HUSAR, R. B., et al.	Sulfur in the Atmosphere
Volume 6	JØRGENSEN, S. E.	Handbook of Environmental Data and Ecological Parameters
Volume 7	JØRGENSEN, S. E.	State-of-the-Art of Ecological Modelling
Volume 8	HINCKLEY, A. D.	Renewable Resources in Our Future
Volume 9	HOLÝ, M.	Erosion and Environment
Volume 10	BEALE, J. G.	The Manager and the Environment
Volume 12	BISWAS, M. R. and BISWAS, A. K.	Desertification

NUCLEAR ENERGY AND THE ENVIRONMENT

Edited by

ESSAM E. EL-HINNAWI

United Nations Environment Programme
Nairobi, Kenya

PERGAMON PRESS

OXFORD · NEW YORK · TORONTO · SYDNEY · PARIS · FRANKFURT

U.K.	Pergamon Press Ltd., Headington Hill Hall, Oxford OX3 0BW, England
U.S.A.	Pergamon Press Inc., Maxwell House, Fairview Park, Elmsford, New York 10523, U.S.A.
CANADA	Pergamon of Canada, Suite 104, 150 Consumers Road, Willowdale, Ontario M2J 1P9, Canada
AUSTRALIA	Pergamon Press (Aust.) Pty. Ltd., P.O. Box 544, Potts Point, N.S.W. 2011, Australia
FRANCE	Pergamon Press SARL, 24 rue des Ecoles, 75240 Paris, Cedex 05, France
FEDERAL REPUBLIC OF GERMANY	Pergamon Press GmbH, 6242 Kronberg-Taunus, Hammerweg 6, Federal Republic of Germany

First edition 1980

British Library Cataloguing in Publication Data

Nuclear energy and the environment.
- (Environment sciences and applications; 11).
1. Atomic power-plants - Environmental aspects
2. Atomic power industry - Environmental aspects
I. El-Hinnawi, Essam E II. Series
301.31 TD195.E4 80-40365

ISBN 0-08-024472-6

In order to make this volume available as economically and as rapidly as possible the authors' typescripts have been reproduced in their original forms. This method has its typographical limitations but it is hoped that they in no way distract the reader.

Printed in Great Britain by A. Wheaton & Co., Ltd., Exeter

To Susan and the Future Generations

CONTENTS

CONTRIBUTORS

A. M. Aikin
Institute for Environmental Studies
University of Toronto
Toronto, Canada

J. R. Beattie
Safety and Reliability Directorate
United Kingdom Atomic Energy
Authority
Culcheth, Warrington, WA3 4NE
Lancashire, United Kingdom

Asit K. Biswas
Director
Biswas and Associates
3 Valley View Road
Ottawa, Canada

F. C. R. Cattell
Australian Atomic Energy
Commission
P.O. Box 41, Coogee, NSW 2034
Australia

J. E. Cook
Australian Atomic Energy
Commission
P.O. Box 41, Coogee, NSW 2034
Australia

J. M. Costello
Australian Atomic Energy
Commission
Research Establishment
Private Mail Bag
Sutherland, NSW 2232
Australia

D. R. Davy
Australian Atomic Energy
Commission
Research Establishment
Private Mail Bag,
Sutherland, NSW 2232
Australia

Essam E. El-Hinnawi
United Nations Environment
Programme
P.O. Box 47074
Nairobi, Kenya

R. F. Griffiths
Safety and Reliability Directorate
United Kingdom Atomic Energy
Authority
Culcheth, Warrington WA3 4NE
Lancashire, United Kingdom

H. Howells
British Nuclear Fuels Ltd.
Risley, Warrington
Cheshire WA3 6AS
United Kingdom

G. D. Kaiser
Safety and Reliability Directorate
United Kingdom Atomic Energy
Authority
Culcheth, Warrington WA3 4NE
Lancashire, United Kingdom

F. Kenneth Hare
Institute for Environmental Studies
University of Toronto
Toronto, Canada

G. H. Kinchin
Safety and Reliability Directorate
United Kingdom Atomic Energy
Authority
Culcheth, Warrington WA3 4NE
Lancashire, United Kingdom

P. W. Mummery
British Nuclear Fuels Ltd.
Risley, Warrington
Cheshire WA3 6AS
United Kingdom

A. Scriven
British Nuclear Fuels Ltd.
Risley, Warrington
Cheshire WA3 6AS
United Kingdom

Y. Sousselier
Institut de Protection et Sûreté
Nucléaire,
Commissariat à l'Energie Atomique
92260 Fontenay-Aux-Roses
France

PREFACE

Nuclear energy is an important source of power. However, nuclear power development raises a number of complex issues; political, socio-economic, environmental and ethical, as well as technical. The proponents of nuclear power advocate it as a safe, clean source of energy that is indispensable to the future world. They assert that it can generate base-load electricity at lower costs than fossil-fuel alternatives, and that without it the rising demand for electricity cannot be met. They consider that it is less dangerous to the environment and to human health than fossil fuels. They also emphasize that the probability of a serious reactor accident occurring is very small, and that radioactive wastes are hazardous for only a few hundred years and can be handled during this period without causing any harm to man and his environment. On the other hand, nuclear power critics consider it to be an unacceptably dangerous source of energy. They question its economic competitiveness, its safety and its health impacts on present and future generations. In particular, they point to the possibility of catastrophic accidents and to a persistent threat from radioactive wastes, which they maintain could endanger civilizations for thousands of years ahead. They assert that nuclear power will lead to proliferation of nuclear weapons throughout the world, and that nuclear fuel-cycle facilities could become targets for terrorists. The critics cite the failure of many States to become parties to the Non-Proliferation Treaty, and point to the ineffectiveness of safeguards to prevent the transfer of nuclear technology from power production to the acquisition of nuclear weapons. They express concern about the absence of safeguards on source materials and about the practical problem of effectively checking nuclear material inventories. There is also criticism of the ease with which states can withdraw from safeguards agreements and of the absence of effective sanctions to deter diversion of safeguarded material.

The idea of this book originated in the course of the preparation of a comprehensive review of the environmental impacts of nuclear energy, as a part of a series of studies undertaken by the United Nations Environment Programme on the environmental aspects of production, transport and use of different sources of energy. The framework of these studies was developed by the editor of this book. For the study on nuclear energy, a number of leading international authorities were invited to write specific papers on the environmental aspects of the different stages of the nuclear fuel cycle; these papers were used primarily in the preparation of a consolidated UNEP report. It was, however, realized that due to the wide interest in the subject, the publication of some of the original papers is more than warranted.

It is hoped that this book will be found to give a balanced assessment of the environmental impacts of the different stages of the nuclear fuel cycle as seen by the international experts in the field; the views expressed in the different chapters of this book are their own and do not necessarily represent those of the organizations/institutions with which they are affiliated or of the United Nations Environment Programme.

I would like to express my gratitude to the authors who prepared the different chapters for their whole-hearted co-operation, to Dr. Mostafa Kamal Tolba, the Executive Director of UNEP, without whose support and encouragement this book could never have been finished, to Mr. P.S. Thacher, Deputy Executive Director of UNEP and Dr. S. Evteev, Assistant Executive Director of UNEP for their assistance and support and to Dr. Asit K. Biswas, Director of Biswas and Associates, Ottawa for his interest, comments and help throughout the preparation of the book.

Nairobi, Kenya
February, 1980

Essam E. El-Hinnawi

1

INTRODUCTION

Essam E. El-Hinnawi
Chairman of Energy Task Force
United Nations Environment Programme
P.O. Box 47074, Nairobi, Kenya

Energy has long been viewed as an essential ingredient in meeting basic human requirements and in stimulating and supporting economic growth and a rising standard of living. Historically, energy has been thought to be cheap and plentiful. Recently, however, it has been realized that fossil fuels, especially oil, are finite in extent and should be regarded as depleting assets, and the term "energy crisis" has become a common catchword all over the world. However, it is difficult to define this "crisis". Does the problem lie in the scarcity, or the vulnerability, of supply? Or is it the rising cost of energy resources as an important cause of the world-wide inflation and recession? Or is the crisis defined by an "excessive" demand for energy, or by widespread habits of energy waste? Of course, it may be all of these things.

At local, national and in some cases regional levels, the environmental aspects of energy production and use have become the subject of wide-ranging debate. Environmental awareness and anti-pollution campaigns have affected the formation of energy policies in many countries, and it has recently been realized that nations are not isolated in this respect; the actions of one country may affect the environment in a neighbouring one. Environmental objectives should not, however, be seen as being inconsistent with, or as imposing constraints upon, energy policy. A balance should be maintained between the need to preserve and improve the quality of the environment and the socio-economic goals and needs which depend on the availability of energy.

---

* Permanent Address:
Research Professor of Energy and Natural Resources
National Research Centre
Dokki, Cairo
Egypt

The environmental aspects of production and use of energy have become an increasingly important factor in the development of national energy policies and strategies and this has been conspicuously so in the case of nuclear energy. The development and operation of nuclear power facilities have taken place, with the awareness that the technology is associated with a number of safety and environmental hazards. However, nuclear activities at both national and international levels have been controlled by a regulatory framework which has ensured high levels of safety and environmental protection. The design, construction and operation of nuclear facilities have been carried out in accordance with stringent standards and there has been extensive research on the effects of radiation on man. Although research has been directed, more generally, towards identifying the effects of radiation on the earth's biosphere, this is clearly a complex area in which further study is required to enable a more detailed assessment of the impacts of low levels of radiation on the biosphere. Furthermore, it is generally recognized that the environmental issues related to nuclear power and other conventional sources of energy have not yet been fully resolved. Assessment of alternative energy strategies, at either the local, regional or global level, requires a thorough and comprehensive evaluation.

ENVIRONMENTAL IMPACTS

The environment - defined as that whole outer physical and biological system in which man and other organisms live - is a whole, albeit a complicated one with many interacting components. The wise management of that environment depends upon an understanding of those components: of its rocks, minerals and waters, of its soils and their present potential vegetation, of its animal life and of its climate. Good management avoids pollution, erosion, and the wastage of resources by irreversible damage. To prevent such types of environmental degradation is even more challenging and certainly more efficient than to redress them after they had occurred.

Of the many potential environmental impacts associated with any particular energy technology, some would be substantial and others small, some important and others of little consequence, some of short duration and others with long term effects, some might be adverse and others beneficial and they might occur in different geographic areas and might affect different communities in different ways. A distinction should be made between the assessment of the nature, scale and geographic distribution of the impact, and the evaluation which is concerned with its value or importance. For many environmental changes which are identified as impacts, the state of knowledge and technology will often only permit a qualitative assessment. Only in a few cases is it possible to evaluate an impact quantitatively. Decisions must ultimately be made on the basis of combination of cost/benefit analysis, other quantified inputs and qualitative information.

Discussion of the environmental impacts of various energy strategies has, in the past, tended to focus more attention on short-term aspects, such as occupational and public health and direct impacts on the physical environment, than on the long-term socio-economic and environmental consequences. However, there is now a growing disposition to analyse these long-term impacts which may range from those for which substantial data exist and around which there is a fair degree of certainty as to the risks involved, to those which are rather speculative in nature and for which very little data are available.

The biosphere consists of different organisms, plants and animal life supported by a number of physical characteristics such as topography, soils, climate, water supply and drainage. For a given development, these physical characteristics and hence the biosphere may be affected. Whenever pollutants are released, the analysis of the environmental impact of these pollutants requires the knowledge of the:

(a) quantity and types of pollutants released;
(b) dispersion of these pollutants in the environment;
(c) Ecological pathways followed by the pollutants;
(d) relationships between the pollutants and the damage to man and his environment;
(e) the extent of the damage including its cost (where it is possible to make this assessment).

The total impact of some pollutants may depend on positive or negative synergistic effects. Although standards have been formulated for "acceptable" levels of several pollutants, it is prudent to assume that for exposure to radiation, and also to carcinogenic chemicals and those which might have mutagenic or teratogenic effects, there is no threshold and that effects can occur at very low exposures. A further feature is that radioactive and many other carcinogenic materials may remain in the environment (and accessible to the food chain) long after the action releasing them has been discontinued. Attempts to assess the long-term impacts of these pollutants, although difficult, should be made taking into consideration the different pathways, biogeochemical cycles and fate of these substances in the environment.

The assessment of the environmental impacts resulting from the different stages of any fuel cycle is important in relation to policy-making and decisions about energy options or "mixes" to be developed. The conservation of the natural environment is essential to the maintenance and regulation of the food, air and water cycles on which human life depends and to socio-economic development. The most important impacts of any energy technology on the human environment are probably the impacts on health and safety and on social well-being. In the case of nuclear power, public concern focuses on the following:

(a) the effects of radiation on man, both somatic and genetic, which may be associated with some parts of the fuel cycle;
(b) the different nature, and distribution in time and between sectors of national economies of costs and benefits of the various stages of the fuel cycle;
(c) the environmental impacts associated with the disposal of radioactive wastes and/or spent fuel;
(d) the availability of plutonium separated in reprocessing of spent fuel and fear or its theft and misuse by terrorists, or of its diversion to produce nuclear weapons;
(e) the need for continual monitoring and surveillance in certain parts of the nuclear fuel cycle. In the case of waste disposal this could extend until the present concepts of safe disposal are demonstrated and accepted.

The nuclear power industry produces a number of environmental impacts similar to those which are caused by fossil fuels (especially by the coal fuel cycle), e.g. thermal pollution, land use, non-radioactive and radioactive emissions, occupational risks ... etc. However, the principal concern with releases from the nuclear fuel cycle operations has been with radioactivity and its effects on the biosphere, and specially on human health. Although the principal objective of radiation protection is the achievement and maintenance of appropriately safe conditions for activities involving human exposure, the level of safety required for the protection of all human individuals is thought likely to be adequate to protect other species, although not necessarily individual members of those species (ICRP-26, 1977). The ICRP therefore believes that if man is adequately protected then other living things are also likely to be sufficiently protected. However, ecological changes are sometimes unforeseen and there is a case for maintaining a watch on the ecology of areas which receive higher radiation doses (Royal Commission Environmental Protection, 1976).

The release of radioactivity into the environment can cause radiation exposure to the biosphere. This can be affected by a number of possible routes or pathways. These "environmental pathways" are illustrated in Fig. 1. Radiation exposure may be either external, internal or both. External radiation exposure occurs when the source of radiation is external to the body for instance if the radioactivity has been deposited on the ground or passes overhead as a cloud. Internal exposure will occur if some of the radioactivity released gets inside the body. This may occur directly by inhaling some of the released radioactivity or by ingestion of activity through consumption of contaminated food. The latter may occur through a "food chain" which may be a single stage, such as direct contamination of leafy vegetables, or involve a series of stages leading to human food material. The nature of these food chains depends on ecological factors; radioactive materials, because of their chemical nature, may be concentrated by the flora or fauna exposed to the contaminant.

The process is highly site specific and the consequential exposure of man will also depend on food consumption rates and habits. It is for the latter reasons that ICRP introduced the concept of "critical group". For any proposed release of radioactivity to the environment it is usually found that there is one pathway whereby a group of people (the "critical group") may receive a radiation exposure higher than that received by the rest of the population. Adequate control of the exposure of members of this group will then ensure that individual members of the public are safeguarded. The size of the critical group is variable dependent on the particular circumstances and is related to the special dietary and working habits of the members of the group. These food chain pathways are often the most important environmental pathways leading to the intake of radioactivity by man and his subsequent radiation exposure.

The hazardous nature of ionizing radiation first became apparent in connection with exposures from man-made sources. The discovery of X-rays and the identification and separation of radioactive substances at the end of the last century brought, in addition to great benefits, unforseen hazards. The need for protection from radiation became obvious as observations of radiation injury in man and radio-biological experiments in animals increased the knowledge of the gross effects of radiation. The first international action to be taken in this respect was by the second International Congress of Radiology in 1928 through the establishment of an International Commission on Radiological Protection (ICRP) to provide guidance and formulate recommendations within the field of radiation protection. Because of the increase in radioactivity in our environment from man-made sources such as accelerators, nuclear reactors and artifically-produced radio-isotopes, the potential for radiation hazards and the problems of radiation protection have also increased. This has made it necessary for UNSCEAR, ICRP and similar bodies to keep under review the evaluation of radiation levels and effects and formulate recommendations for radiation protection (for details, see UNSCEAR, 1977; ICRP-26, 1977).

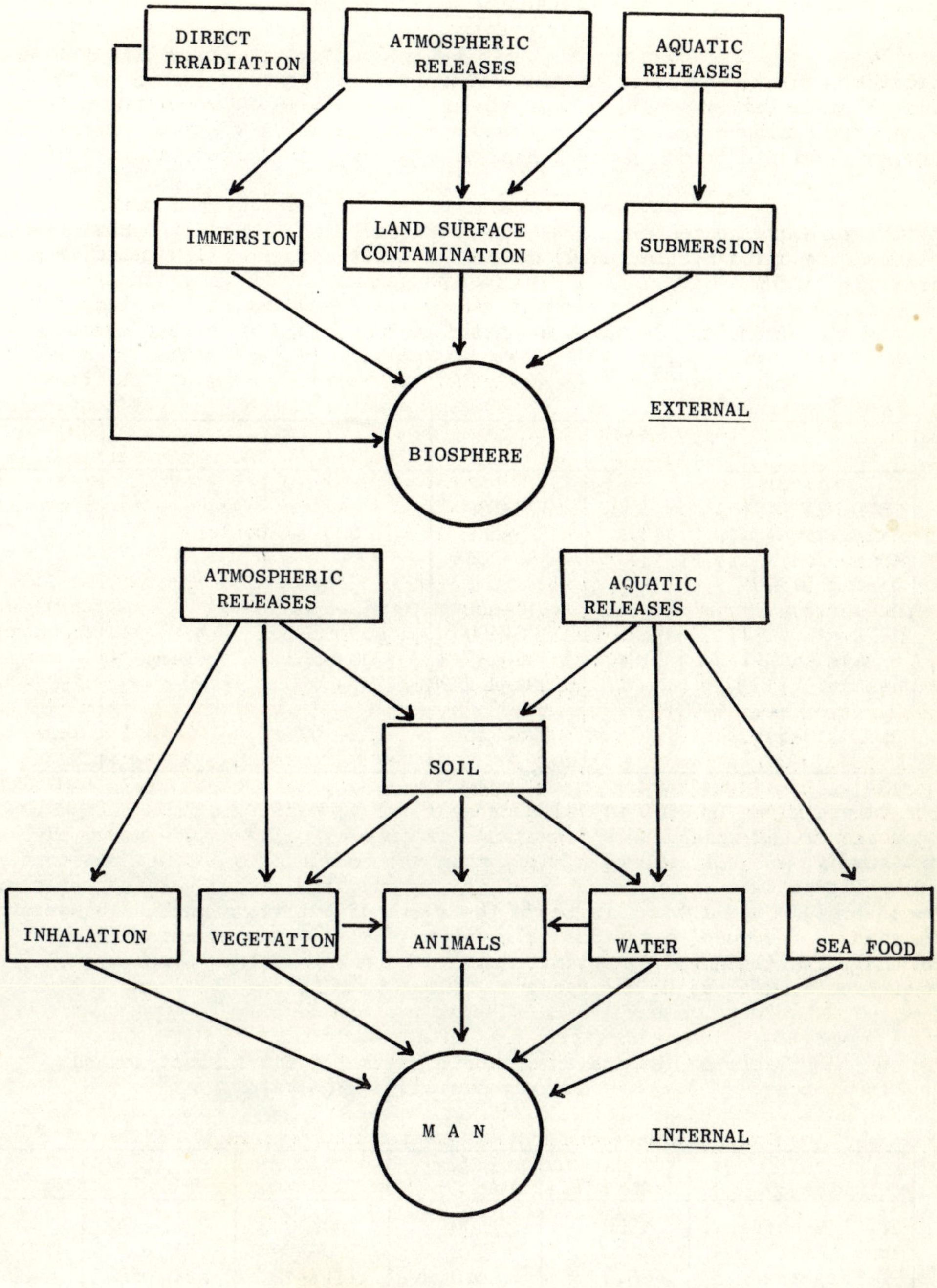

Fig. 1 Radiological Exposure Pathways.

NUCLEAR POWER GROWTH

Nuclear energy has been developed and used commercially for about two decades to meet a fraction of the world electrical energy needs. The total installed nuclear generating capacity in the world in 1978 was 110.8 GW(e) from 227 power reactors operating in 20 countries (IAEA, 1979), constituting about 6% of the world electrical power capacity.

In the past three years, several estimates of the nuclear power growth by the turn of the century have been made (Table 1). The estimates (centrally planned economies not included) demonstrate the wide range of possibilities which are wider after 1985.

Table 1. Comparison of Nuclear Power Growth Estimates (GW(e))

	1985	1990	1995	2000
OECD-NEA (1976)	479-530	-	-	2005-2480
Hanrahan et al. (1976)	358	620	-	1410
Krymm and Woite (1976)	350-400	-	-	1500-1800
Giraud (1976)	420	940	1500	2100
Messer (1977)	310-360	550-650	-	-
Williams (1977)	300-350	500-600	730-870	900-1100
Braatz and Dibbert (1977)	372-456	700-865	1110-1360	-
WAES (1977)	291-412	-	-	913-1772
Duret et al. (1978)	303	-	-	1543
OECD (1978)	278-368	504-700	750-1220	1000-1890

The values given by OECD (1978) represent two trends: the lowest figures are for "present trend" which takes cognizance of current patterns of energy utilization and supply, as well as present delays in the construction of new reactors, and generally assumes a continuation of this trend; the higher values represent an "accelerated" trend which reflects the goals of ambitious nuclear power programmes, planned in response to the possible unavailability of conventional fuels. Recently, the IAEA (1979) estimated that the total nuclear power growth (for the whole world) will be in the range of 1100-1700 GW(e) by the year 2000 (see Table 2).

Table 2. Estimates of World Installed Total Electric and Nuclear Capacity (GW(e))

	1978	1980	1985	1990	2000
Total Electric	1900	2100	2700-3000	3300-3700	5500-6600
Total Nuclear	110	170	300-350	475-600	1100-1700
% Nuclear	5.8	8.0	11.7-12	14-16.2	20-26

After (IAEA, 1979).

It should be noted that estimates of future nuclear energy production, extending over long periods of time, are unavoidably subject to large margins of uncertainty and can provide no more than very general indications, whose validity must be constantly subjected to critical revisions. The numerous energy forecasts (including those given above) rest on multiplicity of different assumptions and different aggregating procedures which make them roughly indicative and subject to changes. The main elements affecting these forecasts are (a) world and regional scenarios of economic development; (b) correlation of economic growth and energy consumption; (c) physical, economical, environmental and political constraints applying to energy production and consumption; (d) future prices of different energy sources; (e) future availability of non-renewable resources and priorities for their use; (f) public reaction and concern about nuclear power development ... etc.

THE NUCLEAR FUEL CYCLE

The "nuclear fuel cycle" refers to the entire programme from the mining and milling of uranium ores, through the manufacture of fuel elements, reactor operation, reprocessing of irradiated fuel (in case of re-cycle option), to the management of wastes produced in all steps of the cycle.

Types of Nuclear Reactors:

Nuclear reactors fall into two broad categories in terms of their basic physical principles - "thermal" for those maintaining the chain reaction with slow neutrons, and "fast" for those relying on fast neutrons. In the first category, fission is induced by slow neutrons that have lost most of their original energy by collision with nuclei of a "moderator". Such slow neutrons readily fission U-235, whilst avoiding capture by the large excess of U-238 in the fuel. Under these conditions a chain reaction can be induced in natural uranium with a heavy water or graphite moderator. If ordinary water (light water) is used as moderator, as in general practice, the fuel must be enriched in U-235 content from the 0.7% value found in nature to about 2-3%. In "fast" reactors, the concentration of fissile nuclei is high enough that the chain reaction can be sustained by the fast neutrons emitted in fission and no moderator is needed. "Fast" reactors can operate on about 20% enriched uranium but normally will use an equivalent amount of Pu-239 produced initially in thermal neutron reactors, but ultimately bred within the fast reactors themselves.

Gas-cooled Thermal Reactors include the Magnox Reactor, the Advanced Gas-cooled Reactor (AGR) and the High Temperature Gas-cooled Reactor (HTGR). In the first the fuel rods of natural uranium metal encased in "magnox", a magnesium alloy, are inserted in a structure of graphite blocks which constitutes the moderator. Pressurized carbon dioxide is the coolant The AGR is a development of the Magnox reactor in which the fuel is slightly enriched (to about 2 per cent U-235) and is in the form of uranium dioxide (UO_2). The HTGR is a different type of reactor. The fuel is U-233 (dioxide or carbide) mixed with a thorium oxide or carbide fertile component, with graphite as moderator. The thorium captures neutrons to become U-233, which is fissile and makes an increasing contribution to the reactor's fuel supply. Helium is used as the coolant gas.

In Water-cooled Reactors, either light or heavy water is used as coolant. Most of the world's commercial power reactors use light water and are referred to as LWRs. The simplest of all types of reactors is the Boiling Water Reactor (BWR) in which water serves as coolant and moderator and directly provides the steam that drives the turbine. The fuel is slightly enriched (typically 2.4% U-235) uranium dioxide. In the Pressurized Water reactor (PWR) pressurized water serves as coolant and moderator. The hot water coolant is used to produce steam in a second circuit at lower pressure. The fuel is also slightly enriched uranium dioxide (about 3% U-235). The Pressurized Heavy Water Reactor (PHWR) uses pressurized heavy water as both moderator and coolant and produces steam in a a second circuit in the same way as the PWR. The uranium dioxide fuel in the PHWR is, however, not enriched. The most common type of these reactors is the Canadian Deuterium Uranium Reactor (CANDU). The Steam-Generating Heavy Water Reactor (SGHWR) would use heavy water as moderator and light water as coolant. The latter boils, as in the BWR, to provide steam directly for the turbine. In this type of reactor, the fuel is uranium dioxide enriched to about 2% U-235.

Fast Reactors as built or proposed to date are also breeders. The Fast Breeder Reactor (FBR) uses a fuel consisting, typically of a mixture of about 20 per cent plutonium oxide with natural or depleted uranium oxide. In the leading version of the FBR, as proposed to date, neutrons released by the reactor fuel convert fertile material (for example, U-238, located mainly in a blanket around the reactor core) to fissile material (Pu-239), hence the name "breeder". The common type of the FBR operating as prototype is the Liquid Metal Fast Breeder Reactor (LMFBR) in which liquid sodium is used as the coolant.

The LWR Nuclear Fuel Cycle:

The most common type of reactor, at least up to the year 2000, is the LWR (see Table 3).

Table 3. Distribution of reactor types up to the year 2000

Year	Total Nuclear Power GW(e)	LWR %	AGR %	HWR %	GCR %	FBR %	HTGR %
1977	87	85.1	3.5	4.6	6.9	-	-
1980	146	86.3	4.1	4.8	4.1	0.7	-
1985	278	90.2	2.2	4.7	2.2	0.4	0.4
1990	504	91.2	1.2	5.6	1.2	0.4	0.4
1995	750	91.1	0.8	6.7	0.5	0.5	0.4
2000	1000	90.1	0.6	7.5	0.2	1.0	0.6

- After OECD/NEA (1977); data do not include USSR, Eastern Europe and China.
- LWR - Light Water Reactor (the ratio of PWR's to BWR's is about 60:40).
  AGR - Advanced Gas-Cooled Reactor.
  HWR - Heavy Water Reactor.
  GCR - Gas-Graphite Reactor.
  FBR - Fast Breeder Reactor.
  HTGR - High Temperature Gas-Cooled Reactor.

Present day fuel for LWR's consist of uranium that has had its content of fissile U-235 enriched to 2.5-3%. Other fissile materials such as Pu-239 or U-233 may also be used. The energy produced in a LWR comes from the fission of U-235 by thermal neutrons. However, thermal neutrons also interact with U-238 to produce fissile plutonium which in turn contributes to the production of energy by the fission reaction. The average conversion ratio for a LWR is about 0.6, which means that 6 atoms of plutonium are produced in the fuel for every 10 atoms of U-235 that are fissioned. At the time the spent fuel is discharged, more than half the fissions occurring in the fuel are from self-produced plutonium rather than U-235.

Fuel is considered spent when it can no longer sustain the neutron chain reaction at economic power levels because of the depletion of fissile materials and the accumulation of neutron-absorbing by-products in the fuel and fuel hardware. At that point, however, the fuel still contains appreciable quantities of fissile isotopes, typically about 260 kg of fissile plutonium and 350 kg of unused U-235 in 44 tonnes of spent fuel, the approximate amount per GW(e)y. After the spent LWR fuel is removed from the core, several options are possible: (a) The spent fuel rods can be disposed of without reprocessing. This is the non-recycle option, with two variations depending on whether the disposition is regarded as permanent (throwaway) or recoverable (stowaway).
(b) Uranium can be recovered by reprocessing the spent fuel. Such uranium has a slightly higher U-235 content that natural uranium and can be further enriched for making new fuel. This option is called Uranium-only recycle. At this stage, there are two variations with respect to the plutonium contained in the spent fuel: permanent disposal (throwaway) or recoverable storage (stowaway).
(c) Plutonium can be recovered from spent fuel and combined with uranium having a low U-235 content or with U-238 to form plutonium-uranium mixed-oxide or "mixed-oxide" fuel. In this case the plutonium had been "recycled".

These options of the nuclear fuel cycle of LWR are illustrated in Figs. 2, 3 and 4.

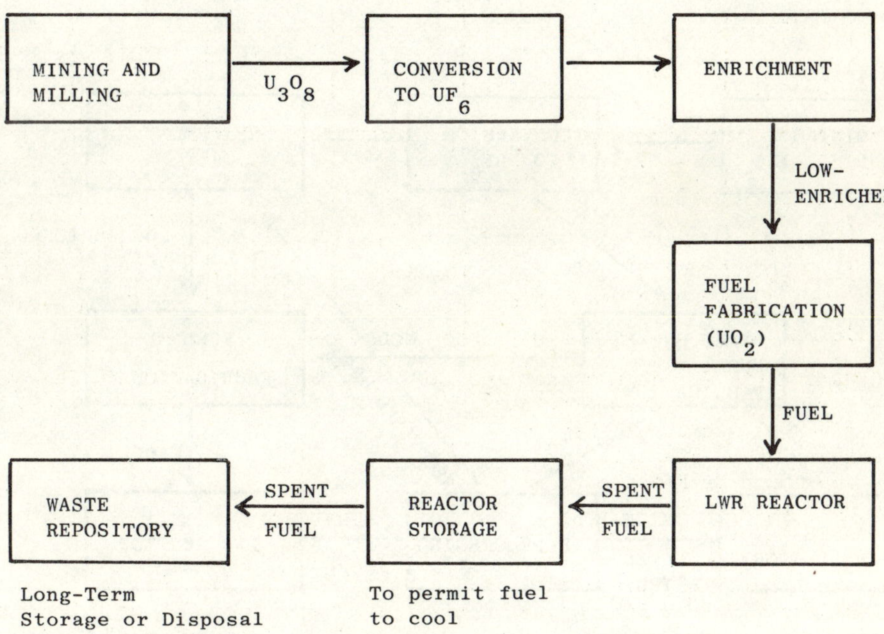

Fig. 2 LWR Fuel Cycle: No re-cycle (APS, 1978).

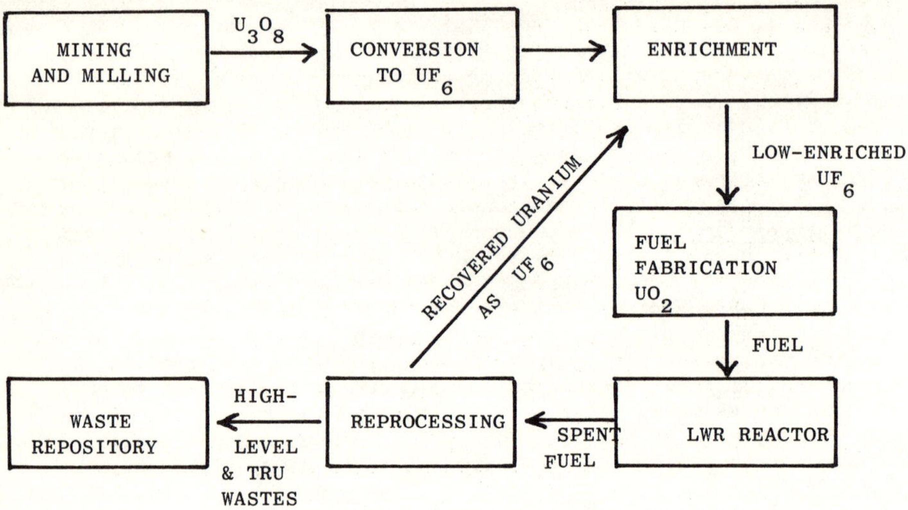

Fig. 3 LWR Fuel Cycle : Uranium Recycle Only (APS, 1978).

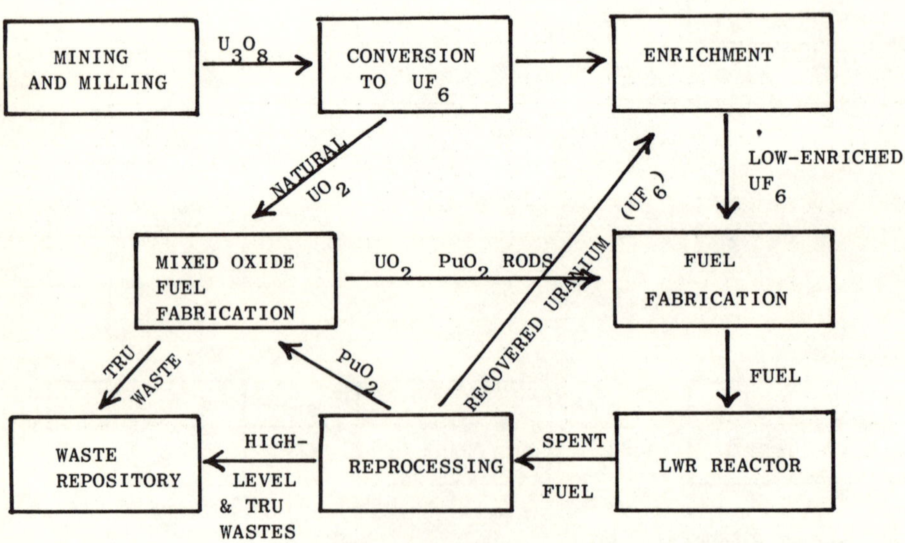

Fig. 4 LWR Fuel Cycle : Uranium and Plutonium Recycle (APS, 1978).

At present, the recycling of LWR fuel cycle has been indefinitely suspended in the U.S.A., and only a limited amount of recycling is underway elsewhere. Several studies have compared the different options (e.g. GESMO, 1976; APS, 1978) and it has been found that:

(a) When both plutonium and uranium are recycled, enriched uranium supply requirements for the year 2000 are reduced by about 20% compared to the uranium-only-recycle option. U_3O_8 and natural UF_6 requirements are reduced by about 30% compared with the no-recycle option.

(b) The recycle of plutonium and uranium introduces a commercial trade in purified plutonium, with additional safeguards requirements.

(c) If all spent fuel is reprocessed and the plutonium recycled, the quantity of spent fuel plutonium discarded in various nuclear wastes is expected to be about 1-2% of what it would be without recycle. However, recycling the plutonium requires additional operations like reprocessing, mixed oxide fabrication, and management of various additional types of wastes. These will raise additional questions of economics, health and safety, and environmental matters, which differ from those encountered in the no-recycle option.

The LWR fuel cycle requirements for the production of 1 GW(e)y are summarized in Table 4.

Table 4. LWR Fuel Cycle Requirements for 1 GW(e)y - no recycle*

Fuel Cycle Component	Unit	Requirement
Mining**	tonne U_3O_8	340
Milling	tonne U_3O_8	307
UF_6 Conversion	tonne U	266
Enrichment	tonne (SWU)***	132
UO_2 Fuel Fabrication	tonne U	43
Irradiated fuel storage	tonne/y (heavy metal)	41.2
Transportation	km	9.4×10^4
Waste management	ha	5

* After GESMO, Vo. 3, p. IV J(E)-17.

** The amount of uranium ore required varies according to ore grade. The average ore grade for GESMO scenario is about 0.1%; the ore mined is 340,000 tonne/GW(e)y.

*** A separative work unit (SWU) is a measure of the work expended to separate a quantity of uranium of a given assay into two components, one having a higher percentage of U-235 and one having a lower percentage.

Other Fuel Cycles:

The substitution of thorium for U-238 in the conventional PWR and BWR technologies offers a real but marginal advantage for resource extension alone. Uranium-Thorium mixed oxide fuels have been fabricated and satisfactory performance has already been demonstrated in LWR's on an experimental basis. As mentioned earlier, the recycle option for LWR's leads to significant reduction in the 30 year reactor lifetime ore requirements (the net lifetime gain from U-Pu recycle versus no recycle is about 30%). The LWR-Th reactor offers a further saving in uranium ore of approximately 16%. From the use of thorium, benefits are possible with the redesign of the reactor core or of the moderator-coolant system to improve the conversion ratio.

The uranium ore requirements for CANDU reactors are less than for the light water fuel recycle. Substantial savings (about 50% can be achieved in CANDU reactors with self-generated plutonium recycle. The fueling of CANDU with U-235, thorium and recycled uranium leads to further savings in the consumption of uranium ore (39 to 45% less for such cycle than for uranium fueling with self-generated plutonium recycle).

The High Temperature Gas Reactor (HTGR) is a helium cooled graphite moderated reactor using natural Th-232 as a fertile material and highly enriched uranium as fissile material. The reactor is fueled with thorium, make-up U-235, and uranium (U-233, U-234, U-235, U-236) recovered from the discharged fuel and recycled. An alternative means of realizing the fuel value of plutonium recovered from discharged LWR fuel would be to recycle this plutonium in the HTGR in lieu of highly enriched U-235 make-up normally used. The uranium produced as a result of neutron absorption in thorium is recycled, along with the plutonium remaining in the discharged fuel.

The LMFBR programme is aimed towards the development of first generation breeders fueled with uranium and plutonium. In the Pu-U fast breeder the core consists of plutonium-uranium mixed oxide fuel surrounded by a blanket of depleted uranium which absorbs neutrons leaking from the reactor core to produce additional plutonium. Cooling is accomplished by means of liquid sodium in the case of the LMFBR or helium gas in proposed gas-cooled designs. The only ore requirements attributable to the breeder are those associated with the production of plutonium for start-up loadings of the first-generation breeders. This plutonium must be obtained from thermal reactors; these reactors will then require more ore because they are thereby deprived of the benefits of plutonium recycle. When operating without Pu recycle, a 1 GW(e)y of LWR operation produces about 260 kg of fissile plutonium. LWR's must operate for 30 GW(e)y without Pu recycle in order to produce about 7,500 kg of fissile Pu required to start-up a 1 GW(e) LMFBR (3000 kg Pu are required for the initial core and 4500 kg for replacement loadings before Pu in discharged fuel is recycled).

Since light water reactors will continue to be the main type of reactors in use at least until the year 2000, the papers presented in the following chapters are mainly devoted to review the environmental impacts of the different stages of the fuel cycle of LWR's.

ACKNOWLEDGEMENTS

The author is grateful to Dr. Y. Ahmed, International Atomic Energy Agency, Dr. J.R. Beattie, Safety and Reliability Directorate, U.K. Atomic Energy Authority; Dr. K.H. Buob, Institut Féderal des Recherches en matiere réacteurs, Würenlingen, Dr. M. Carriera Vasquez, Junta de Energia Nuclear, Madrid; Dr. L.D. Hamilton, Brookhaven National Lab., New York; Dr. H. Howells, British Nuclear Fuels Ltd.; Dr. H.P. Jammet, Institut Curie, Paris, Dr. J.L. Weeks, White Shell Nuclear Research Establishment, Manitoba; Dr. A. Biswas, Biswas and Associates, Ottawa; and Dr. G. Gorrie, Department of Environment, Canberra for their comments and assistance.

REFERENCES

APS: American Physical Society; Study Group on Nuclear Fuel Cycles and Waste Management. Review of Modern Physics 50, No. 1 Part II (1978).

Braatz, U. and Dibbert, H.: Kernbrennstoffversorgung, Jahr. der Atomwirtschaft, (1977).

Duret, M.F. et al.: The contribution of Nuclear Power to World Energy Supply 1975-2000. World Energy Resources World Energy Conference, IPC Sci. Techn. Press, London, (1978).

GESMO: Final Generic Environmental Statement on the use of Recycle Plutonium in Mixed Oxide Fuel in Light Water Cooled Reactors. Vol. 3. U.S. Nucl. Regulatory Comm., NUREG-0002, (1976).

Giraud, A.: World Energy Resources. Conference on World Nuclear Power. Atomic Ind. Forum November, (1976).

Hanrahan, E.J. et al.: Demand for Uranium. Atomic Ind. Forum, Phoenix, USA, (1976).

IAEA: Private Communication, (1979).

ICRP-26: International Commission on Radiological Protection Publ. No. 26, Pergamon Press, Oxford, (1977).

Krymm, R. and Woite, G.: IAEA Bulletin, 18. p. 6; (1976).

Messer, K.P.: Uranium Demand as Judged by Electric Utilities. International Symp. Uranium Supply and Demand; Uranium Institute, London, June, (1977).

OECD: Nuclear Fuel Cycle Requirements, OECD, Paris, (1978).

OECD-NEA: Uranium Resources, Production and Demand, OECD, Paris, (1976).

OECD-NEA: Uranium Resources, Production and Demand, OECD, Paris, (1977).

Royal Commission Environmental Pollution: Sixth Report: Nuclear Power and the Environment. HMSO, London, (1976).

UNSCEAR: Sources and Effects of Ionizing Radiation. United Nations Scientific Committee on the Effects of Atomic Radiation. United Nations, New York, (1977).

WAES: Workshop on Alternative Energy Strategies; Energy Global Prospects 1985-2000. McGraw-Hill, New York, (1977).

Williams, K.R.: Projected Energy Requirements up to the year 2000. International Symp. Uranium Supply and Demand; Uranium Institute, London, June, (1977).

2

A REVIEW OF THE ENVIRONMENTAL IMPACT OF MINING AND MILLING OF RADIOACTIVE ORES, UPGRADING PROCESSES, AND FABRICATION OF NUCLEAR FUELS

J. M. Costello*, D. R. Davy*, F. C. R. Cattell**, J. E. Cook**

*Australian Atomic Energy Commission
Research Establishment
Private Mail Bag, Sutherland, NSW. 2232
**Australian Atomic Energy Commission
P.O. Box 41, Coogee, NSW. 2034

INTRODUCTION

This paper surveys published assessments of the environmental impact of operations in some nuclear fuel cycles from mining of the ores through to the fabrication of nuclear reactor fuel. It was prepared as a contribution to a detailed review by the United Nations Environment Program of the environmental impact of the production and use of all sources of energy.

Comparison of different energy sources requires that the estimation of environmental impact be as realistic as possible. Conservative estimates are inappropriate for comparison purposes. Even when impacts are calculated realistically there are problems of comparison due to the wide spectrum of concerns and the wide variety of methods for analysing and weighing impacts and costs. This review concentrates on the use of natural resources and on the effects of pollutants, particularly as they affect human health. The review is generic in nature. Actual impacts will vary from country to country depending on, among other things, the standards and regulations applied. Within the framework of these standards and regulations impacts will also vary from site to site and on the types of technology used. A weakness of a generic review is that these variations are largely ignored.

Predictions of nuclear power plants to be operational in OECD countries by 2000 shown in Table 1 (OECD/NEA, 1977) indicate that light water reactors will be the predominant reactor type, accounting for about 90 per cent of nuclear power generation. Although alternative reactor strategies are being studied in the International Nuclear Fuel Cycle Evaluation against the criteria of potential for weapons proliferation, environmental impact, economics and institutional arrangements, this report has concentrated on those fuel cycle operations required by light water reactors.

Fuel cycles for gas-graphite reactors (GGRs) and heavy water moderated reactors (HWRs), which are fuelled with natural uranium metal and oxide respectively,

Table 1

REACTOR DISTRIBUTION - LOW POWER GROWTH ESTIMATE*

Year	Total Nuclear Power GWe	LWR %	AGR %	HWR %	GGR %	FBR %	HTR %
1977	87	85.1	3.5	4.6	6.9	-	-
1980	146	86.3	4.1	4.8	4.1	0.7	-
1985	278	90.2	2.2	4.7	2.2	0.4	0.4
1990	504	91.2	1.2	5.6	1.2	0.4	0.4
1995	750	91.1	0.8	6.7	0.5	0.5	0.4
2000	1000	90.1	0.6	7.5	0.2	1.0	0.6

* From OECD/NEA (1977).

Legend: LWR Light Water Reactor
AGR Advanced Gas Cooled Reactor
HWR Heavy Water Reactor
GGR Gas-graphite Reactor
FBR Fast Breeder Reactor
HTR High Temperature Reactor

are similar in the mining, milling and uranium purification stages, but differ in fuel manufacture and fabrication operations; light water reactors (LWRs) and advanced gas-cooled reactors (AGRs), which require fuel enriched in uranium-235, involve the production and enrichment of uranium hexafluoride. Techniques for disposal of waste from uranium production (Tsivoglou and O'Connell, 1962) have depended on the country of origin, local environmental conditions and national attitudes towards the environment. Practices for waste management in mining and processing of radioactive ores are continually under review and international recommendations and national codes of practice on radiation protection have been written (IAEA, 1976, Commonwealth Department of Health, 1975).

Environmental assessments of the impact of nuclear fuel cycles have been made in the US. Early generic studies on the uranium fuel cycle (USAEC, 1974, USEPA, 1973), for support of LWRs derived data on land and water requirements, made quantitative assessments of thermal, chemical and radiological effluents discharged to the biosphere (USAEC, 1974), and estimated health and radiological effects (USEPA, 1973). These generic studies have subsequently been extended to assess the environmental impacts of uranium and plutonium recycle in LWRs (USNRC, 1976) and of some alternatives such as the thorium-uranium breeder reactor fuel cycle (USERDA, 1976). Internationally UNSCEAR periodically reviews radiation doses due to nuclear power production but does not assess the implied public health impact (UNSCEAR, 1977).

This review examines in some detail the fuel cycle of the LWR using uranium dioxide fuel without recycle. Environmental impacts of other nuclear fuel cycles are compared briefly with this fuel cycle.

THE FIRST STAGES OF THE URANIUM FUEL CYCLE

Uranium is mined as a low-grade ore, usually containing greater than 0.1 wt per cent uranium. Normally it is processed at a mine site to concentrate the uranium into an economically transportable form. Commercial concentrates, called yellowcake, typically contain over 90 per cent of oxides of uranium. Natural uranium contains only 0.7 per cent of the uranium-235 isotope, which is capable of generating power in a thermal reactor through thermal nuclear fission. The remaining 99.3 per cent is the isotope uranium-238, which is effectively non-fissile by thermal neutrons but contributes to power generation through production of fissile plutonium.

Use of uranium enriched in the isotope uranium-235 rather than natural uranium, allows a wider range of design options and a more compact reactor core, which may result in lower capital costs. LWRs and AGRs require enriched uranium. GGRs and HWRs operate on natural uranium metal and oxide fuels respectively, and do not need enriched uranium.

If enrichment is required the crude yellowcake is converted into pure uranium hexafluoride (UF_6), which is readily vaporised by heating. The uranium isotopes in the UF_6 vapour are separated in an enrichment plant, typically into a product stream containing about 2-4 per cent uranium-235 and a reject 'tails' stream depleted to about 0.2-0.3 per cent uranium-235.

Fuel fabrication involves processing of the enriched UF_6 into a ceramic oxide (UO_2), which is pressed and sintered into fuel pellets. The pellets are loaded and sealed into metallic tubes, which are assembled into a precise mechanical array to form the fuel element assembly ready for loading into the reactor.

Uranium Mining

Mining operations. Uranium may be mined by three methods:

(a) Opencut mining, which is employed when the orebody lies close to the surface under an overburden which can be removed easily and economically. The operation involves the mining, removal and stockpiling of all the material associated with the ore deposit.

(b) Underground mining, which is generally used for ore bodies at depths greater than 100 m, or when the ore lies under rock strata which would otherwise require substantial blasting for excavation. This operation is more selective for ore and produces less waste rock than opencut mining, but has a lower extraction efficiency.

(c) Solution mining, in which the ore is leached in situ from the underground deposit and the product solution recovered by pumping to a treatment plant. It may be employed where an orebody is horizontal and lies on relatively impervious strata below the static water table. Extreme care must be taken to avoid contamination of ground water (USEPA, 1974). Overall recovery efficiencies of 65-70 per cent have been claimed for solution mining (Frank, 1976).

Uranium can also be recovered as a by-product of other mining operations. For example, in South Africa uranium is extracted from gold mine tailings (Poole, 1975).

Open cut and underground mining accounted for nearly 60 and 40 per cent respectively of US uranium ore production in 1975, but these proportions are expected to be reversed by 2000 (USNRC, 1975; USERDA, 1976). Solution mining accounted for less than 2 per cent of US uranium production in 1974 (USERDA, 1976). Open cut mining has been proposed for deposits in the Australian Northern Territory (Ranger Uranium Mines Pty. Ltd., 1974, Pancontinental Mining, 1977), but some are also suited to underground mining operations.

Milling of Uranium Ores

In the milling operation the ore is processed mechanically and chemically to extract the bulk of the uranium content and to produce an impure concentrate of uranium oxides, called 'yellowcake'. The uranium content of yellowcake is generally expressed in terms of the oxide U_3O_8.

There are essentially two alternative methods of uranium milling; uranium is leached from the finely ground ore either by sulphuric acid or by sodium carbonate, the choice being determined mainly by the acidity or alkalinity of the host rock (Merrit, 1971). The acid leach process is suitable for the majority of ores, while carbonate leaching is applicable to some uranium deposits in Western Australia, Canada and the US.

Description of Processes

Acid leach process (Fig. 1). Ores from the mine are stockpiled into grades based on uranium content and leaching characteristics. Ores are blended and reduced in size by coarse and fine crushing, followed by wet grinding to produce a slurry of finely divided ore. The ore is usually heated with sulphuric acid and an oxidant to leach more than 90 per cent of the uranium. The leach liquor is washed from the insoluble residue (tailings) by counter-current decantation (CCD) and transferred to a purification system. The tailings residue is transferred to a retention system, sometimes with neutralisation. In the purification step, the uranium is extracted from the bulk of other soluble minerals by solvent extraction or ion exchange, and concentrated into aqueous solution. Part of the acid raffinate may be recycled to wash the tailings, and the remainder neutralised and transferred with the tailings to a retention system. The uranium is precipitated with an alkali, (e.g. ammonia), and the diuranate slurry thickened and centrifuged to separate the solid concentrate. Ammonium diuranate is decomposed into U_3O_8 by calcination, and the product pulverised and packaged into drums for transport.

Carbonate leach process (Fig. 2). Ores are stockpiled into grades and blended, crushed and wet ground to a slurry, which is normally finer than for acid leaching. The ore is heated with sodium carbonate solution containing some sodium bicarbonate to leach more than 90 per cent of the uranium and retain it in soluble form. Air or oxygen may be fed to the solution as an oxidant. The leached liquor is decanted or filtered from the undissolved solids, which are washed and transferred to a retention system. Generally, the carbonate solution of uranium does not require purification; uranium may be precipitated selectively by addition of sodium hydroxide with close control of pH to retain contaminant ions in solution. The precipitated slurry of sodium diuranate is thickened and centrifuged. The barren solution is treated with carbon dioxide to regenerate sodium carbonate for re-use in the leaching step. The sodium diuranate may either be dried and packaged as crude yellowcake or treated to reduce its sodium content by dissolution in sulphuric acid and precipitation with ammonia, followed by separation and calcination of the ammonium diuranate product.

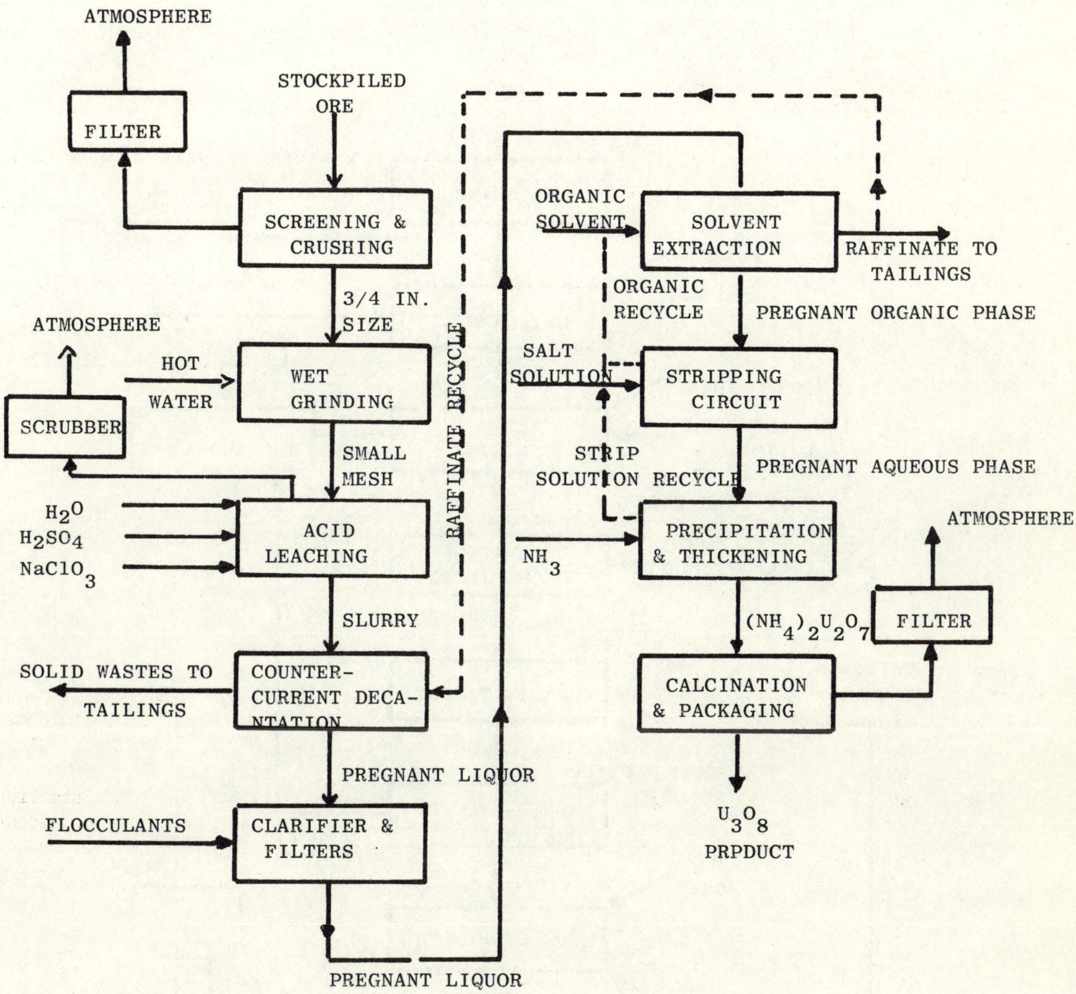

Fig. 1 Simplified Flow Diagram for Uranium Milling Acid Leach Process (After USERDA (1976))

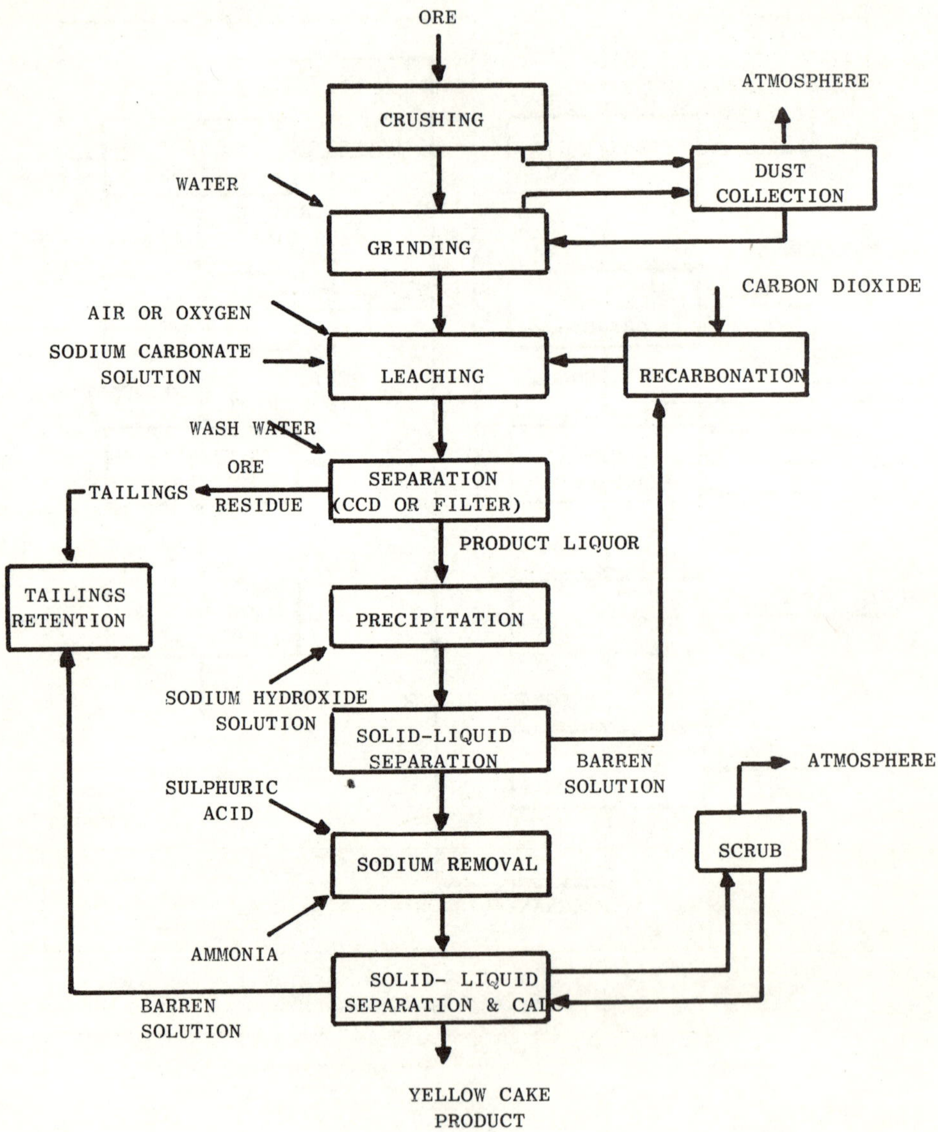

Fig. 2 Uranium Mill- Carbonate Leach Process, Simplified Block Flow Diagram (after Merritt, 1971).

Manufacture of Uranium Hexafluoride

The crude uranium ore concentrates are converted into pure uranium hexafluoride (UF_6), the feed stock for uranium enrichment plants. Two major process types are in current use for UF_6 manufacture: so called wet processes in which the uranium is purified before conversion to UF_6, and a dry hydrofluor process in which complete purification is not achieved until after the UF_6 has been produced. Alternative process technologies have been reviewed (USAEC, 1974, USEPA, 1973, USERDA, 1976) and an environmental impact statement compiled for a wet process (USNRC, 1976). Simplified block flow diagrams for these processes are given in Figure 3 and Figure 4.

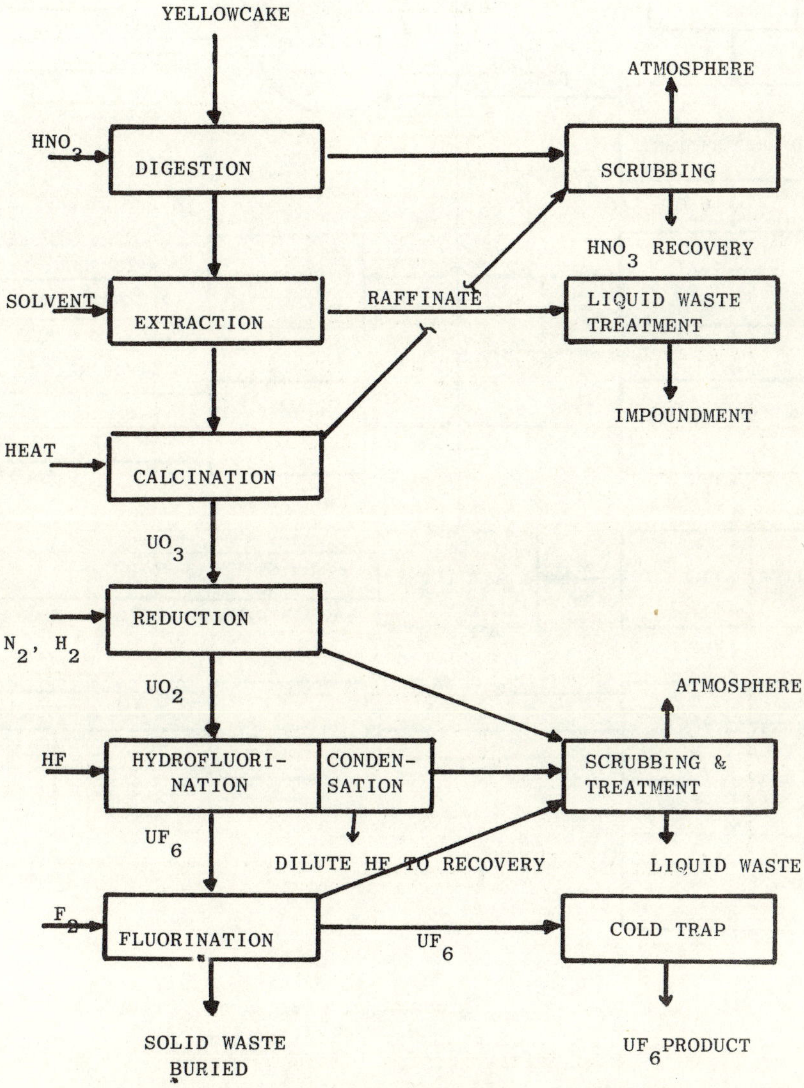

Fig. 3 Schematic Diagram of Wet Process for Uranium Hexafluoride Production (after USERDA, 1976)

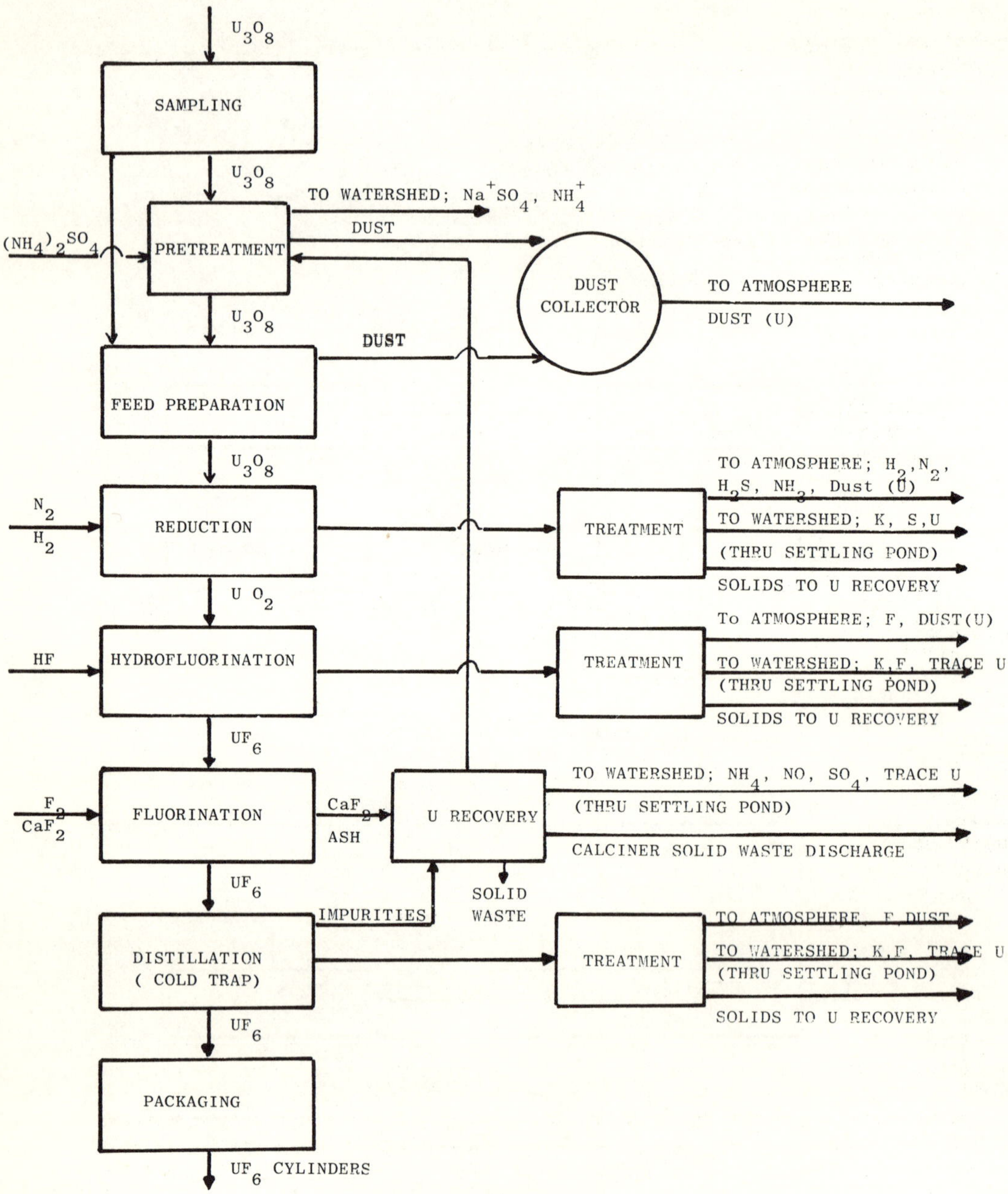

Fig. 4 Schematic Diagram of Hydrofluor Process for Uranium Hexafluoride Production (from USERDA, 1976).

Both process types can achieve virtually complete recovery of uranium, total consumption of fluorine, and high utilisation of a number of other process chemicals, such as hydrogen, hydrogen fluoride, and ammonia (USAEC, 1974).

Operations in a typical wet process (Fig. 3) are:

- sampling, analysis and storage of received yellowcake;
- dissolution of yellowcake in nitric acid;
- purification of the uranyl nitrate solution by solvent extraction;
- concentration and thermal decomposition of uranyl nitrate to uranium trioxide (UO_3), with recovery of nitric acid;
- reduction with hydrogen to form UO_2;
- reaction with hydrogen fluoride, HF, to form uranium terafluoride UF_4; and
- fluorination to UF_6 with elemental fluorine.

The alternative dry process (Fig. 4) involves the following stages:

- sampling, analysis and storage of yellowcake;
- treatment of the yellowcake for processing in a fluidised bed (treatment processes may include crushing and pelletisation of the yellowcake and, possibly, removal of excessive sodium impurity);
- reduction of the U_3O_8 to UO_2 with hydrogen;
- reaction with hydrogen fluorine to UF_4;
- fluorination to UF_6 with elemental fluorine; and
- fractional distillation to separate the crude UF_6 product from other volatile fluorides.

Uranium Enrichment

LWRs require uranium enriched to 2-4 per cent in uranium-235. Commercial enrichment technologies are based on gaseous diffusion or centrifugation of uranium isotopes in the form of UF_6. Environmental impact statements have been prepared for gaseous diffusion plants (USERDA, 1976, 1977) and for a gaseous centrifuge plant with an associated centrifuge manufacturing facility (USERDA, 1976). The impacts of the two technologies have been compared (USNRC, 1976, USERDA, 1976).

Enrichment by gaseous diffusion. In the diffusion process, gaseous UF_6 is compressed and passed over a porous membrane. Molecules of UF_6 containing the lighter isotope, uranium-235, diffuse through the membrane more rapidly than those with the heavier uranium-238 isotope; consequently, the UF_6 passing the membrane has a slightly greater proportion of molecules containing uranium-235. The degree of enrichment for one membrane is minute and over a thousand successive diffusion stages are necessary to change the proportion of uranium-235 from the naturally occurring level of 0.71 per cent in the feed stream to the approximately 3 per cent required in the product stream for LWR fuel with about 0.25 per cent in a reject 'tails' stream. Each stage requires recompression of the gaseous hexafluoride. Uranium enrichment by gaseous diffusion requires large quantities of electrical energy. About 36.5 MWe y of electricity would be needed by a gaseous diffusion plant to enrich the uranium for the generation of 1 GWe y of electricity in a LWR (USNRC, 1976).

Enrichment by gas centrifugation. During centrifugation molecules of UF_6 containing the heavier isotope uranium-238 migrate preferentially to the wall of a rapidly rotating cylinder. There is a consequent enrichment in the lighter uranium-235 isotope in the gas near the tube axis. The separation factor is greater than in the diffusion process, and the two streams removed from the tube axis and wall require only tens of stages arranged in a cascade to produce the required percentages of uranium-235 in the product and reject tails. High centrifugal stresses limit the size of the equipment, and many parallel cascades involving hundreds of thousands of centrifuges are required to achieve the separative capacity of a commercial enrichment plant. However, a gas centrifuge plant uses less than one tenth of the electrical energy required by a diffusion plant of similar capacity. The nuclear reactor could therefore produce at least 280 times the electrical energy required to enrich its uranium fuel by the gas centrifuge process.

Laser enrichment processes. Uranium enrichment with lassers involves the selective dissociation of volatile uranium compounds or selective ionisation of uranium atoms to a degree controlled by differences in physical properties of uranium-235 and -238 isotopes. Laser enrichment might require only 10 per cent of the electricity for gas centrifuge processes (USERDA, 1976), but the technology is still at the laboratory stage and unlikely to influence the environmental impact of the fuel cycle in the near term. It has not been considered further in this review.

Fuel Manufacture and Fabrication

The feed material for the manufacture and fabrication of fuel for a LWR is uranium hexafluoride (UF_6) enriched to about 3 per cent in uranium-235. The UF_6 is converted to uranium dioxide powder (UO_2) which is formed into pellets, sintered to achieve the desired density and ground to the required dimensions. Fuel pellets are loaded into tubes of Ziracloy or stainless steel, which are sealed at both ends. These fuel rods are spaced in fixed parallel arrays to form reactor fuel assemblies.

The process steps involved in making UO_2 powder, shown in Figure 5, illustrates the conventional route by which UF_6 is converted first to ammonium diurante (ADU) and then to UO_2:

- the UF_6, a solid at normal temperatures, is received in sealed cylinders from the enrichment plant;
- the UF_6 is vapourised by heating electrically or with steam;
- the gaseous UF_6 is hydrolysed to form a solution of uranyl fluoride (UO_2F_2);
- ammonia is added to precipitate ammonium diurante (ADU);
- the slurry of ADU is centrifuged or filtered; and
- the ADU is decomposed by heating, pyrohydrolysed with steam to remove traces of fluoride, and reduced to UO_2 powder with hydrogen.

The fabrication operations (Fig. 6) involve the following stages:

- preparation of UO_2 powder of desired size distribution by comminution, compaction and granulation;
- manufacture of UO_2 pellets;
- sintering of the pellets in hydrogen gas;

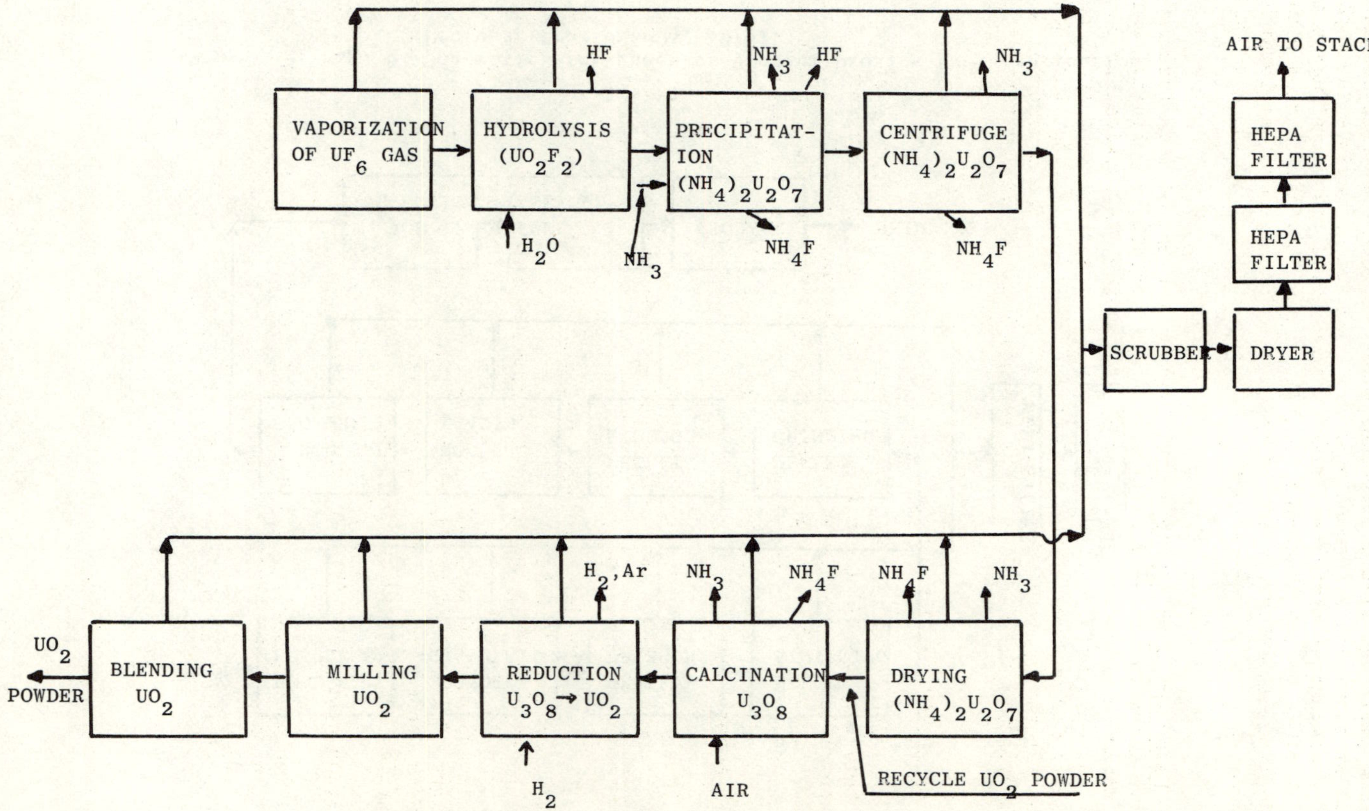

Fig. 5 Schematic Flow Sheet of Uranium Dioxide Conversion Process (from USERDA, 1976).

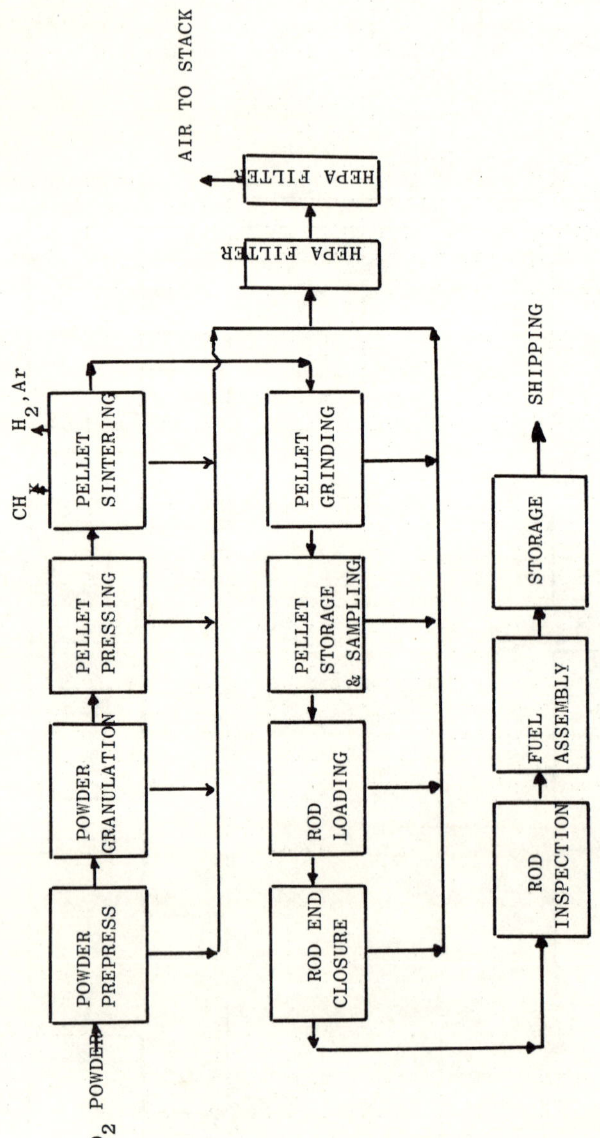

Fig. 6 Schematic Flow Sheet of Uranium Dioxide Fuel Fabrication Process (from USERDA, 1976).

- grinding of the sintered pellets to the required size;
- washing, drying and loading the fuel pellets into tubes of Zircaloy or stainless steel and sealing the ends with welded caps; and
- locating fuel rods in fixed parallel arrays forming the reactor fuel assemblies.

Off-specification material produced during fabrication is fed into a scrap recovery cycle in which uranium oxide is dissolved in nitric acid, the resultant uranyl nitrate is purified by solvent extraction, and the uranium reconverted into UO_2.

ENVIRONMENTAL IMPACT

The environmental and social costs of energy production encompass a wide spectrum of concerns, and a wide variety of methods for analysing impacts. In this review discussion of environmental impacts is limited to a survey of the use and degradation of resources including the effects of pollutants which are released as a result of fuel cycle operations.

In general, for energy production, three phases of fuel cycle operations should be considered:

1. Construction (and maintenance) for which accidents and the effects of routine emission of pollutants due to the preparation of construction materials may be important.
2. Routine operation and accidents.
3. Emissions and impacts which occur after the plant has ceased operating.

In this review impacts are examined briefly in terms of their effects on the non-human environment but mainly in terms of their effects on occupational and public health. The effects of the construction phase are not considered, except to the extent that accidents at the construction site are included in occupational health impacts. In general impacts from construction are expected to be less than those from routine operation. For the early stages of the uranium fuel cycle reviewed here consequences and probabilities of accidents seem likely to result in impacts which are small compared with those from routine operation.

In general analysis of the environmental impact of pollutants requires knowledge:

1. of the quantity of effluent released;
2. of the dispersion of material in the environment;
3. of factors for calculating the transfer of pollutants to sensitive species;
4. of the relationship between concentration of the pollutant and damage to the species; and
5. of the relationship between damage to species and environmental cost.

The total impact of some pollutants depends on synergistic or antagonistic reactions. For some pollutants there is a threshold concentration for each species below which essentially no effect occurs. Provided concentrations are kept below this threshold for the most sensitive indicator the environmental impact is negligible. However, for effects caused by a number of pollutants there appear to be no threshold. It is prudent to assume that for exposure to radiation, and also carcinogenic chemicals, there is no threshold and that effects occurring at low doses are linear with dose to the lowest dose. The total impact of these pollutants therefore requires integration over all pollutant concentrations. A further feature is that radioactive and many other carcinogenic materials may remain in the environment long after the action releasing them has been discontinued. Attempts to assess the future impacts of these pollutants, although difficult, should therefore be made. When different activities are being compared it is important that common ground rules are used to assess impacts.

Several reviews, relevant to the early sections of the uranium fuel cycle, have been published, particularly relating to practices in the US (USAEC, 1974; USEPA, 1973; USERDA, 1976; USNRC 1976; Pochin, 1976). The recent Final Generic Environmental Statement on the Use of Recycle Plutonium in Mixed Oxide Fuel in Light Water Reactors (GESMO) (USNRC, 1976), considered implications of the requirements for natural resources and the releases of chemical, radioactive and thermal effluents for several fuel cycle options appropriate to the US over the period 1975 to 2000. This study was developed as a more comprehensive generic assessment than its predecessors and is based on recent estimates of pollutant releases. The GESMO study is used extensively as a source document for this review and its findings are compared with UNSCEAR and other sources.

The GESMO study contains a generic assessment of each segment of the LWR fuel cycle industry. Model facilities reflecting current technology and which were a composite of typical facilities were assumed for this assessment. No allowance was made for improvements in technology as the industry matures over the period 1975 to 2000.

Ore production from underground and opencut uranium mines was assumed to be in the proportion of 3 to 2, with an ore grade of about 0.085 per cent uranium which is considered likely for future mining in the US. This is about a 50 per cent reduction in the grade considered in earlier studies (USAEC, 1974; USEPA, 1973). This grade is substantially lower than those at some large uranium discoveries in Australia and elsewhere.

Mining and milling operations in semi-arid areas of the US remote from centres of population were assumed. Ores were processed with a 90.5 per cent recovery through a postulated composite model mill processing 85 per cent of ores by acid leach and solvent extraction and 15 per cent through an alkaline leach circuit.

Wet and dry hydrofluor processes installed in approximately equal proportions were assumed for uranium hexafluoride conversion.

Enrichment services consisting of gaseous diffusion and gas centrifugation in the ratio 4 to 1, reflecting uprated existing US gaseous diffusion plants augmented by centrifuge plants, were proposed.

Production of uranium dioxide from enriched hexafluoride was assumed in six plants using conventional processes involving precipitation and calcination of ammonium diuranate (ADU), augmented later by three plants using a dry production route (Westinghouse, 1970). Environmental impact assessment was based on the ADU process.

Use of Natural Resources

Part of the environmental cost of any energy technology arises from the use of physical resources other than the fuels themselves. Land, water and minerals are probably the three most important of these resources. Quantifying the significance of resource use is a difficult problem. The economic value of the resources is a very imperfect measure of environmental cost. There is clearly a difference between water which is evaporated through use and water which is polluted and returned to surface use. With land use it is useful to distinguish between permanent and temporary use and measures of environmental cost, such as the loss of food production, are valuable.

Table 2 shows GESMO predictions of the use of land, water and energy in the first stages of the uranium fuel cycle. Actual quantities will depend on the particular site. For instance, the use of many natural resources in mining and milling will be smaller for Australian deposits which have a higher percentage uranium than the 'typical' deposits assumed for the GESMO study.

Table 2

USE OF NATURAL RESOURCES PER GWe y

Resource	Mining	Milling	Conversion	Enrichment	UO_2 Fuel Fabrication
Land Use:					
Disturbed ha	34	3.8	0.13	0.15	0.023
Committed ha	1.1	3.2	9.3×10^{-3}	-	-
Water m^3					
Discharged to air	-	1.1×10^6	1.2×10^4	9.4×10^5	-
Discharged to water	-	-	1.1×10^5	5.3×10^7	3.0×10^4
Discharged to ground	2.8×10^6	-	-	-	-
Total discharged	2.8×10^6	1.1×10^6	1.2×10^5	5.3×10^7	3.0×10^4
Energy Use					
Coal tonne	-	-	-	840	-
J equiv.	-	-	-	2.3×10^{13}	-
Gas J equiv.	-	3.2×10^{14}	3.2×10^{13}	-	5.6×10^{12}
Fuel Oil m^3	1.9×10^3	-	-	12	-
J equiv.	7.0×10^{13}	-	-	4.0×10^{11}	-
Electricity GWe y	7.3×10^{-4}	1.7×10^{-3}	3.3×10^{-4}	0.035	3.0×10^{-4}
J equiv.	2.3×10^{13}	5.4×10^{13}	1.0×10^{13}	1.1×10^{15}	9.5×10^{12}

Mining. For openpit mines about one third of the total land involved is disturbed temporarily by the actual mining operation while the remaining two thirds remain idle (USAEC, 1973). The effect of underground mining on surface land use varies considerably. Most of the land can be reclaimed after the mining operation and piles of overburden and other wastes graded and contoured. For an openpit mine the final pit area may be converted to a lake.

Large quantities of water may have to be pumped from the mine, depending on its location. This can temporarily lower the water table. The water will form part of the mine water management scheme and may be used as process water. Diesel fuel is required for mine equipment and a modest amount of electricity is used, mainly for ventilation fans and dewatering pumps.

Milling. Most of the land area required for milling is devoted to a pond for the permanent disposal of mill tailings. Although tailings piles should be rehabilitated after the milling operation ceases the land is not likely to be released for unrestricted use in the short term. The Ranger Uranium Environmental Inquiry (1976) recommended that tailings be returned to the pit at the completion of operations.

Process water is also pumped to the tailings dam. Most of this water may be lost by evaporation although there will be some seepage. In the GESMO study natural gas is assumed to be the source of process heat.

Hexafluoride conversion. Of the land commitment of hexafluoride production approximately 10 per cent is disturbed for roads, fills and plant structures. About 1 per cent is permanently committed for waste burial. Over 90 per cent of the water is used for process cooling. Some water is used as process water in the wet solvent extraction process.

Enrichment. Essentially none of the land required for enrichment is committed permanently. Water is required for the operation of cooling towers associated with the gaseous diffusion process and also for the large amounts of electricity which need to be generated. For 80 per cent of the enrichment by gaseous diffusion, assumed in GESMO, 92 per cent of all the electricity required in the fuel cycle up to fuel fabrication was used in enrichment. A gaseous centrifuge plant requires only about one tenth the electricity of a gaseous diffusion plant. However, even with extensive use of gaseous centrifugation, enrichment would still be the most electricity consumptive stage of all the fuel cycle operations considered in this review.

Fuel Fabrication. All of the land required for fuel fabrication is capable of reclamation. (The small amount used for holding ponds or lagoons would present most problems.) Most of the water is used for cooling of process plant and to dilute liquid process wastes prior to release.

Effluents from Fuel Cycle Operations

GESMO predictions of effluents released by the fuel cycle operations are given in Table 3.

Mining Wastes and effluents. Solid waste produced in opencut mining operations is the earth overburden above the orebody and the barren rock in which the ore is dispersed. An average waste to ore ratio of about 30 has been reported (USAEC, 1974) for US mines; ratios of about 4 and 14.5 have been estimated for two Australian uranium deposits (Ranger Uranium Mines Pty Ltd, 1974; Pancontinental Mining Ltd., 1977).
Initially, waste rock from excavation, if free from potential contaminants such as pyritic material, may be used for construction of earthworks, foundations and roads. Excess rock is transported to a dump near the mine. Other solid waste material, separately stockpiled, consists of rock which bears traces of uranium below the economic cutoff grade.

Liquid effluent from uranium mining consists of surface water runoff from the waste rock and ore stockpiles, water seepage through the waste rock and stockpiled

Table 3

EFFLUENTS PRODUCED PER GWe y

Effluent	Mining	Milling	Conversion	Enrichment	UO_2 Fuel Fabrication
Heat dissipated (J)	1.4×10^{14}	4.7×10^{14}	6.1×10^{13}	3.1×10^{15}	3.3×10^{13}
Effluents from Coal Equivalent * of 2/3 Electricity					
Sludge (tonne)	210	500	100	1.0×10^{4}	84
SO_x to atmosphere (tonne)	19	45	8.8	900	7.8
NO_x to atmosphere (tonne)	15	35	7.0	730	6.0
CO to atmosphere (tonne)	0.30	0.70	0.14	15	0.12
Particulates to atmosphere (tonne)	0.93	2.1	0.43	42	0.38
Hydrocarbons (tonne)	0.15	0.35	0.070	7.3	0.060
Plant Effluent to Atmosphere (tonne)					
SO_x	21	0.14	8.5	12	-
NO_x	17	28	4.5	9.3	-
CO	-	-	-	0.19	-
Particulates	2.5	1.4	0.14	0.55	8.8
NH_3	-	-	0.12	-	-
Fluorides	-	-	0.070	0.035	7.8×10^{-3}
Hydrocarbons	1.2	5.5	0.63	0.093	-
Plant Effluent to Atmosphere (T Bq)					
Radon-222	220	41	3.1×10^{-7}	-	-
Radium-226	-	1.2×10^{-4}	9.3×10^{-8}	-	-
Uranium	-	4.9×10^{-3}	6.6×10^{-5}	3.6×10^{-5}	1.1×10^{-5}

* The GESMO study (USNRC 1976) assumed 2/3 of electrical requirements would be derived from coal; 1/3 of requirements from nuclear power.

Table 3 (Cont'd.)

Effluent	Mining	Milling	Conversion	Enrichment	UO_2 Fuel Fabrication
Plant Effluents to Water (tonne)					
sulphate	-	-	14	0.80	-
nitrate	-	-	0.20	0.90	1.5
chloride	-	-	4.0	0.38	-
fluoride	-	-	0.24	9.3×10^{-3}	0.045
sodium	-	-	13	-	-
calcium	-	-	0.98	-	2.2×10^{-4}
ammonia	-	-	4.3	-	0.017
iron	-	-	0.058	-	-
Plant Effluents to Water (T Bq)					
uranium	-	-	2.6×10^{-3}	2.0×10^{-6}	2.5×10^{-3}
thorium-230	-	-	3.9×10^{-4}	-	-
radium-226	-	-	1.3×10^{-5}	-	-
Plant Waste (m^3)					
chemical compounds	-	-	45	0.17	38
mill tailings	-	2.0×10^{5}	-	-	-
radioactive solids	-	-	22	16	1

ore, and water seepage into the mine. This effluent may contain suspended solids and dissolved minerals, including uranium and its decay products. Treatment practices can involve impoundment in retention ponds for settling of solids and surface evaporation of some water, use of mine drainage water as a process feed to the uranium mill, and controlled dilution and discharge of some impounded water during periods of heavy rainfall, after treatment for removal of radium-226 if necessary.

Airborne effluents from uranium mining consist of dusts containing uranium and its decay products, radon-222 released to atmosphere during exposure and breakup of the orebody, and the combustion products of petroleum fuels used by the mining equipment. Radon emanation from the mine and solid wastes will continue after the mining operation has ceased.

Milling effluents. The tailings slurry is the major chemical and radiological waste from the milling process. This stream is a slurry of leached solid ore and waste solutions from the grinding, leaching, uranium purification, precipitation and washing circuits of the mill. The solution from an acid leach process contains predominantly sulphate ions with quantities of soluble metal ions and traces of organic solvents. The sodium removal step in a carbonate leach process also produces a sulphate waste solution. The neutralised solid tailings consist of unleached rock and precipitated mineral hydroxides.

With little of the parent uranium remaining, the shortlived daughters thorium-234 (24.1 d), protoactinium-234m (1.17m) and thorium-231 (25.5 h) are lost by decay from the mill wastes. The remaining 70 per cent of the activity originally present in the ore remains in the waste stream (UNSCEAR, 1977). Yellowcake produced by an acid leach process contains about 5 per cent of the thorium and about 0.1 per cent of the radium originally present in the ore. With a carbonate leach process approximately 2 per cent of the radium may appear in the yellowcake. These elements appear in the waste stream from uranium purification.

The tailings slurry is pumped into a system designed to retain all solids. Management of liquid wastes is very dependent on climate and ranges from treatment prior to discharge to almost total retention.

Radioactive airborne effluents from milling include dusts and radon gas released into the air from ore stockpiles, the crushing, grinding and leaching operations and from the tailings retention system. The releases of dusts produced in such processing operations as crushing and grinding of ore and calcining and packaging of yellowcake are reduced by ventilation extract scrubbers. Tailings piles are a a continuing source of radon and radioactive dust after the milling operation has ceased.

Airborne chemical contaminants released to the environment include combustion products (oxides of carbon, nitrogen and sulphur) from the process steam boilers, sulphuric acid fumes in small concentrations from the leach tanks, oxides of sulphur, sulphuric acid fumes from the sulphuric acid plant, and traces of vaporised organic reagents from the solvent extraction ventilation system.

Hexafluoride process effluents. The effluents from the wet and dry processes differ substantially. Most of the impurities entering with the yellowcake are rejected in the wet process raffinate solution from solvent extraction, whereas in the dry process, most of the yellowcake impurities are contained in solid wastes from the fluorination and distillation stages.

Effluents from a typical wet process (Fig. 3) consist of:

- neutralised aqueous raffinate from solvent extraction,
- caustic effluents and residual fumes from recovery of nitric acid, hydrogen fluoride and treatment of general off-gas streams, and
- a small quantity of solid calcium fluoride from the fluorination step (assuming fluorination in a fluidised bed).

The raffinate stream amounts to about 5 m^3 t^{-1} of uranium processed (USAEC, 1974). It may contain substantial dissolved solids, radium and thorium-230 entering with the yellowcake feed, and about 0.2 per cent of the uranium processed. This stream is neutralised and impounded in a retention pond. Disposal of this effluent is a major problem associated with the wet process (USERDA, 1976, USNRC, 1975). Proposals in the US call for evaporation of the supernatant water and either burial of the sludge, or its transfer to a uranium mill retention system. Some scrubber effluents are treated with lime to precipitate fluoride ion in settling ponds for packaging and burial as calcium fluoride.

Most of the chemical solid effluent from the dry process (Fig. 4) occurs as non-volatile ash containing iron,calcium, magnesium, copper and other fluorides. This residue can amount to about 0.1 tonne per tonne of product UF_6 (USAEC, 1974; Sears et al, 1977); it is reacted with fresh fluorine to recover uranium and the final residue packaged and buried as low activity solid waste. Scrubber effluents arising from treatment of the hydrofluorination off-gas stream are treated similarly to those in the wet process.

<u>Effluents from enrichment processes</u>. The primary source of environmental impact associated with the enrichment of uranium is the emission of particulates and oxides of nitrogen and sulphur from the generation of electrical energy through fossil fuel combustion. An early US study estimated (USAEC,1974) that about 6000 tonne of waste gas emissions (including particulates) may be associated with the production of the annual fuel requirement for a 1000 MWe LWR by the diffusion process. Substantially lower quantities have been reported in more recent assessments (USNRC, 1976, USERDA - 1543, 1976).

The enrichment plant itself generates small quantities of airborne fluorides and oxides of nitrogen and sulphur from the process cooling systems, process cleanup operations, on-site steam plant and auxiliary production facilities. These may contain low concentrations of sulphate, chloride, fluoride and nitrate, and metallic ions such as sodium, calcium, chromium and iron. Uranium losses in the effluents are very low - typically about 0.005 per cent of throughput, which is less than 8 kg per annual fuel requirement of an LWR (USAEC, 1974). Effluents are disposed of by discharge and dilution. Some sludges arising from cleanup operations are retained on site.

<u>Effluents from fuel manufacture and fabrication</u>. The effluents from fuel manufacture and fabrication with greatest potential environmental impact are chemical in nature (USAEC, 1974). Hydrogen fluoride is the most significant airborne chemical effluent from fuel fabrication. The fluorine used for manufacture of UF_6 becomes a waste product during production of enriched UO_2 powder. The gaseous fluoride wastes generated are removed from the air stream by scrubbing and filtration systems.

Liquid effluents from fuel manufacture contain nitrogen compounds formed from ammonia in the production of UO_2 powder and by nitric acid in the scrap recovery operations. Water from the air scubber systems is combined with liquid wastes and the fluoride content precipitated with lime as calcium fluoride (CaF_2) which is filtered and stored on site. The annual fuel requirement for an LWR results in the production of about 26 tonnes of solid CaF_2, occupying a volume of about 8 m^3; this contains a total of about 9 GBq of uranium (USAEC, 1974). The small percentage of fluorine not removed by lime treatment is diluted and discharged. Very small quantities of uranium are released with the effluent gases and liquids.

An alternative dry route for conversion of UF_6 into UO_2 powder has been developed but is not in major use. This process involves fluid bed hydrolysis-reduction techniques and could substantially reduce, gaseous liquid and solid effluents by elimination of the need for ammonia and by recovery of fluoride in a marketable form (USNRC, 1976). However, the environmental impact of fuel fabrication is at present minor compared with that from mining and milling.

The Non-Human Environment

The principal concern with releases from uranium fuel cycle operations has been with radioactivity and its effects on human health. The effects of radioactivity on other species in the biosphere is expected to be much less significant. Other effluent releases occurring at various stages of the fuel cycle may affect significantly not only human health but also animals, plants and materials. The total costs from these impacts should be considered also.

The major non-radiological impact due to emissions from the fuel cycle arises from the use of electricity, most of which is consumed during the enrichment stage. According to the GESMO study the amount of electrical energy required by the fuel cycle stages from mining through fuel fabrication is 38 MWe y per GWe y generated. In the near future a large fraction of this energy will be produced by coal-fired stations. (The GESMO study assumes that two thirds of the electricity generated from 1975 to 2000 will be from coal-fired stations).

This review is not an appropriate place to describe in detail the environmental impact of coal-fired power stations. However, if electricity production by an LWR has a much lower health and environmental cost than does coal, as is suggested by Lave and Silverman (1976), the generation of electricity by coal-fired stations will be a significant source of the total environmental cost of the whole fuel cycle.

Other pollutants which are released as a result of chemical processes in the fuel cycle are potentially of local significance. Their release must be examined on an individual case basis where local environmental conditions will be most important. Waste heat released to water bodies arises mainly from the enrichment process and amounts to approximately 5 per cent of that from power production.

Regions bearing commercial grade uranium ore often contain other toxic heavy metals which can leach and produce environmental degradation (Davy, 1975). The release of toxic metals is a problem common to most metal mines.

The Human Environment

Usually the most restrictive measure of environmental impacts due to releases of radioactivity is the effect on individual human health. Effects on other organisms are judged in terms of damage to whole populations rather than risk to individuals.

Health effects may be transient of permanent, localised or global in their impact, trivial or fatal, occurring immediately or delayed perhaps for many generations after the operation has ceased. Health effects can be expressed in absolute terms, such as estimated numbers of premature deaths or of serious disability. Premature deaths and serious disabilities are however not all equivalent. For example, the average age at death from occupational accidents is less than that from occupational disease. Ideally, when health effects are summed, each should be weighted according to its environmental cost. However, weighting factors are controversial and in this review health effects are restricted to numbers of premature deaths.

Premature deaths or disability may occur in small populations during the present generation or they may be inferred to result over many centuries in the global population. For the same number of effects the risk to individuals is much smaller in the second case than in the first. Although they are not equivalent both estimates have their uses in comparative studies. The calculation of future health effects requires assumptions not only of pollutant releases but also future population sizes, medical practices, and use of the environment.

The assessment of radioactive releases from electricity production can be expressed in terms of annual dose rate, dose commitment or collective dose commitment per unit of electrical energy generated. The dose commitment from a given source is the integral over infinite time of the <u>per caput</u> dose rate. The collective dose commitment is the integral over the population at risk as well as over infinite time. In addition, incomplete dose commitments, resulting from integrals over a finite time, have been used (USEPA, 1973; UNSCEAR, 1977).

The estimates of radiation exposure can be converted to estimates of risk of cancer or genetic damage through the use of the assumption that risk is proportional to exposure (the linear hypothesis). In this review, unless otherwise stated, the proportionality factors (risk coefficients) recommended by the International Commission on Radiological Protection are used (ICRP, 1977). The ICRP factors are consistent with the available information reviewed by UNSCEAR (1977). ICRP warns that in the choice of alternative practices, radiation risk estimates should be used only with great caution and with explicit recognition of the possibility that the actual risk at low doses may be lower than that implied by a deliberately cautious assumption of proportionality. This warning is even more applicable where comparisons involve detriments from different pollutants.

Since the hereditary detriment is likely to be less than the detriment due to somatic injury (ICRP, 1977), it is sufficient for present purposes to consider somatic injury only in this review.

<u>Occupational Health</u>

Workers in any industry are at risk of death or disability from accident or from disease caused by exposure to noxious agents. Fatal accident rates vary markedly from industry to industry, and to a lesser extent, from country to country. An average over all industries is of the order of 10^{-4} per person per year (Pochin, 1974). In some industries death rates from occupational disease exceed those from fatal accidents. Pochin (1976) has reviewed available information on risks to workers from all sources, including radiation for the various stages of the nuclear fuel cycle, while UNSCEAR (1977) has reviewed the information on the radiation exposures. The GESMO study has presented information on occupational dose commitments for their model industry over the period 1975 to 2000. These dose commitments and the inferred number of cancer deaths based on ICRP proportionality factors are given in Table 4.

Table 4

OCCUPATIONAL DOSE COMMITMENTS AND INFERRED CANCER MORTALITY PER GWe y

	Mining	Milling	Conversion	Enrichment	UO_2 Fuel Fabrication
	Person sieverts per GWe y*				
Total Body	2.6	1	0.01	0.007	0.1
G.I. Tract	2.6	0.5	0.007	0.004	0.1
Bone	3.7	4.8	0.1	0.06	0.2
Liver	2.6	0.5	0.01	0.007	0.1
Kidney	3.7	0.6	0.03	0.02	0.1
Thyroid	2.6	0.5	0.01	0.007	0.1
Lung	13.7	10.2	0.008	0.1	4.5
Skin	2.6	0.5	0.03	0.03	0.1
	Inferred cancer mortality deaths per GWe y**				
	0.08	0.04	0.0004	0.0004	0.01

* (After NUREG-0002, 1976).
** Estimated from ICRP risk coefficients: (ICRP, 1977).

The ratio of disabilities of fatalities is smaller for radiation induced than for most occupational health effects. The use of fatalities only as a measure of health cost for comparative purposes therefore overestimates the importance of radiation.

Mining. Pochin estimated the annual average dose to uranium miners from external sources of radiation to be 5 mSv per year, while UNSCEAR used 10 mSv per year. These estimates correspond to a risk of death from radiation induced cancer of $(0.5\text{-}1) .10^{-4}$ per person per year. Pochin assessed the lung cancer risk from exposure to radon and daughters in underground uranium mines as 1.5×10^{-4} per person per year, while UNSCEAR gave a figure for the annual average exposure of underground uranium miners in France and the US as 1.3 and 1.9 WLM per year, respectively. (The WLM, or working level month, is a unit of exposure to radon decay products in air.) UNSCEAR, reviewing data on underground miners, indicates that a lung cancer mortality risk of 2.10^{-4} per WLM per person is a representative figure for miners. Use of this number with the annual average exposure of 1.3-1.9 WLM per year gives a figure for lung cancer mortality of about 3.10^{-4} per year. The total inferred cancer mortality risk from external radiation exposure and from radon decay product exposure for underground miners is in the range $(2\text{-}4).10^{-4}$ per person per year.

Pochin assessed the risks of death from accidents and from ill health resulting from exposure to dust and silica in underground mines as 1.5×10^{-3} and 5×10^{-4} per person per year, respectively. These figures are based on past experience and it is to be expected that attention to industrial safety can reduce these risks. Taken together the figures are significantly larger than the inferred radiation induced cancer mortality. The total mortality risk from all sources for underground miners is estimated to be of the order of $(2-3).10^{-3}$ per person per year. Fatal accident rates in open cut mining are less than in underground mines. However, actual estimates of total risks in open cut mining are not available.

In order to estimate the contribution of occupational deaths to the total detriment per unit of electricity produced, Pochin assumes 200 miner years per GWe y, while UNSCEAR assumes 50, which is more realistic for current mines. The Pochin assumption applied to the estimate above of individual risk gives a figure of 0.4–0.5 deaths per GWe y, while the UNSCEAR assumption gives about 0.1 deaths per GWe y. The contributions from radiation exposure in these numbers are 0.04 and 0.02 deaths per GWe y respectively, compared with 0.08 estimated from the exposure data in GESMO.

Milling. Data on fatality rates from accidents or disease in uranium milling plants are not readily available, but indications are that they are less than in underground mining (Archer et al., 1973). Neither Pochin nor UNSCEAR gives estimates of radiation doses from milling operations; both assume that the contributions to the total detriment per unit of electricity generated are small compared with those from mining. GESMO estimates of the radiation dose from milling indicate an inferred cancer mortality of 0.04 deaths per GWe y. Pochin suggests a figure for the contribution from fatal accidents in milling operations of 0.1 deaths per GWe y.

Hexafluoride Production, Enrichment and Fuel Fabrication. UNSCEAR gives figures for whole body doses in the UK and US for uranium upgrading and fuel fabrication. These average around 3 mSv per person per year and give a collective dose per unit of electricity produced of 1.5 person sieverts per GWe y. From these figures the inferred cancer mortality risk is 3.10^{-5} per person per year and the detriment 0.015 deaths per GWe y. The inferred number of deaths per GWe y based on the GESMO estimates of dose are 0.0004, 0.0004 and 0.01 for hexafluoride conversion, enrichment and fuel fabrication, respectively. If the fatal accident rate averaged over all industries of 10^{-4} per person per year applies to these stages of the fuel cycle, the detriment per unit of electricity from fatal accidents is around 0.05 deaths per GWe y.

Enrichment in particular requires large quantities of electricity, which may be provided by coal-fired power stations. If two thirds of the electricity is provided by coal-fired power stations and if the occupational risk of producing this electricity is 0.5–5 deaths per GWe y (Comar and Sagan, 1976), it follows that electricity use, primarily in enrichment results in another 0.01 to 0.1 deaths per GWe y. Although occupational exposure to radiation may be increased through the use of gas centrifuge technology (USNRC 1976) this will be more than counteracted by the smaller number of occupational fatalities which result from reduced electricity consumption.

Construction and Transportation. Pochin, using a fatal accident rate of 8.10^{-4} per person per year for construction workers deduces a detriment from construction work for the whole nuclear fuel cycle of 0.25 accidental deaths per GWe y. This review arbitrarily assigns 40 per cent of these deaths to the construction operations required for the fuel cycle stages under consideration, i.e. 0.1 deaths per GWe y in total. Transport presents risks of fatal accidents and from radiation exposure. Pochin assesses the detriment from fatal accidents in transport for the whole nuclear fuel cycle as 0.003 deaths per GWe y, and the collective dose as 0.3 person sieverts per GWe y, i.e. an inferred cancer mortality of 0.003 deaths per GWe y. UNSCEAR gives two estimates of the collective dose from the transport of unirradiated fuel, the figures being 10^{-4} and $3x10^{-3}$ person sieverts y, i.e. an inferred cancer mortality of less than 10^{-4} deaths per GWe y. Pochin's figures cover more of the fuel cycle than is being considered here and the UNSCEAR figures less. This review adopts detriments of 0.001 fatal accidents and 0.001 inferred cancer deaths per GWe y resulting from transport operations associated with the stages from mining to fuel fabrication.

Summary of Occupational Detriment. The total detriment associated with the stages from mining to fuel fabrication per unit of electricity produced is in the range 0.4 to 0.9 deaths per GWe y, with 0.04 to 0.13 of these being inferred deaths from radiation induced cancer. Underground uranium mining is the largest single contributor, the total detriment is estimated to be in the range 0.1 to 0.5 per GWe y, with an inferred radiation induced cancer component of 0.02 to 0.08 deaths per GWe y.

Public Health.

The calculation of health effects due to the release of radioactive effluent from fuel cycle operations requires a model relating effluent release to exposure of humans. The GESMO study, based on the US nuclear industry from 1975 to 2000, considered the release of effluents from model facilities which were composites of a number of technologies. Routes leading to exposure from airborne dispersion which were considered by GESMO were inhalation, shortly after release and after resuspension of deposited material, dietary intake of deposited material and external irradiation from submersion in a radioactive plume and from material deposited on the ground. Airborne radon was assumed to be liberated free from its daughters which were allowed to grow in with time. Doses were also computed for hydrologic pathways; drinking water, consumption of fish, and also vegetable and animal products produced on irrigated ground.

Dose commitments to the general (US) public presented in GESMO are given in Table 5. The dose commitments represent a 50 year dose commitment received by the population from mining to fabrication per GWe y of nuclear power. Isotopic concentrations used in the GESMO calculations were those which would exist in the final year of a 40 year plant lifetime. The inferred cancer mortalities in Table 5 are based on ICRP risk coefficients (ICRP, 1977).

An alternative approach for the naturally occurring isotopes which are released in those sections of the fuel cycle being reviewed is to estimate dose commitments by assuming that the isotopes released behave similarly to their naturally occurring counterparts. This method has been used by UNSCEAR. Results based on the UNSCEAR approach are compared with those reported in GESMO.

Mining. Early assessments of the impact from mining either under-estimated or ignored radon releases (USAEC, 1974; US EPA, 1973). The airborne wastes are considered by GESMO and UNSCEAR to represent the major exposure route, and result from radon-222 and, to a lesser extent, dust which are liberated as a result of mining operations. UNSCEAR, citing US data (Sears et al., 1975), reports a radon release for mining and milling of 6.3 TBq per GWe y. The GESMO study, using more recent data, estimated that the radon release during mining would be about 220 TBq per GWe y. This figure was calculated from measurements of radon concentrations in underground mines. A release of 220 TBq per GWe y is used here as the best estimate for radon emanation although it is higher than other estimates (USAEC, 1974; UNSCEAR, 1977) including recent upper limit

Table 5

COLLECTIVE DOSE COMMITMENTS TO PUBLIC AND INFERRED CANCER MORTALITY PER GWe y

	Mining	Milling	Conversion	Enrichment	UO_2 Fuel Fabrication
	Person sieverts per GWe y*				
Total Body	6.3	1.2	0.1	2×10^{-4}	0.006
G.I. Tract	0.3	0.05	0.01	0.01	0.006
Bone	20	4	0.2	0.002	0.1
Liver	5	1	0.002	4×10^{-5}	1×10^{-5}
Kidney	24	4.5	0.3	0.001	0.02
Thyroid	0.02	0.003	1×10^{-4}	5×10^{-5}	1×10^{-5}
Lung	2	0.4	0.002	0.001	4×10^{-4}
Skin	0.02	0.003	0.08	5×10^{-5}	1×10^{-5}
	Inferred cancer mortality deaths per GWe y**				
	0.09	0.02	0.001	3×10^{-5}	1×10^{-4}

* After NUREG-0002 (1976)
** Estimated from ICRP risk coefficients: (ICRP, 1977).

calculations for the Ranger mine (Ranger Inquiry, 1976).

In calculating dose commitments UNSCEAR assumed a realistic dispersion model. Table 6 shows the collective dose commitment for radon release calculated by UNSCEAR but modified for a release of 220 TBq per GWe y and for an assumed quality factor for alpha radiation of 20 (ICRP, 1977).

Table 6

COLLECTIVE DOSE COMMITMENT TO PUBLIC AND INFERRED CANCER MORTALITY RESULTING FROM MINING (PER GWe y)

	TBq Release	Person sieverts per GWe y: Segmental Bronchioles	Lung (Whole)	Bone Marrow	Bone Lining Cells	Gonads Whole Body	Cancer Mortality
During Mining							
Radon-222	220	30	7	0.05	0.05	0.05	0.01
Lead-210	0.11		1	0.6	2	0.5	0.01
After Mining*							
Radon-222	$4\ y^{-1}$	50	10	0.09	0.09	0.09	0.02
Lead-210	$0.002\ y^{-1}$		2	1	4	0.8	0.02

* For continuing release after mining an incomplete (100 year) dose commitment has been calculated.

The release of 220 TBq of radon-222 corresponds to the release of 0.11 TBq of lead-210. UNSCEAR estimates the collective dose commitment due to the deposition of released particulates by assuming that the released nuclides, once deposited, behave similarly to their naturally occurring counterparts. The estimated collective dose commitment from the release of 0.11 TBq of lead-210 is also shown in Table 6.

The number of early fatalities estimated from the UNSCEAR based figures is approximately one fifth of those based on the GESMO model. The Nuclear Energy Policy Study Group (1977) has observed that the calculational model in GESMO appears to be conservative and therefore overestimates exposure.

After the mining operation has been completed radon will continue to be released from an openpit mine unless remedial action is taken or the pit fills with water. Radon releases from an underground mine would be very much smaller since there is no longer forced ventilation. The annual release of radon from a worked out openpit mine may be of the order of 4TBq per GWe y, (Wilde, 1978). There will also be a radon release from the waste rock. Davy (1978) has estimated that for some mines in the Northern Territory radon emissions from waste piles may be comparable to those from inactive tailings piles, but that these emissions are only a few per cent of the above figure. It is impossible to estimate how this release and its significance will vary with time. Because of the uncertainties of projections into the future UNSCEAR is followed and an incomplete dose commitment is calculated. This incomplete dose commitment is calculated for 100 years during which time if there were no abatement 400 TBq of radon would be released. The incomplete dose commitment based on UNSCEAR calculations is included in Table 6.

Table 6 also shows the inferred health effects both from the initial release and from the continuing release. The inferred cancer mortality from the initial release of radioactivity is 0.02 deaths per GWe y compared with an estimate based on GESMO data of 0.09. If radon release continued unabated the annual death rate is estimated to be 4×10^{-4} deaths per year per GWe y. The dose commitment from dust released during the mining operation will be very site specific and is not included.

Global dose commitments from the release of airborne mine and milling effluents have been ignored by UNSCEAR and GESMO. Because of the 3.8 day half-life of radon-222 and the 10 day atmospheric half-life of the particles on which the radon daughters attach themselves the global dose commitment is not expected to be greater than 20 per cent of the national commitment given in Tables 6 and 7. Collective dose commitments resulting from inhalation depend on population density whereas those from ingestion depend on agricultural production. For the Northern Territory in Australia inferred health effects will be very much smaller than for the US because of the lower population densities and limited agricultural production.

Milling. If available technology for the optimum in management of liquid wastes is used, collective dose commitments from milling will depend almost entirely on airborne releases. During milling radon-222 is released as a result of the crushing, grinding and chemical operations and there is also a continuing release from the tailings. This release from tailings depends very much on their physical state. In particular the release from tailings which are covered with water is very much less than from those which have dried. After mining and milling have ceased radon will continue to be generated and released from the tailings as a result of the decay of radium-226.

According to UNSCEAR the radon release from a uranium mill and active tailings area is 6.3 TBq per GWe y. Table 7 summarises estimates of radon releases from a number of mines and mills. These values are less than the estimate in the GESMO study of 41 TBq per GWe y which has been supported in recent testimony (Magno, 1978) and is accepted for the present review. Table 8 shows the collective dose commitment and inferred health effects (calculated in the same way as those for mining) which result from this release.

Table 7

CONTRIBUTION OF URANIUM MINING AND MILLING OPERATIONS TO DAILY RADON DISCHARGE RATES

	Ranger (Davy, 1976)	Highland (USAEC, 1973)	Shirley Basin (USAEC, 1974)	Bear Creek (USAEC, 1977)	USAEC (1974)	ERDA-1541 (1976)
Plant t U/d	7.6	3.4	2.5-3.1	1.2	~2.6	2.5
Capacity t ore/d	3000	2000	1500-1800	1000	1450	1600
	Radon Discharge G Bq d^{-1}					
URANIUM MINING	270			750	Nil assumed	560
URANIUM MILLING						
Ore dumps and stockpiles	96	*	96	11	)	17
Crushing plant	81	*	19	*	) 48	27
Treatment plant	63	*	*	*	)	*
Tailings retention system	140	27-44	48	160	)	37
Milling Total	380	27-44	160	170		81

* Not stated.

If the tailings become dry after milling has ceased it is estimated that provided no remedial action is taken to reduce radon emanation the continuing radon release would be about 4 TBq per year per GWe y. Incomplete, 100 year dose commitments, for this release are included in Table 8. Despite the inherent uncertainties inferred health effects in some instances have been integrated over the lifetime of thorium-230 (UNSCEAR, 1977; Pohl, 1976; Fry and Cook, 1976). There is a difference between the values in Table 8 and those estimated from GESMO. The inferred cancer mortality from the release of radon during the mine life given in Table 8 is 0.004 deaths per GWe y compared with the GESMO value of 0.02. If radon emanation occurred unabated the inferred cancer death rate would be 4×10^{-4} deaths per year per GWey. (In the US an NRC objective (Magno, 1978) is that radon releases from inactive tailings piles should be no more than twice the background flux in which case the inferred death rate would be approximately one hundredth of this rate). Radioactive dust is also released during milling. Releases given in GESMO are somewhat larger

Table 8

COLLECTIVE DOSE COMMITMENT TO PUBLIC AND INFERRED CANCER MORTALITY RESULTING FROM MILLING (PER GWe y)

		PERSON SIEVERT PER GWe y					
	TBq Release	Segmental Bronchioles	Lung (Whole)	Bone Marrow	Bone Lining Cells	Gonads Whole Body	Cancer Mortality
During Milling Radon-222 (+ Lead-210)	41	6	1	0.1	0.5	0.1	0.004
Dust*:							3×10^{-4}
Uranium-238	4.8×10^{-3}		0.03	0.005	0.018		
Thorium-230	1.2×10^{-4}		0.01	0.007	0.13		
Radium-226	1.2×10^{-4}		0.007	0.007	0.007	0.007	
Lead-210	1.2×10^{-4}		0.001	7×10^{-4}	0.003		
After Milling* (no radon abatement)	$4 y^{-1}$	50	10	1	4	0.9	0.04

* Incomplete (100 year dose commitment)

than those in UNSCEAR. The uranium-238 release is 6 times larger than that estimated for the Ranger mine and the radium release about half. The estimated incomplete collective dose commitments from dust releases are also given in Table 8. The inferred health effects due to dust releases are much less than those from radon releases.

Hexafluoride production, enrichment and fuel fabrication. Doses incurred by the general public through hexafluoride production, enrichment and fuel fabrication are smaller than those from mining and milling. Estimated dose commitments calculated by GESMO are given in Table 5. These correspond to inferred cancer mortalities per GWe y of 0.001, 3×10^{-5} and 1×10^{-4} for hexafluoride production, enrichment and fuel fabrication respectively, approximately 5 times greater than would be estimated from use of UNSCEAR methods.

Since enrichment in particular requires large quantities of electricity the public health effects resulting from the use of this electricity also need to be considered. If two thirds of the electricity requirement is provided by coal fired power stations (USNRC, 1976) and if the public health risk consequences of producing this electricity are between 2 and 120 deaths per GWe y (Comar and Sagan, 1976) it follows that electricity use, mostly in enrichment, results in another 0.05 to 3 deaths per GWe y. (This estimate depends on the interpretation of observations of the impact of air pollution on human health and also on the location of generating plant relative to population centres. An estimate from papers presented to UNEP on the generation of electricity from coal should be used for these premature deaths. However, premature deaths from air pollution due to coal-fired generating plant are not directly comparable with those caused by radiation).

ALTERNATIVE FUEL CYCLES

Recycling Uranium and Plutonium

This review has considered the environmental impact of an LWR fuel cycle in which neither the uranium nor the plutonium is recycled. If fuel is reprocessed and uranium is recycled the GESMO study predicts that for the uranium fuel cycle the environmental impact from mining and milling would be reduced by 10.5 per cent, from hexafluoride production would be reduced by 10.6 per cent, from enrichment would be increased by 0.9 per cent and from uranium dioxide fuel fabrication would be unchanged. If both plutonium and uranium are recycled in LWRs the environmental impact from mining and milling would be reduced by 22.5 per cent, from hexafluoride production by 24.3 per cent, from enrichment by 14.0 per cent and from uranium dioxide fuel fabrication by 13.8 per cent. Recycling of uranium and plutonium would, however, result in new environmental impacts due to reprocessing and mixed oxide fuel fabrication.

The HWR Fuel Cycle

Heavy water reactors have been predicted to contribute between 4 and 8 per cent of nuclear power generated between 1977 and 2000 (Table 1). Heavy water reactors are fuelled with natural uranium dioxide. Generation of one GWe y of electricity by an HWR requires about 85 per cent of the uranium mining and milling requirement and about 400 per cent of the fuel fabrication requirement of an LWR without recycle. Manufacture of uranium hexafluoride and uranium enrichment operations are not required. The fuel fabrication operation will however contain additional effluents from purification of uranium which normally is effected at the hexafluoride stage. The radiological impact of the HWR cycle to fuel fabrication would be controlled by the uranium mining milling and fuel fabrication components and appears to be about equal to that for an LWR without recycle.

The GGR Fuel Cycle

Table 1 shows that the fraction of nuclear power generated in the western world by gas graphite reactors is predicted to fall from nearly 7 per cent in 1977 to about 0.2 per cent in 2000. GGRs are fuelled with natural uranium metal. Generation of one GWe y of electricity in a GGR has about a 30 per cent greater uranium mining and milling requirement and about nine times the fuel fabrication requirement of an LWR. The greater requirements for uranium mining, milling and fuel fabrication for the GGR suggests that the GGR fuel cycle to fuel fabrication would have a radiological impact about 80 per cent greater than that for an LWR without recycle for the same amount of electricity generated.

The Thorium-Uranium Fuel Cycle

A partial environmental assessment has been made(USERDA-1541, 1976) of the individual stages of a conceptual light water converter and breeder (LWBR) fuel cycle. The use of natural resources and the quantities of effluents produced and their concentrations in atmospheric and surface water discharges from these operations were assessed, but radiological and health effects were not reported. The environmental impact of the mining to fuel fabrication stages in the thorium-uranium fuel cycle would be determined by the impacts of two complementary reactor systems, the thorium converter power reactor and the thorium breeder power reactor. Converter reactors would require a uranium fuel cycle similar to that of the LWR, together with the mining and milling of thorium ores and their fabrication into breeder elements. The thorium breeder reactors would be fuelled by uranium-233 recovered by reprocessing spent fuel from converter reactors. No mining of thorium ores would be required by LWBRs for many decades after the converter reactors were decommissioned, since thorium would be recovered and stockpiled from the spent converter fuel. In the short term the thorium-uranium fuel cycle does not seem to offer environmental advantages over that of the LWR. The long term environmental advantages in fuel sufficiency, claimed for the LWBRs are inherent in any breeder concept and could be compared with the potential reduction in uranium mining from use of the western world's stockpile of depleted uranium in fast breeder reactors operating on the plutonium-uranium cycle.

ADDITIONAL WASTE TREATMENT

A number of studies correlating the cost of waste treatment alternatives and the environmental impacts of effluents for the early stages of the uranium fuel cycle have been completed (Sears et al., 1977; 1975; Pechin et al., 1975). A cost-benefit study (Sears et al., 1975) regarding the effectiveness of additional waste treatment equipment at milling plants for the reduction of radioactive releases and of cost commitment of the discharges to the environment reviewed both acid leach and carbonate leach processes and considered short and long term radiological releases. This study concluded that the greatest dose reduction and most effective cost benefit, expressed as $US per sievert reduction in dose, could be obtained by the following actions:

1. Minimising airborne dispersal of tailings materials by covering exposed tailings during the life of the mill and rehabilitating the tailings retention system with a permanent stabilised earth cover after cessation of milling operations;

2. use of more efficient dust collectors on ore dust streams at mills processing ores containing less than 6 per cent of moisture;

3. use of more efficient dust collectors on yellowcake dust streams at alkaline leach mills. (The alkaline leach process can produce a yellowcake product containing up to 2 per cent of the radium-226 originally present, a factor of 20 greater than in the acid leach method);

4. minimising the long term release of radon-222 from stored tailings by covering them with thick layers of earth.

The report noted that potentially up to 99 per cent of radon in milling off-gasses could be removed on beds of activated charcoal, but more laboratory and operating data on charcoal absorbers would be needed for design of large scale systems. The quantity of radon released from the mill is small by comparison with that from stored tailings. There was scope for improvements in scrubbers and filters for reduction in discharges of ore and yellowcake dusts. The moisture content of ores is a controlling influence in dust release; an eight fold reduction in dust release occurred when moisture content was increased from 6 to 8 per cent.

Disposal of the solid tailings and liquid mill effluent constitutes the main radiological impact from uranium milling. These contain about 70 per cent of the radioactivity present in the original ore (UNSCEAR, 1977). Significant impacts from stored tailings can be caused by emission of radon-222, and discharge of radium-226 by wind dispersal of solid tailings or seepage of effluent water, containing dissolved radium and other radionucleds from tailings storage structures. During the operating life of the mill, the rate of emission of radon-222 from tailings is estimated to be reduced by about 90 per cent by coverage with about 60 cm of water (Fry, 1975). After conclusion of operations, coverage of the tailings with an earth layer 3-7 m thick could reduce the rate of radon-222 emission by 80 to 98 per cent (Sears et al.,1975) in addition to stabilising the pile against wind erosion. Asphalt has been suggested as an alternative diffusion barrier for radon: a layer 8 mm thick is equivalent in efficiency to nearly 7 m of earth; however the integrity of such a layer in the long term needs appraisal. Reduction in the impact of seepage from tailings systems can be achieved by lining the base and walls of storage embankments with bentonite clays of low permeability and by chemical treatment of effluents to precipitate the majority of radium and other radionuclides. Neutralisation of acidic tailings effectively reduces concentrations of dissolved minerals, some radioactivity and amine solvent which could otherwise be present in seepage water. Tailings neutralisation has been compulsory in Ontario, Canada, since 1960 because of the wet environment but is not widely practised in the US where mills retain tailings at pH 2 and rely on sandstone to neutralise and precipitate potential contaminants in seepage. Barium chloride has been reported to be effective for removal of radium-226 and is being used by two mills in the US and all mills in Ontario for treatment of overflow waters from tailings sites (Sears et al., 1975).

The dose commitments from emission of radon-222 and seepage of dissolved radium-226 from tailings can be reduced in the first few thousand years by removal of the radium by leaching with sodium chloride solution or by successive leaching with nitric acid. Sodium chloride leaching was inefficient for Ambrosia Lake, US, tailings. After leaching the extracted radium would be concentrated and converted into an immobile form possibly for deep geologic burial. For this treatment to have a lasting effect may require removal of thorium-230 which has a half life of over 80,000 years, and which is the precursor of radium-226. It seems impracticable to apply radium and thorium removal processes to existing abandoned tailings.

The success of tailings stabilisation by revegetation has been reviewed (Sears et al., 1975). The long term erosion characteristics of an above ground retention system are uncertain, and it was recommended (Ranger Inquiry, 1976) that tailings from development of the Ranger Uranium deposit in the Australian Northern Territory should be temporarily stored above ground in an engineered structure and returned to the opencut for coverage and revegetation when mining had ceased. The feasibility of this has not yet been established.

The effectiveness of alternative waste treatment practices in uranium hexafluoride production on reduction of population radiological dose commitments has been assessed (USEPA, 1973, Sears et al., 1977). Treatment practices considered included improvements to extract dust scrubbers and filters and treatment of liquid effluents to remove radium and to isolate solid wastes by burial, possibly by return to mill tailings sites.

The radiological dose commitment from enrichment operations is a very small fraction of that for the fuel cycle (Table 5) and no cost benefit study has been published of the effectiveness of alternative waste treatment practices.

SUMMARY

The environmental effects of the uranium fuel cycle from mining to the fabrication of uranium dioxide fuel have been examined. Actual effects will depend on conditions specific to the actual facilities. However a generic estimate of health effects, typical of the US industry, has been given and is summarised in Table 9.

Table 9

INFERRED PREMATURE FATALITIES PER GWe y

	Cause of Fatalities			Total
	Radiation	Accidents	Other	
OCCUPATIONAL				
Mining	0.02-0.08	0.08-0.3	0.03-0.1 (a)	0.1-0.5
Milling	<0.01-0.04	0.1		0.1-0.15
Hexafluoride conversion, Enrichment, UO_2 Fuel fabrication	0.011	0.05	0.01-0.1 (b)	0.07-0.16
Construction		0.1		0.1
Transport	0.001	0.001		0.002
Total	0.04-0.13	0.33-0.55	0.04-0.2	0.4-0.9
PUBLIC (c)				
Mining (d)	0.02-0.09			0.02-0.09
Milling (d)	0.004-0.02			0.004-0.02
Hexafluoride conversion	0.001			0.001
Enrichment	3×10^{-5}		0.05-3 (b)	0.05-3
Fuel fabrication	1×10^{-4}			1×10^{-4}
Total	0.025-0.11		0.05-3	0.075-3.1

(a) Diseases not related to radioactivity.
(b) From electricity production by coal-fired power stations.
(c) Lower values based on UNSCEAR, higher values on GESMO.
(d) Continuing release after operation finished is assessed in text.

The number of premature deaths from occupational causes is estimated to be between 0.4 and 0.09 per GWe y of electrical energy produced. Of this number between 0.04 and 0.13 arise from cancer associated with radiation exposure and between 0.3 and 0.6 from accidents at fuel cycle facilities. The reminder result from diseases and accidents associated with the generation of electricity from coal (0.01 to 0.1) and from diseases due to uranium mining but not associated with radiation exposure (0.03 to 0.1).

The number of premature deaths among the general public per GWe y is estimated to be between 0.025 and 0.11 from cancer associated with radiation exposure and between 0.05 and 3 from the production of electricity from coal, mainly for enrichment.

Mining and milling are, for the section of the fuel cycle reviewed, the main sources of radiation exposure, and by inference, of cancer for both occupational workers and the general public. If no attention is given to reducing emanation of the radioactive gas radon from mines and mills which have finished production and continuing number of inferred cancer mortalities from this cause is 8×10^{-4} deaths per year per GWe y.

Radiological health effects for the early stages of other non-breeder fuel cycles will be comparable to those of the LWR but will be less for breeder fuel cycles because of the reduced impact of mining and milling.

REFERENCES

Archer, V.E., et al. - Cancer mortality among uranium mill workers, J. Occup. Med. 15, 11, (1973).

Comar, C.L. and L.A. Sagan, Health effects of energy production and conversion, Annual Review of Energy 1,581 (1976).

Commonwealth Department of Health: Code of practice on radiation protection in the mining and milling of radioactive ores, Canberra, ACT, (1975).

Davy, D.R. - Environmental aspects of uranium mining and milling in Australia. Pacific Basin Conference 'Nuclear Power Development and the Fuel Cycle', ANS Honolulu, October 1976.

Davy, D.R. (Ed.) - Rum Jungle environmental studies, AAEC/E365, (1975).

Davy, D.R. Unpublished results 1978.

Frank, J.N. - Cost model for solution mining of uranium, US Energy Research and Development Administration, (1976).

Fry, R.M. and J.E. Cook - Comment on paper by Pohl, R.O., Search, 7, 351 (1976).

Fry, R.M. - Radiation hazards in uranium mining and milling, Atomic Energy in Australia, 18, October (1975).

IAEA: Management of wastes from the mining and milling of uranium and thorium ores, Safety Series No.44, (1976).

ICRP - Recommendations of the International Commission on Radiological Protection, ICRP Publication 26, Pergamon, Oxford (1977).

Lave L.B. and L.P. Silverman: Economic costs of energy-related environmental pollution, Annual Review of Energy 1, 601 (1976).

Levins, D.M., et al. - Leaching of radium from tailings. OECD Seminar on Management, Stabilisation and Environmental Impact of Uranium Mill Tailings, Albuquerque, New Mexico, 24-28 July 1978.

Magno, P.J. Affidavit presented to Perkins (Duke Power) Hearings (1978).

Merritt,R.C. - The extractive metallurgy of uranium. Colorado School of Mines Research Institute (1971).

The Nuclear Energy Policy Study Group - Nuclear power, issues and choices, report sponsored by the Ford Foundation and administered by the MITRE Corporation, Ballinger Publishing Co., Cambridge, Mass.,(1977).

OECD/NEA- Uranium resources, production and demand, Paris December, (1977).

Pancontinental Mining Limited - The Jabiluka project, draft environmental impact statement, (1977).

Pechin, W.M. et al. - Correlation of radioactive waste treatment costs and environmental impact of waste effluents in the nuclear fuel cycle for use in establishing 'as low as practicable' guides - fabrication of light water reactor fuel from enriched uranium dioxide, ORNL-TM-4902, (1975).

Pochin, E.E. - Estimated population exposure from nuclear power production and other radiation sources, NEA/OECD, Paris (1976).

Pochin, E.E. - Occupational and other fatality rates, Community Health 6, 2, (1974).

Pohl, R.O. - Health effects of radon-222 from uranium mining, Search, 7, 345 (1976).

Poole, L.G. - World uranium resources. Nuclear Engineering International, 20, 95 (1975).

Ranger Uranium Mines Pty Ltd - Environmental impact statement, (1974).

Ranger Uranium Environmental Inquiry - Second Report, Australian Government Publishing Service, Canberra, ACT (1976).

Ryan, R.K. and P.G. Alfredson - Laboratory studies of the treatment of liquid waste streams from uranium milling operations. Proc. Aust. Inst. Min. Met., No. 253, March (1975).

Sears, M.B., et al. - Correlation of Radioactive waste treatment costs and the environmental impact of waste effluents in the nuclear fuel cycle for use in establishing 'as low as practicable' guides - milling of uranium ores, ORNL-TM-4903, Vol. 1, (1975).

Sears, M.B.,et al. - Correlation of radioactive waste treatment costs and the environmental impact of waste effluents in the nuclear fuel cycle - conversion of yellowcake to uranium hexafluoride, Part 1, the fluorination-fractionation process, ORNL/NUREG/TM-7, (1977).

Tsivoglou, E.C. and R.L. O'Connell: Waste guide for the uranium milling industry. US Department of Health, Education and Welfare, Public Health Service § Division of Water Supply § Pollution Control (1962).

United Nations Scientific Committee on the Effects of Atomic Radiation - Sources and effects of ionizing radiation, United Nations, New York (1977).

USAEC - Final environmental statement related to operation of the Highland uranium mill, Docket No. 40-8102, (1973).

USAEC - Final environmental statement related to operation of Shirley Basin uranium mill, Docket No. 40-6622, (1974).

USAEC - Final environmental statement related to operation of Bear Creek uranium mill, Docket No. 40-8452, (1977).

USAEC - Fuels § Materials Directorate of Licensing - Environmental Survey of the uranium fuel cycle, WASH-1248 (1974).

US Energy Research § Development Administration - Final environmental statement, light water breeder reactor program, environmental assessment of a conceptual thorium-uranium fuel cycle, ERDA-1541, Vol.4 (1976).

US Energy Research and Development Administration - Final environmental statement, expansion of US uranium enrichment capacity, ERDA-1543, (1976).

US Energy Research and Development Administration - Final environmental impact statement, Portsmouth gaseous diffusion plant site, Piketon, Ohio, ERDA-1555, (1977).

US Environmental Protection Agency - Environmental analysis of the uranium fuel cycle, Part 1, fuel supply, EPA-520/9-73-003-B (1973).

US Environmental Protection Agency - State of the art - uranium mining, milling and refining industry, National Environmental Research Centre, Office of Research and Development (1974).

US Nuclear Regulatory Commission - Final environmental statement related to the Sequoyah uranium hexafluoride plant, Kerr-McGee Corp., NUREG-75/007, (1975).

US Nuclear Regulatory Commission - Final generic environmental statement on the use of recycle plutonium in mixed oxide fuel in light water cooled reactors, NUREG-0002, Vol.3, Health, Safety § Environment, (1976).

Westinghouse Electric Corporation - Applications and supplements, USAEC Docket No. 70-1151.

Wilde, R.M. - Testimony Perkins (Duke Power) Hearings (1978).

3

THE ENVIRONMENTAL IMPACTS OF NUCLEAR POWER PLANTS

Essam E. El-Hinnawi
Chairman of Energy Task Force
United Nations Environment Programme
P.O. Box 47074, Nairobi, Kenya

INTRODUCTION

Like any other power station, the interaction between nuclear power plants and their surroundings has recently received considerable attention. The impact that a power station may have on the environment depends to a large extent on its location with respect to the load centre, populated areas, open water, agricultural land, etc.

For nuclear power plants, siting considerations are stringent because of the release of radioactive substances into the environment during normal operation, potential release in accident situation, and post-operational phase when decommissioning may be undertaken. The selection process must satisfy the technical, socio-economic and environmental safety requirements which are most important of the various factors considered for site studies. The methodology for site selection involves an iterative approach to test various alternatives as to location, type of power plant; cooling system, land requirements and exposure of the public to radiation from normal operation and potential exposure in the case of accidents, or radioactive effluents arising therefrom. After the preliminary screening, there may remain only a few alternative sites which must then be evaluated using weighting factors (Site Selection Criteria - see Fischer and Ahmed - 1974). The "best available" site will rarely be ideal in all aspects and in most cases the result of balancing the benefits against the cost and detriments the society will be prepared to accept, will determine the suitable site. A reactor site should, therefore, be seismically acceptable, have an adequate supply of cooling water available and be easily accessible.

---

* Permanent address:

Research Professor of Energy and Natural Resources
National Research Centre
Dokki, Cairo
Egypt

NON-RADIOLOGICAL IMPACTS*

Land Requirements:

For nuclear power plants, land requirements may vary from plant to plant depending on the situation of the plant with regard to open water, population areas, the type and size of reactor, and meteorological conditions. Typical values for site requirements have been 120-240 ha for 2000 MW plants without cooling ponds comprising two units each; small plants do not necessarily require proportionally less land. The average requirement for nuclear power plants in the U.S.A. has recently been estimated at 41 ha per 1000 MW(e) (NUREG-0002, 1976). Additional land is needed for the switchyard and transmission lines.

Not all land in an exclusion area is necessarily unproductive. Access limitations at some plants have turned land into wildlife refuges, in other cases limited farming and cattle grazing have been possible, though adequate radiation monitoring was needed. Such farming uses are likely to expand as the number of plants increase and land values go up. Similarly, at shore locations the availability of heated water will enable use of some of the sequestered shore for fish farming. A long-range land-use problem may arise from the need to control population patterns near large nuclear power plants. Environmental impact assessments attempt to allow for long-term population trends; however, nuclear site selection usually assumes that low population density areas near a plant will retain their current character even though the availability of large amounts of power and the availability of waste heat may attract additional industries and their employees to the neighbourhood.

Thermal Discharges:

The heat produced by nuclear fission in the reactor core is extracted through suitable coolants which in turn runs a turbine to operate the generator to produce electricity. At the exhaust of the turbine the steam is condensed to water to maximize the energy conversion and then is returned to the boiler or reactor to repeat the cycle. A large amount of heat is rejected in the condensing process, and the rejected heat is substantially greater than the heat equivalent of the electric energy generated. The thermal efficiency of light-water reactors at present is approximately 33%** which means that almost two-thirds of the heat energy generated in a reactor core has to be rejected to the environment, in the vicinity of the power station. It should be noted that thermal discharges are not restricted to nuclear power plants, they are also produced by fossil-fueled power plants.

The bulk of the waste heat is transferred from the steam to the cooling water in the condensers. Water is commonly used as the absorbent because of its general abundance, low cost, high specific heat, and ability to dissipate heat in the evaporation process. A LWR nuclear power plant of 1,000 MW(e) with a thermal

---

* See also A. Biswas (this volume).

** The thermal efficiency of heavy-water reactors is about 30%, of Magnox reactors about 33%, and of AGR's, HTGCR's and LMFBR's about 40%, the latter is equal to that of a modern fossil-fueled power plant.

efficiency of 33%, discharges in the condenser about 2,000 MW(t). If the cooling water flow rate is 50 m^3/s, the water temperature at the condenser outlet will have an increase of about 10°C. The cooling water is extracted from some suitable source (river or lake ... etc.), passed through the condenser where its temperature is increased by about 10°C, and returned to the source body of water. Eventually the warmed sink gives up this extra heat to the atmosphere. Such a system, which is referred to as "once-through" cooling, may cause unacceptable environmental changes. In such cases, it is necessary to eliminate thermal discharges to the water source by passing the heated cooling water through a separate "cooling pond" or "cooling tower" system and then returning it to the condenser for re-use.

A cooling pond is a large, shallow body of water that achieves its cooling by natural evaporation. Warm water from the condenser is pumped into one end of the pond and cooler water is extracted from the other end. A source of make-up water is required to replace the water lost by evaporation (usually about 3-5% of the water throughput). Cooling ponds are relatively inexpensive but require a rather large land area. A 1,000 MW(e) plant may require about 400-1,000 ha. Sometimes the water is mechanically sprayed into the air to enhance the evaporative cooling, in which case it is called a "spray pond".

Cooling towers can be classified as wet or dry. Wet cooling towers achieve cooling by evaporation, and so, like cooling ponds, require a source of make-up water. Dry cooling towers are closed systems and achieve cooling by conduction and convection. In such cases there is no loss of water. For economic reasons, dry cooling towers are very seldom used for power plants. For wet cooling towers, an outside source of make-up is required to compensate for the water lost to evaporation (about 3%), as well as water lost during "blowdown". Blowdown is the continuous or periodic flushing of the cooling system to remove solids and chemicals which accumulate in the circulating cooling water. This flushing can become a water pollution problem unless special treatment is provided. These towers also create huge plumes of water vapour which can contribute to local fogging and icing problems.

A water body provides the environment for many species of organisms, and changes in its temperature, chemical composition and flow rate may affect the number and kind of such organisms. Thermal discharges affect the water-based ecosystem to various degrees (Hutchison, 1976; Hochachka and Somero, 1973, Coutant, 1975, Biswas, 1974, Gibbons, 1976). Heat influences all biological activity, ranging from feeding habits and reproduction rates of fish via metabolism to changes in nutrient levels, photosynthesis, eutrophication and degradation rate of organic material. It is also necessary to distinguish between thermal effects in temperate and tropical or sub-tropical habitats. In temperate habitats, as water temperature starts to decrease at the end of the summer, the ecosystem activity is reduced as well. Addition of heated water from power plants may then maintain the level of ecosystem development. In contrast, in the tropics and sub-tropics, the water temperature would be high in summer, and additional thermal input could be detrimental. This could be especially important for water bodies that become shallow in the summer, and under such circumstances, the siting of power plant becomes an important consideration. In any case, it should be noted that whereas many studies are available on the effects of thermal discharges in temperate climates, information is more limited on tropical and sub-tropical climates, where the effects could be more pronounced. Seasonal changes could be critical to certain species at specific stages of their life cycle (Hutchison, 1976). Although studies have been conducted on a variety of organisms, predictions as to the effects of temperature changes and maximum temperatures still cannot be made with certainty especially if the changes are gradual, occur for only a short time, or affect limited areas or volumes of large bodies of water.

When properly controlled, thermal discharges have resulted in an increase in the ability of certain commercially valuable aquatic species to multiply, while at the same time decreasing the time for the species to reach maturity. Experience has shown that at a number of plant locations the discharge of waste heat to a stream or reservoir has improved fishing in the vicinity of the discharge during the cooler months of the year. This increase in fish populations in heated discharge regions due to higher availability of nutrients has led to the apparent paradox that most fish kills have occurred not during the release of heated effluents but whenever a plant shutdown occurs suddenly due to load dumping or equipment failure. This phenomenon, known as "cold shock", results from the sudden cooling of the water plume to whose temperature the fish have become acclimatised, leaving them with a suddenly reduced body temperature, in a state of stupor, and suffering a loss of equilibrium (Coutant, 1977). Unless the temperature is raised again quickly, the fish may die in a matter of minutes or a few hours. The effect is most pronounced during the cold season when ambient water temperatures are low. It can be minimized either by lowering the temperature slowly to give the fish a chance to adjust to the new temperature or to seek other pastures, by diffusing the heated water well, or by combining the discharge mixing zones of several plants so that the shutting down of any one plant has a less drastic effect on the temperature of the receiving water.

Another major cause of fish kills is the entrainment of small fish in the intake water at velocities too high for them to escape, leading to death either by impingement on the trash screens or by thermal shock as they are carried through the condenser pipes. Thus, the use of water for cooling purposes at steam-electric plants may have effects on aquatic organisms other than those resulting from purely thermal discharges. The destructive effects of passing fish and their larvae or eggs through pumps and condensers may indicate the need for intake screens, preferably with travelling screens having little or no impingement velocity. The most visible environmental impact of some power plants has been due to fish kills resulting from improperly designed intake structures, leading to impingement and mutilation of passing fish on the trash screens. Most recent designs incorporate wide-intake, low-velocity systems with trash racks whose main function is to remove driftwood and other solids, with finer screens to remove algae and other entrained plants and animals.

Before entering the condenser system the water has to be purified softened and demineralized, so that the water leaving the plant will be cleaner than the inlet stream. In addition, chemicals used intermittently for defouling the condensers could adversly affect fish and fish food organisms. Thermal discharges enhance the solubility of chemicals and the rate of biochemical reactions, and this may be significant in view of the wide range of plant chemicals, such as detergents, algicides, corrosion inhibitors, that may be contained in the plant effluents (IAEA, 1975).

Much has been written and experience acquired about the possible beneficial uses of thermal discharges (Belter, 1975; IAEA, 1975, 1977; Biswas and Cook, 1974; Lee and Sengupta, 1977) from power plants. Of all beneficial uses, the most promising appears to be the employment of waste heat for residential and industrial space heating in the winter and absorption-type air conditioning in the summer. Such waste heat utilization for district heating has been demonstrated successfully in several countries, for example in Sweden (Josefsson and Thunell, 1967), West Germany, Finland, France and the U.S.A. Agriculture is also a potential user of waste heat. Irrigation with heated water could promote winter seed germination and growth and extend the growing season. Hot houses are used to grow tropical or sub-tropical crops in more temperate regions. However, a number of problems need to be solved before large-scale use of heated water for irrigation could become common practice. Also, the effects of any plant shut-down on such uses of warm water has to be explored. An obvious advantage of raising the temperature of the receiving body of water is in the provision of ice-free shipping lanes and ports. However, the range of waterway affected may be limited to 16-30 km of ice-free water, and the adverse effects on the ecology would be most pronounced during the summer months. Another potential use of condenser discharge water is in aquaculture. Marine and freshwater organisms may be cultured and grown in channels and ponds fed with heated water. For example, it may be possible to grow commercially valuable oysters in the area where they cannot normally reproduce or survive due to low water temperatures. Culture experiments with shrimp, eel, white fish and other species have been carried out with thermal discharges in some countries. In Japan, a large culture programme involving shrimp, red snapper and abalone has been in progress at the Tokai reactor plant and it has been reported that growth rates measured for shrimp and snapper are much higher in the 7°-8°C warmer discharge water than at natural temperatures and no radioactive release or contamination has been detected (Belter, 1975). In the U.S.A., catfish production operations are being carried out in conjunction with steam-electric power plant operations in Tennessee and Texas. Aquaculture is also practiced in France and other countries.

Transmission Lines:

The environmental impact of high-voltage transmission lines* may be considered with regard to six aspects: aesthetic considerations, land requirements, communications, hazards, ozone, and habitat effects. Since underground lines at this time are limited to 250 kV and to short-range distribution systems, high-voltage transmission lines will remain part of our environmental scene for a considerable time. Efforts have been made to design more attractive towers and to select cables permitting longer free distances between towers. Routing can take advantage of topographical features to reduce visibility, and use of higher transmission voltages will reduce the number of lines radiating from major power centres. Land requirements will vary with the number of lines and the height of towers, but the right-of-way for major transmission lines may be 30-120 m wide. In general, use of the land below the lines is restricted to pasturing or low-intensity farming, though in most cases the utility may be sole owner of the right-of-way, may keep it fenced and restricted. In any case, it is evident that the land demand for easements associated with any new power station that may be 60 kms or more for major load centres will greatly exceed the land area of the plant itself and of the exclusion area. Regarding communication problems, high tension lines may cause some interference with nearby radio and television reception and may introduce fluctuation on signal strengths on windy days or under icy conditions. Tall towers and multiple lines may pose a hazard to air traffic, particularly under conditions of poor visibility, and air traffic terminals need to be suitably located.

Ozone may be generated by any corona or electrical discharge in air or other oxygen medium. The quantities produced depend on the severity of the discharge and the quantity of oxygen in the affected volume. Corona discharges can increase as a result of abrasions, corrosion effects, foreign particles or sharp points on electrical conductors, or incorrect design that produces excessively high potential gradients. Experimental work has been conducted to determine the added ozone production around a 1,000 kV high-voltage test line. The results showed an insignificant increase in ozone concentration over that produced by sunlight. Consequently, current types of high tension lines are not expected to pro-

---

* High-voltage transmission lines are common to all electric power stations: nuclear, fossil-fueled or hydro-electric.

duce harmful ozone concentrations.

Power line right-of-way has great potential as a wildlife habitat. Shear clearing through heavily forested areas is not inconsistent with good forestry and wildlife management practices. A common management practice in large sections of unbroken forest land is to open the tract by means of small, evenly spaced clearings. The rationale for this practice is to provide diversity and food in the forest environment. Wildfires originally provided this type of habitat. Power line right-of-way creates long linear forest openings that are indefinitely maintained to prevent power outages. The sunlight penetrating the forest via the right-of-way stimulates understorey growth adjacent to the power line. Periodic line maintenance may perpetuate these beneficial wildlife habitat conditions.

Transmission lines can have adverse impacts when crossing wetland areas. There is some evidence to indicate the behavioural modifications may occur for waterfowls, which could result in the absence of birds covering an area within 1 km of the transmission lines. The swaying of the lines in the wind, their reflective properties, and the humming of the lines could explain the abnormal behaviour of the birds. Direct current transmission could also have effects on migratory birds using magnetic homing. Electric fields associated with a transmission line can produce a charge on animals or human beings, within the range of its influence (Janes, 1977). Questions have been raised on the resultant effects of this displacement of currents on biological systems, but this is normally well below the values generally accepted as "safe" levels. Comprehensive reviews of existing studies on biological effects of high voltate electric fields are available (Kaufman and Michaelson, 1974; Bridges, 1975, 1977). Kaufman and Michaelson (1974) conclude that "research to date has failed to provide convincing evidence that human exposure to stationary or low-frequency electric fields has any harmful biological effect". According to Bridges (1975), "although the great bulk of evidence suggest that there are no significant effects of electric fields encountered under extra-high voltage lines, further research is needed".

RADIOLOGICAL IMPACTS*

During the operation of nuclear power reactors, radionuclides are formed by fission of the nuclear fuel and by neutron activation of structural materials, corrosion products, and impurities in reactor coolant water. Most of the fission products remain in the fuel elements, but a fraction can escape through the fuel cladding fissures into the coolant. Most of the radioactive-isotopes formed either by fission or activation process and released into the coolant or moderator are removed by gaseous and liquid water processing systems. Part of the radioactive material, nonetheless, may eventually be released into the environment.

From the enormous variety of fission and activation products generated by reactor operations, some are particularly relevant, either because of the ease with which they appear in effluent streams or because of their radiological significance. In this paper , emphasis will be given to the environmental impacts of fission noble gases, particularly krypton-85, some activation gases, iodines, tritium, carbon-14, radionuclides in particulate forms released into the

---

* Discussion in this paper centres around LWR's. For data on effluents and on occupational and population exposure from gas-cooled and other types of reactors, see UNSCEAR (1977).

atmosphere and water bodies. Special consideration will be given to tritium and krypton-85 as both radionuclides are long-lived and are distributed worldwide. Krypton-85, particularly, deserves special attention because of the inherent difficulty of its control and its essentially non-reactive and mobile nature in the environment (USEPA, 1977). Because of its long-half-life, the radiation exposure commitments resulting from the environmental build-up of Carbon-14 are considerably larger than those from noble gases and tritium and, therefore, proper consideration of this nuclide is also necessary. Several radioiodines, particularly I-131, are radiologically significant in the local environs. Iodine-129 is also relevant as far as the global effects are concerned.

Airborne Effluents:

Fission Noble Gases: Noble gases (Krypton and Xenon) may escape from fuel-cladding fissures and are difficult to retain. Noble gases escape out of the reactor primary coolant boundary either by leakage from both coolant and/or moderator system; from effluents of the side stream of the coolant and/or moderator which is continually diverted into a sub-system for control and purification; from the main condenser exhaust; from secondary coolant blow-down; or by continuous removal of non-condensable gases, in the steam flow of BWRs via the main condenser air-ejector system and eventually other leakages through the turbine gland-seal system, the condenser mechanical vacuum pump ... etc.

A minor fraction of noble gases escaping the primary coolant boundary directly released into the environment; most of this fraction passes through the reactor stack although an insignificant amount may reach the environment via building leakages. The major amount of radioactivity is, however, treated by a gaseous radioactive waste treatment system. Cryogenic systems, charcoal-adsorption systems and chemical methods can be used for separation and containment of noble gases. While charcoal beds provide a wide range of delay times to permit decay of the short-lived rare gases, cryogenic systems and absorption in liquid fluorocarbons minimize the volume of long-lived separated rare gases for long term storage. The fraction of noble gases which do not decay during the delay time imposed by the treatment system are released into the environment through the reactor stack.

Table 1 gives the noble gases discharged in airborne effluents from LWR's. The composition of the release from PWR's is primarily Xe-133 (half-life time 5.3 d). In BWR's, the important components released are Xe-135, Xe-138, Kr-88, Kr-87 and Kr-85m.

Tritium: Ternary fission in reactor fuels and neutron reaction with various isotopes dissolved in the coolant or moderator and with the coolant or moderator itself, gives rise to tritium. Most of the tritium activity released into the environment is converted to tritiated water (HTO) and participates in the normal water cycle. Most of the current tritium world inventory will be in the oceans as a result of transport by rain and by direct exchange of water vapour between air and sea water; once in the ocean, tritium will be dispersed through mixing processes. The average normalized release of tritium in airborne effluents is about 200 Ci/GW(e)y and 50 Ci/GW(e)y for PWR's and BWR's respectively (UNSCEAR, 1977, p. 179).

Table 1. Noble gases discharged in airborne effluents from LWR's (after UNSCEAR, 1977, pp. 176-177) normalized to 1 GW(e)y

	Half-Life	PWR Ci/GW(e)y	BWR Ci/GW(e)y
Xe-133	5.27 d	1.5×10^4	2×10^5
Xe-135	9.2 hr	540	3.6×10^5
Xe-138	14.2 min	190	1.9×10^5
Kr-85	10.57 y	170	3×10^4
Kr-88	2.79 h	150	2×10^5
Xe-135 m	15.7 min	130	6×10^4
Kr-85 m	4.4 h	100	8×10^4
Xe-133 m	2.26 d	90	8×10^3
Kr-87	76 min	70	1.7×10^5
Kr-83 m	1.86 h	–	1×10^5
Xe-131 m	11.96 d	160	–
Ar-41	1.86 h	50	–
Xe-137	3.82 min	–	1×10^4
Others		–	4×10^3

Carbon-14: Carbon-14 is produced in LWR's by (n, α) reactions with oxygen-17 present in the oxide fuel and in the moderator, by (n, p) reactions with nitrogen-14 present in the fuel as impurities, and by ternary fission. Ternary fission production per unit electrical energy generated is practically independent of reactor design, while the normalized production of C-14 by the other reactions depends on the enrichment of the fuel, the relative masses of the fuel and moderator, the concentration of nitrogen impurities in the fuel and the temperature of the fuel and moderator. The average production of C-14 in LWR's has been estimated to be about 20 Ci per GW(e)y. The amount which can be assumed to be released at the reactor is about 30% of the total production, i.e. about 6 Ci per GW(e)y. (UNSCEAR, 1977, p. 181). However, higher releases have been reported: 13 Ci/GW(e)y for the Yankee Rowe PWR in U.S.A. and up to 200 and 460 Ci/GW(e)y for PWR and BWR for the USSR (UNSCEAR, 1977, p. 181).

Iodine: A number of iodine isotopes are produced in reactors by fission and by decay of other fission products. Iodine-131, with a half-life of 8 d, is the main concern from an environmental point of view. I-129 (half-life: 1.6×10^7 y) has not been identified in power reactor environs, and its significance, in terms of possible release activities, is much less than that of the other iodine isotopes. The amount of I-131 released in airborne effluents has been estimated to be about 2-5 Ci/GW(e)y from BWRs and $5\text{-}50 \times 10^{-2}$ Ci/GW(e)y from PWRs (UNSCEAR, 1977, p.184).

Particulates: Most of the fission and activation products can form aerosols, which may be released with the airborne effluents. Particulates may arise either directly or as decay products of fission noble gases. Aerosols originate particularly from primary coolant leaks, but they can also be generated by working on contaminated components or cleaning contaminated surfaces. High efficiency particulate air filters (HEPA) retain 99.99 per cent of the particles and, hence, particulate activities are very low. The average total activity of particulates has been estimated to be 500 m Ci/GW(e)y from PWR's and 1,100 m Ci/GW(e)y from BWR's. Over 80% of the average PWR release activity is Rb-88. Other small contributions are Cs-134, Cs-137, Co-58, Co-60, and Mn-54. For the BWRs, Ba-140,

La-140 and Sr-89 each account for about 30 per cent of the total activity, with I-131, Co-58, Co-60, Cs-134 and Cs-137 each contributing 10 per cent or less.

Liquid Effluents:

Radioactive liquids containing several radionuclides may be released from reactor operations. Such liquid effluents are produced not only by leakages from the reactor boundary but also due to various discharges, e.g. from laboratories, laundry, showers ... etc. In some cases, a substantial amount of activity may arise from spent fuel storage pools. The liquid effluents are normally treated by a liquid radioactive waste treatment system; residual radioactivity arising after treatment is ultimately discharged mixed with the cooling water stream, into the aquatic environment.

Tritium in liquid effluents has been estimated to be about 2,000 Ci/GW(e)y for PWRs and about 100 Ci/GW(e)y for BWRs (UNSCEAR, 1977, p. 179). Besides tritium, a number of other radionuclides are encountered in liquid effluents. These are summarized in Table 2. The average normalized releases excluding tritium are approximately 8 Ci/GW(e)y for PWRs and 60 Ci/GW(e)y for BWRs. The radionuclides Cs-137 and Cs-134 account for about 30% of the total activity discharged by PWRs and for about 70 per cent of that from BWRs. I-131 and I-133 make up 10-40% and Co-58 and Co-60 about 15% of the activity in liquid effluents from LWRs.

Table 2. Radionuclides other than tritium discharged in liquid effluents (after UNSCEAR, 1977, pp. 189-190; normalized to 1 GW(e)y)

	Half-life	PWR Ci/GW(e)y	BWR Ci/GW(e)y
I-131	8 d	2.3	3.6
Cs-137	30.3 y	1.7	25.0
I-133	20.5 h	1.3	0.15
Cs-134	2.1 y	1.0	17.3
Co-60	5.2 y	0.42	6.1
Cr-51	27.7 d	0.18	0.58
Mn-54	313 d	0.12	1.9
Ce-144	290 d	0.08	0.34
Ru-103	41 d	0.02	
Ag-110 m	270 d	0.02	0.006
Sr-89	51 d	0.01	0.80
I-132	2.4 h	0.01	
Fe-59	45 d	0.01	0.046
Nb-95	35 d	0.006	
Zr-95	63 d	0.004	0.012
Mo-99	68 h	0.003	0.16
Sr-90	28 y	0.002	0.098
Zn-65	244 d		0.23
Ce-141	32.8 d		0.009

Solid Wastes:

Solid radioactive wastes may be produced in significant quantities by reactor operation; the nature and level of radioactivity in such wastes depend on the sources of production. In the case of no-recycle option the spent fuel elements which contain large quantities of fission products, together with trans-uranium elements and unspent uranium, will be categorized as solid wastes. Filters and resin columns used to remove radioactivity from liquid and gaseous effluent streams will form solid wastes with rather high concentration of radio-nuclides. Activated and contaminated reactor parts which are removed during repair and maintenance operations fall into the category of high level solid radioactive wastes. A less significant source of solid radioactive waste is the contaminated material such as gloves, laboratory ware ... etc. which are used in operations involving radioactive substances. The yearly cumulative amount of solid radioactive wastes generated by reactor operations excluding spent fuel elements has been estimated to be about 2,800 m^3 containing about 100 Ci of activity (Kibbey, 1975). The environmental impacts of such solid radioactive wastes, is not considered to be significant in the short term.

Occupational Exposure at Nuclear Power Plants:

Occupational radiation exposure at LWRs is due primarily to external exposure to gamma radiations from fission and activation products. Less information is available than desirable on the breakdown of the exposure by type of operation, but much of it appears to be associated with refueling, maintenance, inspection, and waste treatment operations. The size of plants on-line at present is increasing, and most new plants are in the 1,100 to 1,400 MW(e) range. There is insufficient operating experience with these larger plants as yet to accurately predict the collective doses incurred with operating them during their total design lives. The best estimate of expected performance may be taken from the record of current operational experience The exposures for 1973 to 1975 from the collective U.S. licensed plants are presented in Table 3. At this time there seems to be no clear trend of increase or decrease in exposure with size or age of the units. During 1973-1975 a total occupational dose of about 400 to 500 man-rem resulted from operating a LWR for a year. Assuming the ratio of generated power to rated capacity was 0.7 for each year, the estimated average occupational exposures would be 1,200 man-rem/GW(e)y for the period 1973-1975.

Table 3. Estimated collective dose to LWR operating and maintenance personnel (NUREG 75/032, 75/108)

Year	No. BWRs	No. PWRs	No. LWRs	Avg. Rated Capacity MW(e)	Total Annual Electric Energy Produced* MW(e)	Annual Average Collective Dose man-rem/ reactor	Average man-rem/ GW(e)y
1973	14	12	26	546	9932	542	1400
1974	14	18	32	581	13014	448	1100
1975			44	640	19712	482	1100

* Estimated using an average capacity factor of 0.7.

Exposures to individuals at reactor plants are generally kept below 5 rem/y (whole body dose) with exposures exceeding that level rare. Since the usual practice in quoting average dose is to include many workers receiving much less than 5 rem/y, an estimate of the average occupational dose is not significant without information on the dose frequency distribution in the population of workers.

Population Exposure:

Radionuclides released in airborne or liquid effluents during reactor operation undergo a series of complex physical, chemical and biological processes before reaching man. Such processes depend on the location of the reactor, meteorological conditions, and the different exposure pathways.

The noble gases released from reactors usually make the largest contribution to the dose to the local population. Using the average release composition based on operating experience in a dispersion model incorporating average weather conditions and assuming discharge from a 30 m effective stack height (Bryant and Jones, 1973), it is possible to estimate the dose per unit release as a function of distance from the stack. With the further assumption of a uniform population density of $100/km^2$ in the region from 1 to 100 km and $25/km^2$ at greater distances, the corresponding collective doses can also be estimated. For the average BWR total normalized release of 1100 x 10^3 Ci/GW(e)y, the local collective dose would be 500 man-rem per GW(e)y (to 100 km), and the additional regional contribution would be 50 man-rem per GW(e)y (100-1500 km), including the contribution from Xe-133 beyond 1000 km (UNSCEAR, 1977, p. 191). For PWRs, the total normalized release of noble gases is about 15 x 10^3 Ci/GW(e)y. The local collective dose would be 1.4 man-rem per GW(e)y to 100 km, with an additional regional contribution of 0.11 man-rem per GW(e)y (100-1500 km), including the extrapolated contribution from Xe-133 beyond 1000 km.

Tritium present in airborne effluents from reactors can contribute to the local collective dose primarily through the inhalation pathway and secondarily through the ingestion pathway. The local collective dose contribution after deposition through the ingestion pathway may be quite variable from site to site owing to differences in local hydrology and water usage. The intake of airborne tritium occurs by inhalation and by passage through skin, in nearly equal proportions. The behaviour of tritium in the environment and its effects on biota are far from being well understood. Extended studies are needed to understand the incorporation of tritium in biological compounds and to investigate the long-term components in tritium elimination by man and other mammals (Farges and Jacobs, 1979). The estimated local collective doses (1-100 km) for tritium are 0.04 man-rem per GW(e)y for PWRs and 0.01 man-rem per GW(e)y for BWRs (UNSCEAR, 1977, p. 194). Tritium in liquid effluent from reactors can contribute most directly to the local collective dose if the discharge is to a river that provides a drinking water supply. The local collective dose has been estimated to be 7 man-rem per GW(e)y for PWRs and 0.7 man-rem per GW(e)y for BWRs.

Carbon-14 makes a very small contribution to the local collective dose. It is more significant as component of the carbon cycle contributing to the global collective dose. For a normalized C-14 release from a light-water reactor of 6 Ci per GW(e)y, the local collective dose commitments are estimated to be 0.7, 0.6, 2.2 and 2.4 man-rem per GW(e)y for the lungs, gonads, bone lining cells and bone marrow, respectively (UNSCEAR, 1977, p. 194). Regional contributions to the collective dose are very small compared to the global collective dose commitment from C-14.

Assessment of local dose from I-131 released from reactors is complicated by the various forms that iodine may take: particulate, elemental, organic or as hypoidous acid. Elemental iodine readily deposits on forage and enters the cow-milk-man pathway. Organic iodine, however, is retained much less efficiently by vegetation, and its deposition velocity is 200-1000 times smaller than that of the elemental form. Particulate-associated iodine will be deposited at rates intermediate between those for the elemental and organic forms. The behaviour of hypoiodous acid uncertain; it may simply decompose to the elemental and organic forms. Physico-chemical transformations occurring during atmospheric transport may also affect the amount of iodine in its various organic forms, since these are not stable in sunlight. The elemental form would be expected to become rapidly associated with airborne aerosols, so that deposition at distances beyond the immediate vicinity of the release would be largely governed by the particulate behaviour. This is also shown by fallout measurements. On the basis of average operating experience, the normalized release of I-131 for BWRs is 5 Ci per GW(e)y (UNSCEAR, 1977, p. 196). The thyroid collective dose commitment is, therefore, estimated to be 110 man-rem per GW(e)y. The normalized collective dose commitments for PWRs are smaller by two orders of magnitude.

The activities of particulates released in airborne effluents may be quite variable but are relatively low. From releases based on average operating experience, estimates may be made of the local collective dose commitments from immersion and from inhalation intake while the radioactive material remains airborne and from the ingestion pathway and the exposure to contaminated ground following deposition. The estimated collective dose commitments are somewhat higher for BWRs than PWRs, ranging from 0.03 to 0.2 man-rem per GW(e)y to the lung, due primarily to Co-60, and 3 to 10 man-rem per GW(e)y to total body, due to Co-60 from external exposure and to Cs-134 and Cs-137 from external exposure and through the ingestion pathway. The dose commitments to bone marrow and bone lining cells are due almost entirely to the small amount of Sr-90 released (UNSCEAR, 1977, p. 196).

Radionuclides discharged in liquid effluents may result in doses to man through the pathways of drinking water and fish consumption for releases to fresh water, and of ocean fish and shellfish consumption for releases to salt water. A proportion of the population may also be exposed on shorelines to external irradiation from sediments. In areas where irrigation of crops is practised, especially when sprinkling methods are utilized, a significant contribution to the collective dose could result. This pathway should then be considered. Ingestion of water fowl and immersion exposure from swimming are other pathways, but these have generally been found to be insignificant contributors to the collective dose. The estimated collective doses of radionuclides other than tritium in liquid effluents are given in Table 4. The collective doses to the total body include contributions from ingestion of Cs-137, Cs-134 and Zn-65, and from external exposure of Cs-137, Cs-134, Co-60, Mn-54, Ce-144, Cr-51, Sb-125, and Ru-106 in sediments.

Table 4. Collective doses to particular organs from radionuclides other than tritium in liquid effluents from reactors (man-rem GW(e)y)* (After UNSCEAR, 1977, p. 200)

Reactor	Total body	Lower large Intestine	Bone	Thyroid
		Release to salt water		
PWR	$2\ 10^{-2}$	$7\ 10^{-2}$	$3\ 10^{-6}$	$1\ 10^{-2}$
BWR	$2\ 10^{-1}$	1	$2\ 10^{-4}$	$3\ 10^{-2}$
		Release to fresh water		
PWR	6	20	$6\ 10^{-2}$	2
BWR	80	3	3	4

* RBE = 1 for conversion to rem.

DECOMMISSIONING OF NUCLEAR FACILITIES

Decommissioning a nuclear facility can be defined as the measures taken at the end of the facility's operating lifetime to assure the continued protection of the public from the residual radioactivity and other potential hazards in the retired facility. Two basic approaches to decommissioning of each facility are generally considered (Smith and Schneider, 1978):

(a) Immediate Dismantlement - Radioactive materials are removed and the facility is decontaminated and disassembled immediately following final shutdown. Upon completion, the property is released for construction of a replacement nuclear facility or for other purposes in which case authorization and special permission would be required.

(b) Safe storage with or without Deferred Dismantlement - Radioactive materials and contaminated areas are secured and structures and equipment are maintained as necessary to assure the protection of the public from the residual radioactivity. During the period of safe storage, the facility remains limited to nuclear uses. Dismantlement is deferred to allow the radioactivity within the facility to decay to lower levels. Upon completion of dismantlement, the property is released for special uses as mentioned above.

Methods for decommissioning of nuclear facilities range from minimal removal and fixation of residual radioactivity with maintenance and surveillance (protective storage or Stage I according to current IAEA terminology) to extensive cleanup, decontamination and entombment (Stage II, according to IAEA). Each of these methods of safe storage requires surveillance and care during the holding period which may vary in length from a few years to decades. Each method ends with the deferred dismantling of the nuclear facility (Stage III, according to IAEA).

Entombment (Stage II) of a nuclear facility requires the encasement of the radioactive material in concrete or other structural material sufficiently strong and structurally long-lived to assure retention of the radioactivity until it has decayed to levels which permit unconditional release of the site. The entombed structure must be designed to outlast any of the confined radiological or chemical hazards. Unless the structure is to be re-entered later and decommissioned further, the potential chemical and radiological hazards should vanish in no more than about 200 years in order to fulfill the bases for entombment. But this cannot be applied in practice due to the presence of long-lived radionuclides. In the United Kingdom, Stage II is regarded as a storage period, either for a defined period or of unspecified duration after which Stage III (total dismantling) is to be undertaken.

It was assumed possible to entomb the reactor pressure vessel and its internal structures within the biological shield since the principal source of radiological dose with Co-60 which decays with relatively short half-life (5.27 years). Thus, within about 100 years, the residual radioactivity will have decayed to levels indistinguishable from normal background, well within the safe structural lifetime of the entombment structure. The amounts of long-lived radionuclides formed in the relatively brief operating life of these early plants were sufficiently small as to present no significant hazard. However, in large power reactors that have operated for 30 to 40 years, the induced Nb-94 and Ni-59 activities in the reactor vessel and its internal components are well above unconditional release levels and, since Ni-59 has an 80,000 year half-life and Nb-94 has a 20,000 year half-life, the radioactivity will not decay to unconditional release levels within the foreseeable lifetime of any man-made surface structure. It should be noted, however, that Ni-59 is a β emitter of low energy and low radiotoxicity and might not pose a serious environmental problem on the long run.

In the case of a fuel reprocessing plant the problems of decommissioning are different from those of nuclear reactors. Such plants are designed to facilitate the decontamination of the plant vessels as an operational requirement, so that the plant can be decontaminated to very low levels. Additionally there is not the problem of neutron activation of the plant vessels.

Entombed facilities are a nuisance, subject to vandalism and intentional (but not necessarily malicious) intrusion by man. Any structure conceived by man can also be compromised by him. Extra hardening of entombed facilities such as filling them with concrete will prevent or retard deliberate unauthorized penetration for a period of time, but the facilities will still require some surveillance. It is also difficult to reverse this type of decommissioning if a decision is made later to remove the facility. Provisions for subsequent retrieval of radioactive material under entombment could be done as an option, but this would sacrifice penetrability protection. Furthermore, implementation of the entombment mode would significantly increase the total number of radioactivity-containing sites that require surveillance. This would contradict the philosophy of long-term protection of the public and non-proliferation of radioactivity-containing sites.

Several studies on decommissioning of nuclear facilities (Smith and Schneider, 1978; Nemec, 1978; IAEA, 1978; Detilleaux and Lennemann, 1977; Gasch et al., 1978; Essmann et al., 1978; Auler et al., 1978 and Martin et al., 1978) have outlined the technical and regulatory aspects of the problem. Table 5 gives a summary of decommissioning data for a LWR (PWR, about 1 GW(e)).

Table 5. Decommissioning Data for LWR (PWR) 1 GW(e) (after Smith and Schneider, 1978)

	Immediate Dismantlement	Passive Safe Storage with Deferred Dismantlement After:		
		10 Years	30 Years	100 Years
Total Cost (Million 1978 $)	42	50	52	51
Time requirements (Years)	4	4.5	4.5	4.5
Occupational dose (man-rem)*	1326	792	470	445
Public dose (man-rem)*	22	8	3.5	3
Serious accidents fatalities*	0.097	0.105	0.107	0.113
Manpower (Cumulative man-years)*	320	334	362	460

* Includes decommissioning operations, interim care, and transportation where applicable.

The volumes of radioactive material for disposal estimated to result from the dismantlement of this LWR are:

Activated material (m^3)	1191
Contaminated material (m^3)	16078
Radioactive wastes (m^3)	618
Spent fuel (m^3)	37
Truck shipments	1363
Rail shipments	28

All non-transuranic material was assumed to be buried in a shallow land burial facility, and all high-level material was assumed to be placed in deep geologic storage. Other estimates of radioactive material resulting from LWR's are lower than the figures given above.

The estimates of radiation doses to workers and the public resulting from decommissioning of LWR are given in Table 6. Nearly all of the radiation dose to the public results from transport of radioactive materials from the facility to the burial site.

Table 6. Estimated dose commitment to workers and the public from decommissioning of LWR (after Smith and Schneider, 1978), in man-rem/GW(e)y

	Immediate Dismantling	Passive Safe Storage with deferred dismantling after: 10 years	30 years	100 years
I. Workers				
Decommissioning	40.8	27.8	14.7	13.9
Transportation	3.4	1.3	0.5	0.5
Safe Storage	-	0.3	0.5	0.5
II. Public				
Decommissioning	0.3×10^{-5}	$< 0.3 \times 10^{-5}$	$< 0.3 \times 10^{-5}$	$< 0.3 \times 10^{-5}$
Transportation	0.7	0.3	0.1	0.1
Safe Storage	-	←——— negligible ———→		

Decommissioning of fuel reprocessing plants was studied by Smith and Schneider (1978) using the Barnwell Nuclear Fuel Plant as a model. Although the facility is not yet operating, it is believed to be representative of commercial fuel reprocessing plants. The main chemical process building, spent fuel receiving and storage station, liquid radioactive waste storage tank farm, and conceptual high-level liquid waste solidification facility were postulated to be decommissioned. The volumes of radioactive material to be disposed of from the plant are:

TRU-material (m^3)	4600
Non-TRU material (m^3)	3100
Truck shipments	160
Rail shipments	180

Table 7 summarizes the decommissioning data for this fuel reprocessing plant and Table 8 gives the estimated radiation doses from the decommissioning operation.

The issues involved in decommissioning nuclear facilities are complex - more so in some respects than in the construction and commissioning of plants. The various aspects of the problem - technical, economical, radiological, environmental and organizational are conflicting in many ways, and will not be resolved until waste disposal routes are defined. Setting aside the uncertainties of waste disposal, it is reasonably clear that the dismantling of nuclear plants is technically feasible.* Reactors are likely to present the greatest problems because of their large numbers, induced radioactivity and the large volumes of radioactive waste which will be produced. However, the technical problems will diminish as experience is gained on the first few plants. The major issue to be resolved is the timescale on which complete dismantling of a plant is to be undertaken. Although there may, in some cases, be pressures for early dismantling, this would result in economic and other penalties. In particular, pressure would in turn be placed on the establishment of suitable waste disposal routes, and a need for early decision on this problem could foreclose some options. Moreover, delaying dismantling for a period of 50-100 years could minimize the need for remotely controlled operations and permit a much greater degree of manual effort without significantly enhanced radiological detriment. The technical problems posed in the dismantling of fuel reprocessing or fabrication plants differ in a number of respects from those of reactors, and the balance between the costs and benefits will generally favour early dismantling.

There is at present some controversy as to whether consideration should be given at the design stage to introducing features that would ease decommissioning problems following a major accident. Clearly some thought must be given to the effects of relatively minor accidents, but major accidents must be very rare events and it would appear to be difficult to justify provisions to remedy or ameliorate the effects of such an accident during the decommissioning process. Rather it is felt that an ad hoc approach is probably the only feasible one as regards decommissioning the plant, though some thought should be given to the possible need to deal with large quantities of very highly contaminated waste.

---

* The ELK River nuclear reactor in the U.S.A. has been dismantled. The most difficult parts of the reactor decommissioning operation, namely the removal to safe storage of the reactor calandria, has already been successfully undertaken (Marchildon, 1971).

Table 7. Decommissioning Data for Fuel Reprocessing Plant
(after Smith and Schneider, 1978)

	Immediate Dismantlement	Custodial Safe Storage With Deferred Dismantlement After			Passive Safe Storage With Deferred Dismantlement After		
		10 Years	30 Years	100 Years	10 Years	30 Years	100 Years
Total Cost (Million 1978 $)	67	78	95	157	74	77	90
Time Requirements (Years)	5.2	2.4	2.4	2.4	2.7	2.7	2.7
Occupational Dose (man-rem)*	532	453	333	179	445	312	137
Public Dose (man-rem)*	19	15	10	4	15	10	4
Accidents Fatalities*	0.021	0.025	0.034	0.060	0.023	0.024	0.030
Manpower (Cumulative man-years)*	423	510	693	1338	481	514	634

* Includes decommissioning operations, interim care, and transportation where applicable.

Table 8. Estimated dose commitments to workers and public from decommissioning a fuel reprocessing plant (after Smith and Schneider, 1978), in man-rem

	Immediate Dismantlement	Custodial Safe Storage With Deferred Dismantlement After			Passive Safe Storage With Deferred Dismantlement After		
		10 Years	30 Years	100 Years	10 Years	30 Years	100 Years
I. Workers							
Decommissioning	512	423	290	113	426	296	124
Transportation	20.2	16.7	11.6	4.7	16.7	11.6	4.7
Safe Storage	-	12.8	31.4	61.4	1.8	4.4	8.6
II. Public							
Decommissioning	10.2	8.2	5.1	2.0	8.2	5.1	2.0
Transportation	8.5	7.1	5.0	2.1	7.1	5.0	2.1
Safe Storage	-	←——— negligible ———→					

ACKNOWLEDGEMENTS

Grateful acknowledgements are made to Dr. A. Gonzalez, Comision Nacional de Energia Atomica, Argentina, Dr. Y. Ahmed, International Atomic Energy Agency; Dr. J.R. Beattie, Safety and Reliability Directorate, U.K. Atomic Energy Authority; Dr. A. Biswas, Biswas and Associates, Ottawa; Dr. M. Carriera Vasquez, Junta de Energia Nuclear, Madrid; Dr. L. Farges, IAEA; Dr. L.D. Hamilton, Brookhaven National Lab., New York; Dr. H. Howells, British Nuclears Ltd., Dr. H.P. Jammet, Institut Curie, Paris; Dr. Z. Jaworovski, Department of Radiation Hygiene, Warsaw; Dr. L. Weeks, White Shell Nuclear Research Establishment, Manitoba; Dr. J.M. Costello, Australian Atomic Energy Commission, Sutherland; Dr. Y. Sousselier, Commissariat à l'Energie Atomique, France and Dr. B. Wachholz, U.S. Department of Energy, for their comments and assistance. The views expressed in this paper, however, are those of the author, and not necessarily of the scientists mentioned above.

REFERENCES

Auler, I. et al.: Decay Behaviour and Structure of the Radioactive Inventory Considering the Decommissioning of a Nuclear Power Plant with a LWR. Intern. Symp. Decommissioning Nuclear Facilities, IAEA, Vienna Paper SM-234/1, (1978).

Belter, W.G.: Management of Waste Heat at Nuclear Power Stations; in "Environmental Effects of Cooling Systems at Nuclear Power Plants", IAEA, Vienna, (1975).

Biswas, A.K.: Energy and the Environment. Planning and Finance Service, Environment Canada, Ottawa, Report No. 1 (1974).

Biswas, A.K. and Cook, B.: Beneficial Uses of Thermal Discharges; Planning and Finance Service, Environment Canada, Ottawa, Report No 2, (1974).

Bridges, J.E.: Biologic Effects of High Voltage Electric Fields. Electric Power Research Institute, Palo Alto, California Report 381-1, (1975).

Bridges, J.E.: Environmental Consideration Concerning the Biological effects of Power Frequency (50 or 60 Hz) Electric Fields. Proc IEEE Paper F77-256-1, (1977).

Bryant, P.M. and Jones, J.A.: Environmental Behaviour of Radionuclides Released in the Nuclear Industry. IAEA Publication STI/PUB/345, Vienna, (1973).

Coutant, C.C.: Temperature Selection by Fish, a Factor in Power Plant Impact Assessment in: Environmental Effects of Cooling Systems at Nuclear Power Plants, IAEA, Vienna p. 575, (1975).

Coutant, C.C.: Cold Shock to Aquatic Organisms; Guidance for Power Plant Siting Design and Operation. Nuclear Safety, 18, 329, (1977).

Detilleaux, E. and Lennemann, W.L.: Criteria, Standards and Policies Regarding Decommissioning of Nuclear Facilities. Nuclear Power and Its Fuel Cycle Vol. 4, IAEA, Vienna, (1977).

Essmann, I. et al.: Provision of Decommissioning of the German Utilities for LWR-power plants. Intern. Symp. Decommissioning Nuclear Facilities, Vienna, IAEA Paper SM-234/2, (1978).

Farges, L. and Jacobs, D.: Symposium on Tritium, At. Energy Rev. 17 (1979).

Gasch, A. et al.: Results of an Analysis of the Quantities of Radioactive Waste which Develop during the Decommissioning of Nuclear Power Plants. Intern. Symp. Decommissioning Nuclear Facilities, IAEA paper SM-234/3, Vienna, (1978).

Gibbons, J.W.: Thermal Alteration and the Enhancement of Species Population; in "Thermal Ecology II", Edited by G.W. Esch and R.W. McFarlane, U.S. ERDA, Washington, D.C., (1976).

Hochachka, P.W. and Somero, G.N.: Strategies of Biochemical Adoption. Saunder Publ. Co. Philadelphia, (1973).

Hutchison, V.H.: Factors Influencing Thermal Tolerances of Individual Organisms in "Thermal Ecology II" Edited by G.W. Esch and R.W. McFarlane. U.S. ERDA, Washington, D.C., (1976).

IAEA: Environmental Effects of Cooling Systems at Nuclear Power Plants, IAEA, Vienna, (1975).

IAEA: Urban District Heating Using Nuclear Heat. Proceed. of a Symposium, Vienna, IAEA Doc. STI/PUB/461, Vienna, (1977).

IAEA: Intern. Symposium Decommissioning Nuclear Facilities. IAEA-205, Vienna, (1978).

Janes, D.E.: Background information on High Voltage Fields. Envir Health Persp. 20, 141, (1977).

Josefssen, L. and Thunell, J.: Nuclear District Heating, a Study for the Town of Lund. in: Containment and Siting of Nuclear Power Plants, IAEA, Vienna, (1967).

Kaufman, G.E. and Michaelson, S.M.: Critical Review of the Biological Effects of Electric and Magnetic Fields; in "Biologic and Clinical Effects of Low Frequency Magnetic and Electric Fields. Thomas Publ. Co. Springfield, III, (1974).

Kibbey, A. and Godbee, H.W.: Solid Radioactive Waste Prechers at Nuclear Power Plants; Nuclear Safety 16, 581, (1975).

Lee, S.S. and Sengupta, S.M.: Waste Heat Management and Utilization. Proceedings of a Ccnference, Miami Beach, 1976. Machan. Engin. Dept. Univ. Miami, (1977).

Marchildon, P.: The NRX Calandria Replacement; Chalf River Nuclear Lab., Canada, (1971).

Martin, A. et al.: Criteria for the Management of Redundant Nuclear Facilities, IAEA, Vienna, Paper SM-234/46, (1978).

Nemec, J.F.: An Engineering Approach to Decommissioning. Intern. Symp. Decommissioning Nuclear Facilities, IAEA, Vienna, Paper IAEA-SM-234/15, (1978).

NUREG-75-032: Occupational Exposure at Light Water Cooled Power Reactors (1969-1974); U.S. Nuclear Reg. Comm., Washington, D.C., (1975).

NUREG-75/108: Seventh Annual Radiation Exposure Report; U.S. Nuclear Reg. Comm.,

Washington, D.C., (1975).

NUREG-002: Final Generic Environmental Statement on the Use of Recycle Plutonium in Mixed Oxide Fuel in Light Water Cooled Reactors. Vol. 3. U.S. Nucl. Regulatory Comm., (1976).

Smith, R.I. and Schneider, K.J.: Analyses of the Decommissioning of a Pressurized Water Reactor and a Fuel Reprocessing Plant. Intern. Symp. Decommissioning of Nuclear Facilities, IAEA, Vienna, Paper IAEA-SM-234/16, (1978).

UNSCEAR: Sources and Effects of Ionizing Radiation. United Nations Scientific Committee on the Effects of Atomic Radiation. United Nations, New York, (1977).

USEPA: Environmental Radiation Protection Standards for Nuclear Power Operations, U.S. Environment. Protection Agency, Washington, D.C., (1977).

4

NON-RADIOLOGICAL ENVIRONMENTAL IMPLICATIONS OF NUCLEAR ENERGY

Asit K. Biswas
Director, Biswas and Associates
3 Valley View Road
Ottawa, Canada

"The optimist proclaims that we live in the best of all possible worlds; and the pessimist fears this is true."

James Branch Cabell

INTRODUCTION

Every stage of energy development and utilization has an impact on the environment. The generation of electricity, whether it is by fossil fuels or by nuclear power, or whether all performance criteria have been met, has environmental implications for air, water, and soil, and also the global climate. The extent of the impacts could range from very small and insignificant ones at the lower end of the spectrum, to the major ones that are difficult to manage and at best very difficult to control. In general, it can be said that the nature of waste disposal practices, regulatory requirements and the status of technological developments define the level of impacts. These interrelationships become clear if the elementary law of conservation of matter is considered in an environmental sense, that is, in the ultimate analysis man does not "consume" any materials: they are simply changed from usables to residuals, which are then discharged into the environment in different forms.

Even though the energy industry is one of the largest and most widely spread industries in the world, the environmental consequences of energy production, conversion and use can be described as relative newcomers as an area of major international concern. Increasing global energy production and use, without concomitant increase in efficiency, means that more residuals are being discharged into the environment. Also our perception of the environmental problems with relation to energy growth, like in many other areas of resource utilization, is due partly to increasing levels of pollution and partly to increasing realization of the problems created by pollution. This gradual change in perception has gradually begun to permeate the political processes, and this is being increasingly reflected in the debate that is currently going on in many parts of the world, especially in the area of nuclear energy. A significant percentage of the current discussions on the pros and cons of nuclear energy is not based on scientific facts - the discussions are often political and emotional.

ENVIRONMENTAL ASPECTS OF NUCLEAR ENERGY

In the area of environmental implications of nuclear energy, most of the discussions so far have been on the radiological aspects: safe disposal of radioactive wastes; safety of nuclear power production, including avoidance of catastrophic accidents; safe handling of plutonium; and problems associated with uranium mining. Since this paper is on the nonradiological aspects of nuclear energy, none of these concerns will be discussed herein. Good summaries of these problems can be found in Biswas (1974, 1977), Duckworth et al. (1977), IAEA (1971, 1973, 1975a, 1975b), Nuclear Energy Policy Study Group (1977), UNEP (1976), and Workshop on Alternative Energy Strategies (1977).

In addition to radiological aspects, the only other environmental implication of nuclear energy that has received considerable scientific and public attention is the question of thermal pollution. Much literature (IAEA, 1974, 1975a, 1975b, Coutant, 1975, Esch and McFarlane, 1975; Cairns, 1976) exists on this topic. In contrast, however, not much work has been done on the land use aspects. Even less has been done in the areas of noise and conventional accidents, but this is not surprising since in nearly all cases these two constitute minor problems. Accordingly, much of the emphasis in the present paper has been on thermal pollution and land use problems

THERMAL POLLUTION

Probably the most extensive man-made production of heated water is by steam-operated electrical power plants, either by the burning of fossil fuels or by the release of nuclear energy (Brock, 1975). Power plants are usually located near adequate sources of water, viz. rivers, lakes or estuaries, and they use this conveniently available water as a medium for cooling. During this cooling process, the water used becomes heated, and is then usually discharged back into the natural environment.

Both nuclear and coal power plants contribute to thermal pollution in natural bodies of water, but the extent of thermal pollution depends on various factors like types of generation and the efficiency of the plant in question. The thermal efficiency of fossil-fired plants has steadily increased over the last few decades. A modern fossil-fired plant, based on the steam cycle with superheat, has a thermal efficiency of around 40 percent (IAEA, 1974), but further technological improvements to increase the thermal efficiency are becoming more and more difficult to achieve. It should be emphasized that in countries having a large number of power plants using fossil fuels, the average thermal efficiency is likely to be less than 35 percent.

As a rule nuclear power plants are thermally less efficient than plants using fossil fuels, and accordingly contribute to somewhat increased temperature differentials, which are usually expressed as the difference between intake and outlet temperatures, t. This temperature differential between fossil fuel and nuclear plants arises mainly due to two principal reasons. Firstly, nuclear power plants convert only 30 to 33 percent of energy to electricity (Nuclear Energy Policy Study Group, 1977). Secondly, a coal-fired plant discharges part of its waste heat into the atmosphere, but in contrast a nuclear plant discharges all of its heat into the receiving water.

Because of these two reasons, a nuclear plant could contribute to "50 percent or more local heating of a body of water than a comparable coal plant" (Nuclear Energy Study Group, 1977).

Different "average" figures are available on the extent of thermal differentials between nuclear and fossil fuel plants. Only four will be mentioned here, but it should be realized that the figures quoted are not all that different. According to Harleman and Stolzenback (1972), typical figures for temperature differentials for 1,000 MW power plants would be 9°C for fossil-fuel-fired plants and 13°C for nuclear plants. The Nuclear Energy Policy Study Group (1977) stated that for a typical 1,000 MWe nuclear plant using once-through cooling, it would usually heat up "a thousand cubic feet of water some 10°C each second". According to IAEA (1973), nuclear plants on average use about 180 litres of cooling water per second per kilowatt-hour, with an average maximum temperature rise across the condenser of about 10°C. Plants using fossil fuels require about 115 to 150 litres per second per kilowatt-hour for a maximum temperature rise of about 8°C. Brock (1975) carried out a study of a fossil fuel plant in Lake Monona, Wisconsin, and found an average temperature increase over an annual cycle of 8°C, although at certain times the increase was considerably higher.

Many advanced industrialized countries have so far been able to locate power plants near major water bodies, and thus have been using these sources for cooling purposes. However, with increasing demand for electricity consumption, it is unlikely that this practice can continue indefinitely, since water temperatures would rise to unacceptable levels. According to IAEA (1974), assuming a thermal efficiency of 33 percent and temperature differential of 10°C, 1 GW(e) of electrical capacity requires about 50 m^3/sec of cooling water. If the temperature of the receiving waters may not exceed some reasonable increment, at least twice this amount of water should be available. If this reference figure of 100 m^3/sec is used, the following situation emerges for the United Kingdom (England and Wales), the Federal Republic of Germany, and the United States (IAEA, 1974):

UK: Average runoff 2,100 m^3/sec. Reference ceiling of 21 GW(e) reached in 1950s. Shift to cooling towers and sea cooling. Has more than 300 wet towers, providing cooling for 50% of total installed generating capacity.

FRG: Average runoff 4,000 m^3/sec. Reference ceiling of 40 GW(e) reached in 1960s. Cooling towers commonplace. Most nuclear plants to be equipped with wet towers.

USA: Average runoff 53,000 m^3/sec. Reference ceiling of 53 GW(e) to be reached in 1980. Due to regional lack of water, 13% of existing thermal plants equipped with cooling towers; 35% of plants under construction or ordered will have cooling towers.

The US Atomic Energy Commission (1974) carried out an overall analysis of the environmental reports submitted as a part of the licensing process of nuclear power plants. At that time it was forecasted that 102,000 MW of nuclear power capacity would be placed in operation by 1980. Approximately 34 percent of the plants proposed once-through cooling from fresh-water sources, some 50 percent would use a closed-cycle and the balance would be located on coastal or estuarine areas. For a 1,000 MW(e) plant, estimates of condenser cooling water ranged from 45 m^3/sec to 60 m^3/sec, depending on the extent of the temperature rise across the condensers (Belter, 1975).

The types of cooling systems for removing waste heat from power stations can be divided into the following four major categories, depending on the mechanism for heat dissipation or their unique characteristics (US Nuclear Regulatory Commission, 1976).

i) Once-Through Systems

Using Natural Water Bodies
Using reservoirs on Navigable Rivers

ii) Evaporative Systems

Natural Draft Cooling Towers
Mechanical Draft Cooling Systems
Spray Cooling Systems
Cooling Lakes, Ponds, Canals, etc.
Combinations of Ponds and Mechanical Systems

iii) Dry Systems

These may include both natural and mechanical draft towers. Costs are likely to be relatively high and one would expect them to be used only in very special circumstances, as in arid regions

iv) Hybrid Wet/Dry Systems

These are systems which may operate in either the dry or the evaporative/wet mode, or simultaneously in both modes, depending upon prevailing meteorological conditions and design specifications

The type of cooling system used mainly depends on site conditions, especially availability of water for dissipating heat.

Thermal pollution affects the water-based ecosystem since heat is an important natural parameter in a water system. Heat influences all biological activity, ranging from feeding habits and reproduction rates of fish via metabolism to changes in nutrient levels, photosynthesis, eutrophication and degradation rate of organic material. Fig 1 shows some factors that may influence the thermal tolerance of organisms (Hutchison, 1976). It should be noted that some of the factors shown in Fig 1 seem to have little effect on aquatic organisms, at least based on scientific evidence currently available.

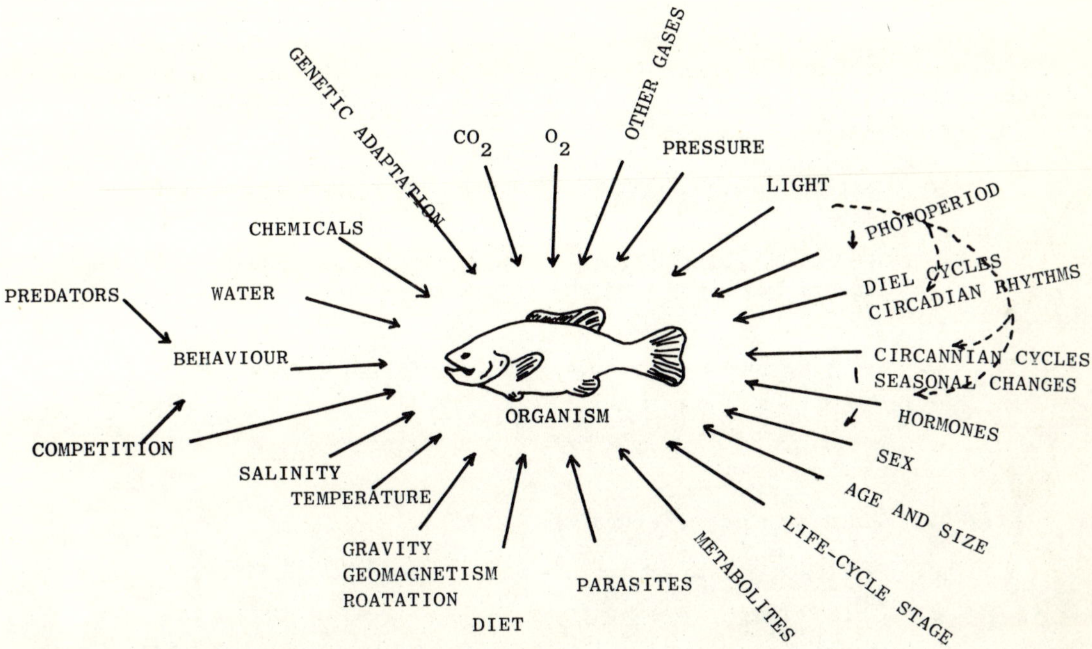

Fig 1. Some factors that may influence the thermal tolerance of organisms. Other influences are quality, quantity, duration and rate of change of environmental factors; methods of acclimation and acclimatization; sublethal exposure to limiting factors; and interrelations between controlling, directing and lethal forces.

The effects of environmental forces on organisms were divided into five categories by Fry (1971). These were lethal, controlling, limiting, masking and directive factors. Hutchison (1976) consolidated these five factors to three, lethal, controlling and directive, for the specific case of consideration of thermal effects on organisms. Lethal effects occur when temperatures kill organisms within a finite time frame. Effects are said to be controlling, if temperature influences metabolic rates and the underlying biochemical reactions, and control of isoenzyme systems uniquely fashioned to operate over the environmental temperature ranges of ecotherms (Hochachka and Somero, 1973). Directive forces can be said to result when temperature affects spontaneous systems.

Depending on specific circumstances, temperature increases could have a number of effects on aquatic organisms (Abrahamson, 1972; Biswas, 1974; Cairns, 1970; Coutant, 1975; Gibbons, 1976; Sylvester, 1972; UNEP, 1976) Among these are:

1. thermal death directly due to increased or decreased temperature;

2. internal functional aberrations, i.e. changes in growth, respiration, etc.;

3. interference with spawning or other critical activities in the life cycle;

4. disruption of normal biological rhythms, including migration patterns;

5. increased susceptibility to chemical toxins, and pathogenic organisms;

6. increased predation rate due to changes in avoidance reactions induced by temperature changes, decrease in swimming speed, stamina, etc.;

7. decreased spawning success and decrease in survival rate of young fry;

8. reduction in dissolved oxygen concentration due to higher BOD requirements;

9. disruption of food supply;

10. increased growth of taste - and odour-producing blue-green algae; and

11. competitive replacement by more thermally tolerant species.

Principles of thermal ecology at the population level are not well defined, but Fig. 2 is an attempt by Gibbons (1976) to provide theoretical responses, in a qualitative fashion, of plant and animal populations exposed to higher temperatures than normally encountered. Increase in temperature initially can be considered for a few specific species, aquatic organisms live in environments having fluctuating temperatures, and they can tolerate reasonable fluctuations, whether they are natural or man-induced. Basically, temperature can affect organisms in two ways. With increases in temperature, the rates of chemical and enzymatic reactions in the cell could increase, and growth may become faster. However, if the rise is high, then at a certain level proteins, nucleic acids and other cellular components that are sensitive

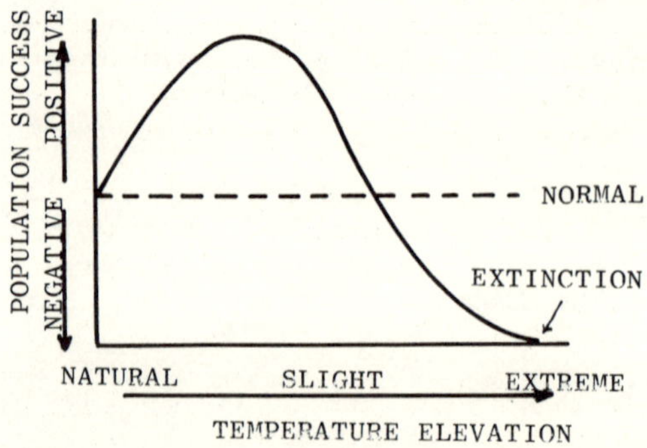

Fig. 2. Theoretical responses of plant or animal populations exposed to environmental temperatures above those naturally encountered.

to high temperatures, may be irreversibly inactivated (Brock, 1975). If temperature increases beyond the point where inactivation reactions set in, cell functions fall sharply to zero. In other words, for every organism, there are minimum and maximum temperatures beyond which growth is no longer possible, and an optimum temperature when the growth is most rapid. These temperatures, for certain common northern fish species, are shown in Table 1 (IAEA, 1974). These three temperatures, known as cardinal temperatures, are characteristics of each organism, but the cardinal temperatures of different organisms could vary widely. The cardinal temperatures can be considered to be genetically fixed properties (Brock, 1975), that is, if the temperature regime of an aquatic habitat is modified, the population structure will probably change primarily due to migration and growth of new species rather than through physiological or genetic changes of existing species.

Table 2 shows the upper temperature limits for different groups of organisms, and Table 3 shows the number of species of water beetles and blue-green algae collected in hot springs at different temperatures (Brock, 1975). From these tables, it is evident that if the temperature exceeds 40°C, species diversity reduces markedly. The ecosystem becomes greatly simplified since fish and other vertebrates disappear completely; the number of species of insects and other invertebrates and higher plants is greatly reduced. The ecosystem tends to become microbial.

Table 2

UPPER TEMPERATURE LIMITS FOR DIFFERENT ORGANISM (AFTER BROCK, 1975)

Group	Temperature in °C
Animals	
Fish and other aquatic vertebrates	38°C
Insects	45-50°C
Ostracods (crustaceans)	49-50°C
Protozoa	50°C
Plants	
Vascular plants	45°C
Mosses	50°C
Eucaryotic algae	56°C
Fungi	60°C
Procaryotic microorganisms	
Blue-green algae	70-73°C
Photosynthetic bacteria	70-73°C
Nonphotosynthetic bacteria	99°C

Table 1

FISH DEVELOPMENT AND GROWTH RELATED TO TEMPERATURE IN CELSIUS (AFTER IAEA, 1974)

Species	Spawning	Development of fry	Maximal growth	Optimal	Lethal
Burbot	1.5-1.0	1.5 } max. for fry			
Whitefish species	0.0-4.0	6.1 }			
Cod	max. 14	10.0			19.8-24.4
Plaice	14				28-31
Flounder					31-34
Char		4-6 }			
Salmon	max. 10	4-8 }	15-18	14-15	23
Sea trout, young					23
Sea trout, adult			16-19	13-17	25-26
Rainbow trout		} natural temperature	14-18	13	23-24
Pike		6-15 }			29
Perch	12-18	}			30-33
Bream		10-18 }			31-38
Roach	10-14	12-19 }		23-24	33
Carp		13-24 }	20-25	27-32	35-37
Herring	max. 5.5	18-24 }			19.5-24.7

Table 3

NUMBER OF SPECIES OF WATER BEETLES AND BLUE-GREEN ALGAE COLLECTED IN HOT SPRINGS AT DIFFERENT TEMPERATURES (AFTER BROCK, 1975)

Water beetles		Blue-green algae	
Temp. (°C)	No. of species	Temp. (°C)	No. of species
30	60	10-15	42
31	58	15-20	54
32	55	20-25	76
33	52	25-30	86
34	47	30-35	90
35	46	35-40	86
36	45	40-45	76
37	35	45-50	60
38	33	50-55	25
39	33	55-60	24
40	22	60-65	2
41	15	65-70	1
42	10	70-75	1
43	6		
44	4		
45	2		

It is difficult to generalize about the effects of thermal pollution on the aquatic environment. However, a few statements can be made. Addition of heat does affect the water quality, most of important of which is the reduction of oxygen. The reduction of oxygen in water decreases with temperature. Thus, oxygen solubility from air into water is 16 percent lower at 30°C compared to 20°C, and is 29 percent lower at 40°C than 20°C (Brock, 1975). Secondly, rate of biological processes increases with temperature, and accordingly biochemical oxygen demand (BOD) increases as well. The net result of these two effects could be significant reduction in the dissolved oxygen content of water, which under some specific site conditions could become anaerobic, thus posing serious consequences for an ecosystem. Sensitive species like trout require higher dissolved oxygen content in water, as shown in Table 4.

Table 4

MINIMUM OXYGEN CONTENT REQUIRED IN THE WATER FOR DIFFERENT TEMPERATURES

Type of Fish	Temperature in °C	Minimum Oxygen Content (mg/1.) Required
Sea trout	21	2.53
Rainbow trout	21	1.82
Salmon	15-25	1.51-2.85
Pike	15-29	0.72-1.40
Roach	15-23	0.60-1.60
Perch	15-25	0.40-1.40

It is also necessary to distinguish between thermal effects in temperature and tropical or subtropical habitats. In temperate habitats, as water temperature starts to decrease at the end of the summer, the ecosystem activity reduces as well. Addition of heated water from power plants could then contribute to extensive ecosystem development, which would otherwise not have been possible. In contrast, in the tropics and subtropics, the water temperature would be high in the summer, and additional thermal input could be detrimental. This could be especially important for water bodies that become shallow in the summer, and under such circumstances, the siting of a power plant becomes an important consideration. In any case, it should be noted that whereas many studies are available on the effects of thermal discharges in temperate climates, not enough research has been carried out on the tropical and subtropical climates, where the effects could be more pronounced.

In addition, seasonal differences in thermal tolerance should also be noted. Seasonal changes could be critical to certain species at specific stages of life cycle. This is shown schematically in Fig. 3 (Hutchison, 1976) for Pacific salmon. It has also been found that critical thermal maximum (CTM) of shore fishes in the Gulf of California is lower than their maximum habitat temperatures in the summer (Heath, 1967). Hutchison (1976) also reports that the "CTM of newts (Notophthalmus viridescens) all collected in the same locality is significantly different at the different times of the year".

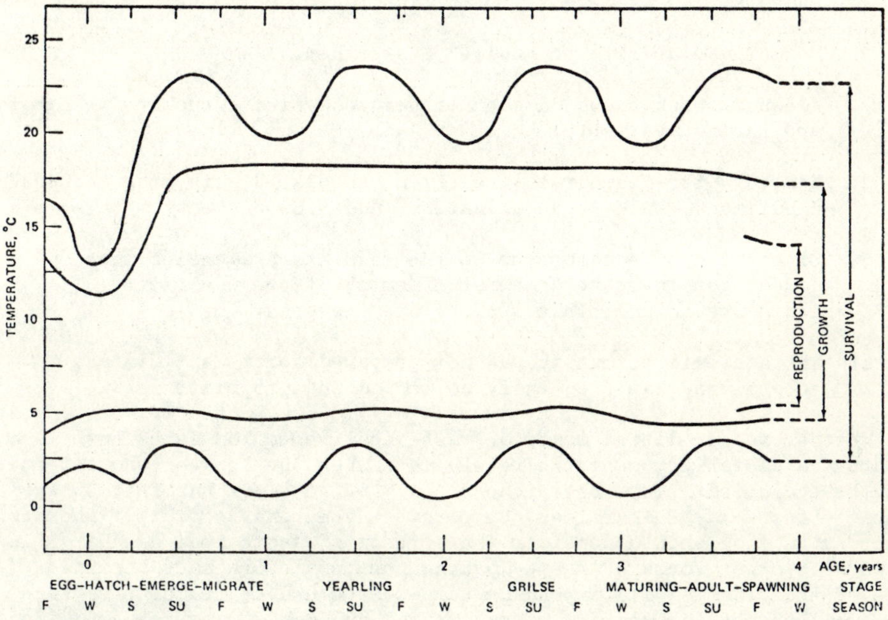

Fig. 3 Schematic representation for minima and maxima of different life processes in Pacific salmon from data compiled by Brett (1969).

From all the experimental studies carried out so far, it does not appearr that discharge of heated effluents have had major impacts on the aquatic ecosystem, viz. no massive fish kills have been reported, which could be directly attributed to thermal discharges. It is, however, important to note that the combined effects of heat and other pollutants could be serious. For example, power plants often treat cooling waters with chemicals to control corrosion, neutralize pH or control biological fouling. These chemicals are often discharged to the receiving waters, along with the thermal discharges. There is very little experimental information available on the combined effects of temperature and other chemicals on different plant or animal species.

Some evidence exists on the effects of temperature on the development of radiation damage in organisms and cells. For example, it has been observed that radiation of effects on Ascaris eggs were greater when they were kept at higher temperatures after irradiation (Strefer, 1975). Studies of temperature effects on the survival rate of mammalian cells after irradiation indicated that the lowest sensitivity was at 37°C. If the temperature was increased to 39°C or 41°C, the recovery from sublethal damage was considerably impaired. Much work, however, remains to be done on these types of possible synergistic effects before any definitive generalized statements can be made.

Cairns (1976) has suggested six reasons for the difficulty of documenting the subtle perturbations of aquatic ecosystems due to thermal discharges. These are the following:

1. They are difficult to distinguish from normal successional changes.

2. Normal variability or "noise" masks them.

3. Adequate background data on either ecosystem structure or function are rarely available.

4. Heated water-discharges are often associated with other stressful conditions, such as chlorination, addition of heavy metals, etc.

5. Since it has been more up to the regulatory agencies than on discharges to prove "harmful effects," there was little incentive to undertake appropriate research.

6. The academic community was not prepared for this task, and accordingly has frequently collected inappropriate data.

Whether the predicted scale of waste heat rejection is potentially harmful (above a socially acceptable level) in either the long or short term, remains to be evaluated. Some researchers claim effects are harmful; others suggest that effects may be harmless, or perhaps even beneficial to fisheries. Any further evaluation should consider the effects of waste heat discharge in conjunction with other forms of waste loading for total drainage basins. The nature of the effects will depend heavily on the unique features of each situation, which poses the problem of evaluating and monitoring the condition of each particular ecosystem. Mathematical models have largely followed the basic DO-BOD relationship with very little attempt at modelling other quality parameters (Biswas, 1979). Development of comprehensive models necessitates large amounts of data which are usually unavailable and often have to be interpolated. Models, may, however, permit the planner to gain a broader perspective on a complex problem. It is evident that considering the costs and time involved, total investigation is seldom practical. Therefore, investigation should perhaps be directed towards identifying and selectively studying the more meaningful parameters such as key species (presence/absence), and species diversity. In extending this, however, to developing generalized indices of environmental quality, there is a danger that these can be easily misused or misinterpreted (Biswas and Cook, 1974).

Much has been written about possible beneficial uses of thermal discharges (Belter, 1975; IAEA, 1977, 1975a, Biswas and Cook, 1974; Lee and Sengupta, 1977) from both fossil fuel and nuclear power stations. Among possible uses cited are agriculture, aquaculture, space heating and cooling, industrial applications - including waste and water treatment, and extension of navigation season. For example, by controlling the water temperature in winter, it is possible to increase and control the growth rates of shrimps, prawns, catfish, oysters and other species. Use of heated irrigation water can extend the growing season, the prevent damages occurring from early or late frosts.

The author has carried out a comprehensive analysis of all possible uses of waste heat around the world (Biswas and Cook, 1974). From this study, the following conclusions emerge:

1. Waste heat from power stations is low temperature heat. Normally its energy is so degraded in temperature that its possible practical uses are severely restricted.

2. Power stations require tremendous amount of cooling water. A A single 1000-MWe nuclear plant can use 46 to 60 m^3/sec of cooling water. To get a beeter perspective, it is approximately equivalent to the water requirements of a city like Chicago. Some existing and proposed sites will have four to ten such units. In 1975, the United States had eighteen sites having three or four such units. It is unlikely that such a tremendous quantity of heated water can be used beneficially.

3. Most of the proposed beneficial uses are for winter, when thermal pollution problems are normally not severe. There is very little need for thermal discharges in summer, when thermal pollution problems are generally acute.

4. From the above three facts, it is evident that the prospect of using thermal discharges to alleviate thermal pollution problems significantly in the future is not very promising.

LAND USE

With regard to possible impacts of nuclear energy on land use, the principal impact is due to mining and milling of uranium ore. Since these impacts have radiological implications, they are not within the scope of this paper. Accordingly, only a very brief account of the extent of the problem will be given. Naturally, land distributed from any mining operation depends on the concentration of the ore. If the current uranium ore concentration of around 0.2 per cent is considered, it has been estimated that some 20 to 50 acres of land has to be disturbed annually in order to obtain enough uranium to support a 1,000 MWe nuclear power plant. Furthermore, another 30 to 70 acres of land will be required for milling operations and for storage of tailings for a 1,000 MWe plant per year for 0.2 per cent uranium (Nuclear Energy Policy Study Group, 1977). When richer and shallower uranium ores become depleted with time, naturally more land area has to be disturbed to obtain the same amount of uranium. Accordingly, at more concentrations of 0.03 to 0.07 per cent, the land requirements for a 1,000 MWe nuclear plant increases to about 100 to 400 acres per year. This figure is comparable to the land requirements for the coal cycle to generate 1,000 MWe per year (Nuclear Energy Study Group, 1977).

Land used by individual nuclear power stations vary considerably. The U.S. Nuclear Regulatory Commission (1976) analysed the land requirements of 72 existing or planned nuclear power stations in the United States. The average value for the 72 stations was 11 megawatt per acre. Sixteen stations used less than 10 MWe/acre, whereas 20 had density higher than 20 MWe/acre. The Indian Point Nuclear Generating Plant 1, 2, and 3, on the Hudson River - some 24 miles north of New York City, occupies a land area of 235 acres for 2,100 MWe, or 9 MWe/acre. However, it should be noted that the plant uses once-through cooling, which requires a minimum site area. Land area used also depends on factors like value of the land and the scarcity of available land.

An indication of the land area necessary for possible future nuclear energy centres can be obtained from a recent report prepared by the U.S. Nuclear Regulatory Commission (1976). Assuming that the reactors are grouped in clusters of four, a nuclear energy centre site will occupy a maximum area ranging from 75 square miles for a 40-reactor centre to about 19 square miles for a 10-reactor centre.

Land requirements for cooling systems for nuclear power plants differ quite widely. For a 1200-MWe nuclear plant, following are typical land areas necessary. For once-through cooling towers will usually take no more than about 16 acres. Thus, land required for a cooling tower is not necessarily a major component of total land occupied by a nuclear plant, since a typical 1,000 MWe nuclear station without a pond requires more than 200 acres. Land requirements for cooling ponds and canals vary widely, even for areas having identical meteorological conditions. As a rule, the larger the pond area, lower is the temperature of cooling water available to the condenser. For a 1,000 MWe plant, the areas could range from 800 to 3,000 acres, and many exceptions to the extremes of this range are not exactly unusual (U.S. Nuclear Regulatory Commission, 1976). Thus, cost of land available is a principal factor determining the land area to be used. An analysis of areas that cooling ponds occupy in the United States, for all nuclear units in use or planned, indicates an average pond utilization area of 2.1 acres per MWe, with minimum and maximum of 0.68 and 3.6 respectively. Minimum land areas required for spray ponds are typically 50 to 100 acres per 1,000 MWe (US Nuclear Regulatory Commission, 1976).

It has been estimated that for pressurized light-water reactors, the land requirements for all operations is about 77 km^2 (ECE, 1978).

The corridors for transmission lines require commitments of large strips of land. Land areas used as rights-of-way for transmission lines are long and narrow, usually ranging from 110 to 200 feet for single lines to as much as 500 to 1200 ft for multi-line corridors. According to the Solandt Commission (1975), one double-circuit 500-kV line requires a right-of-way width of 250 ft, but for two such lines, rights-of-way will at least be 425 ft wide, and may occasionally extend to more than 1,200 ft.

Land area necessary often depends on the voltage being used. In the United States, the highest transmission voltage in 1892 was only 10-kV. By the mid-1950's, transmission began at 345-kV, and by the late 1960's, it increased to 765-kV (Janes, 1977). By 1971, 22 percent of all the U.S. transmission was at 500-kV, and it is estimated that by 1980, 15 percent of transmission capability will be on 765-kV lines (Janes, 1977). Extensive research has already been done at 1500-kV level (EPRI, 1975). Wider corridors may be necessary for higher kV lines, but on the other hand, fewer corridors may be necessary. Depending on length, one 765-kV line could replace two 500-kV or four 345-kV lines (U.S. Nuclear Regulatory Commission, 1976).

While it is unanimously agreed that substantial land is used for transmission lines, estimates differ somewhat for land requirements per mile of right-of-way. A booklet published jointly by the U.S. Department of the Interior and the Department of Agriculture (1970), estimated that approximately 3,600 electric utilities in the United States "operate over 3,000,000 miles of overhead electric transmission lines, which present nearly 4 million acres of land for right-of-way". It was forecast at that time approximately 100,000 miles of new transmission lines will be constructed on 1½ million acres of right-of-way, in each decade for the balance of this century. These figures have been basically accepted by later investigators (Hendrikson, et al., 1974; Jackson, et al., 1978; Kitchings et al., 1974), even though one of the basic assumptions was that the demand for electric power will double every decade.

The above estimate indicates an average land requirement of about 13 acres per mile of right-of-way. This is approximately the same (12 acres per mile or 30,000 m^2/km), as estimated by Hammond et al., (1973) and the Economic Commission for Europe (1975). In contrast, the U.S. Nuclear Regulatory Commission (1976) provides an estimate of 24 to 37 acres per mile. The estimates, however, are not all that widely incompatible as they appear on the first sight. The difference lies in the width of the corridor considered:

110 ft for the USDI-USDA estimate (Hendrikson et al., 1974), and 200 ft to 300 ft by the U.S. Nuclear Regulatory Commission (1976).

One possible way of looking at the environmental implications of high voltage transmission lines is to compare them with other linear systems that compete for land, viz, pipelines, expressways and railways. This is shown in Table 5 (Solandt Commission, 1975). As noted in Table 5, there are secondary uses of right-of-way. It should also be noted that the environmental implications of transmission lines remain the same, irrespective of whether electricity is generated by nuclear or a coal cycle.

OTHER IMPLICATIONS

There are other environmental implications of nuclear energy, which are worth considering. Among these are aesthetic aspects, habitat alterations and atmospheric effects.

Aesthetic Aspects

Normally nuclear power stations are architecturally well-designed. Major aesthetic problems arise with cooling towers and circuit towers for transmission lines. Because of technical requirements, very little can be done to change the design of cooling towers. However, with the increasing recognition of the environmental dimensions of energy production, distribution and use, many regulating agencies and utility companies have retained landscape architects and environmental planners to improve the current situation. Environmental design criteria are being continually developed and updated so as to minimize adverse environmental consequences. One example is the architecturally designed towers, using tubular steel construction, that can replace lattice steel structure. The new tower is more pleasing to the viewer, and may cost 125 percent or more than conventional steel towers.

Extensive public hearings undertaken by the Solandt Commission (1975) in Ontario, Canada, give a good picture of public attitudes to and perception of electric transmission lines. In the case of a 500 kV line, a single circuit tower is 120 ft high and a tower capable of carrying two circuits is 162 ft high. Such 162 ft high towers are all too highly visible under most circumstances. The Solandt Commission came to three important conclusions, so far as public preferences are concerned:

1. Where several circuits are required to achieve the necessary capacity and security, these circuits should be placed on as few towers as possible.

2. Where several lines of towers have to be used between two points, they should be kept side by side throughout the entire route, rather than running separately on different rights-of-way.

3. Where there is more than one line in the same right-of-way, the preference is to have towers located in lines abreast.

There is no doubt that the visual impact of transmission lines can be significantly reduced by putting them underground. Current estimates, however, indicate that even though the present-day costs can be reduced by better design methods, costs would still be approximately 10 to 15 times overhead line costs (U.S. Nuclear Regulatory Commission, 1976).

It should also be emphasized that contrary to popular belief, underground transmission lines will not eliminate land use impacts. Even though the diameter of the cable would be only between 12 to 36 inches, the right-of-way requirements may exceed 100 ft due to the necessity of having access roads, adjacent storage facilities, adequate space to allow earthmoving equipments to operate, and a place to deposit excavated soil. Furthermore, the need to have free and ready access in order to monitor or carry out repair work could, unlike overhead transmission lines, severely restrict the use of the corridor for farming purposes. It has also been suggested that underground high voltage lines may effect some types of surface vegetation (U.S. Nuclear Regulatory Commission, 1976). If this proves to be correct, possible alternative uses of such corridors can be further restricted.

Habitat Alterations

It is important to prevent trees from growing into or falling onto transmission lines, which could interrupt service. Thus, herbicides are often used to control utility rights-of-way. Plants exceeding 6 ft are usually cut before herbicides are applied. One problem that often arises from such spraying of herbicides is the drift that eventually effect non-target foliage. Also, extensive research at Hubbard Brook, New Hampshire, (Hitchings et al., 1974) shows that where herbicide was applied after all trees and shrub were cut and dropped in place, the cut area had an accelerated loss of nutrients, with losses of cations 3 to 20 times as that from undisturbed systems. Increased erosion is often noticed, indicating soil losses.

Transmission lines can also have adverse impacts when crossing wetland areas. There is some evidence to indicate that behavioral modifications may occur for waterfowls, which could result in the absence of birds covering an area within a $\frac{1}{4}$ mile of the transmission lines. The swaying of the lines in the wind, their reflective properties, and the humming of the lines could explain the abnormal behavoir of the birds. Direct current transmission could also have effects on migratory birds using magnetic homing (U.S. Nuclear Regulatory Commission, 1976).

Electric fields associated with a transmission line can produce a charge on animals or human beings, within the range of its influence. For example, at certain locations, ground level field strengths can approach appreciable values like 10-kV/m for a 46-ft high 765-kV line (Janes, 1977). In comparison, the earth's DC electric field is about 0.13-kV/m, and can be as high as 3-kV/m under thunder clouds, even in the absence of local lightning. Questions have been raised on the resultant effects of this displacement of currents on biological systems, but this is normally well below the values generally accepted as "safe let-go" levels. The startling effect of this shock is probably the greatest hazard.

Comprehensive reviews of existing studies on biological effects of high voltage electric fields are available (Kaufman and Michaelson, 1974; Bridges, 1975, 1977). Kaufman and Michaelson (1974) conclude that "research to date has failed to provide convincing evidence that human exposure to stationary or low-frequency electric fields has any harmful biological effect". According to Bridges (1975), "although the great bulk of evidence suggests that there are no significant effects of electric fields encountered under extra-high voltage lines, further research is needed".

Atmospheric Effects

Possible atmospheric effects of nuclear power plants are mainly local.

Table 5

COMPARISON OF FEATURES OF LINEAR LAND USES

	500 kV Transmission Line	Pipeline	4 Lane Controlled Access Expressway	Railway
Right-of-way Width (Typical)	250' - 600'	60'	150' - 300'	60' - 100'
Effect as Physical Barrier	Minimal except at towers (every 900 feet)	Minimal	Yes	Yes
Secondary Uses of Right-of-way A. Farming B. Forestry/Orchard	 Yes Yes, (with 25' height restriction)	 Yes	 No No	 No No
Local Noise	Not serious	None (except at pumping stations)	Considerable (often serious)	Intermittent
Atmospheric Pollution	Negligible (ozone)	No	Yes	Intermitten
Visibility (at a distance)	162' high towers and lines, often visible at great distance	Limited and Local	Limited and Generally Local	Limited and Generally Local

Table 6

ENERGY FLUX OF POWER STATIONS AND CITIES

Process	Area (km^2)	Energy flux density (W/m^2)	Fraction of average solar flux at ground
Waste heat from power stations			
Dresden and Braidwood (over area for city of 1 million	230	35.3	0.22
Summit, Salem, Hope Creek (12 mi x 5 mi)	155	73.8	0.46
Typical cooling pool (Assumed 2 acres/ MWe generation)	-	247	1.54
Typical cooling tower (Chalk Point Design Data)	8.9×10^{-3}	128,000	800
48,000-MWe NEC with area of 48,000 acres	194	495	3.1
48,000 NEC with area of 8,000 acres	33	2,970	18.5
Anthropogenic heat from cities			
Manhattan, New York City	59.8	630	3.94
Moscow	878	127	0.79
Washington, D.C.	173	44	0.28
Los Angeles	3,500	21	0.13
Boston-Washington metropolitan area (projection for year 2000)	31,200	36	0.23
Sheffield, England	48	19.2	0.12

There have been a few reports of localized fogs due to cooling towers. Table 6 shows energy flux densities from power stations, as well as anthropogenic heat from some major cities (U.S. Nuclear Regulatory Commission, 1976). Similarly, studies (Fern and Brabets, 1974) indicate that contributions of high voltage transmission lines to the increase of ozone levels are negligible.

CONCLUSIONS

The different non-radiological implications of nuclear energy are reviewed in this paper. While it indicates that there are some environmental problems that exist at present, and a few more could emerge in the future, all these problems are solvable with the present state of technological development, and the extent of environmentgdegradation can be significantly reduced by appropriate design and planning.

ACKNOWLEDGEMENTS

Grateful acknowledgements are made to Dr. L Farges, International Atomic Energy Agency, Vienna; Dr. W. Osburn, U.S. Department of Energy, Washington, D.C.; Dr. William Ramsey, Resources for the Future, Washington, D.C.; Dr. E. El-Hinnawi and Dr. F Sella, United Nations Environment Programme, Nairobi, Kenya; for their assistance in preparation of this paper. The opinions expressed, however, are those of the author, and not necessarily of the scientists mentioned above.

REFERENCES

Abrahamson, D.E. Ecological Hazards from Nuclear Power Plants, in "Careless Technology," Edited by M.T. Farver and J.P. Milton, Natural History Press, Garden City, New York, pp 798-811 (1972)

Belter, W.G. Management of Waste Heat at Nuclear Power Stations, Its Possible Impact on the Environment, and Possibilities of Its Economic Use, in "Environmental Effects of Cooling Systems at Nuclear Power Plants," IAEA, Vienna, pp. 3-23 (1975)

Biswas, Asit K. Models for Water Quality Management, McGraw-Hill Publishing Co., New York, to be published (1979)

Biswas, Asit K. Environmental Implications of Energy Development, Engineering Issues, American Society of Civil Engineers, Vol, 103, No. EI 1, Proc. Paper 12681, pp. 49-59 (1977)

Biswas, Asit K. Energy and the Environment, Planning and Finance Service, Environment Canada, Ottawa, Report No. 1, 42 p. (1974)

Biswas, Asit K. and Cook, B., Beneficial Uses for Thermal Discharges, Planning and Finance Service, Environment Canada, Ottawa, Report No. 2, 173 p. (1974)

Brett, J.R., Resumé - Temperature and Fishes, Chesapeake Science, Vol. 10, pp. 275-276 (1969)

Bridges, J.E. Environmental Considerations Concerning the Biological Effects of Power Frequency (50 ot 60 Hz) Electric Fields, in Proceedings IEEE PES Winter Meeting, Paper F77-256-1, Institute of Electrical and Electronics Engineers, New York (1977).

Bridges, J.E., Biologic Effects of High Voltage Electric Fields, Report 381-1, Electric Power Research Institute, Palo Alto, California (1975)

Brock, T.D., Predicting the Ecological Consequences of Thermal Pollution from Observations on Geothermal Habitats, in "Environmental Effects of Cooling System Systems at Nuclear Power Plants," IAEA, Vienna, pp. 599-621

Cairns, J., Heated Waste-Water Effects on Aquatic Ecosystems, in "Thermal Ecology II," Edited by G.W. Esch and R.W. McFarlane, ERDA, Washington, D.C., pp. 32-38 (1976)

Cairns, J., Ecological Management Problems Caused by Heated Waste Water Discharge in the Aquatic Environment, Water Resources Bulletin, Vol, 6., pp. 868-878 (1970)

Coutant, C.C., Temperature Selection by Fish - A factor in Power-Plant Impact assessments, in "Environmental Effects of Cooling Systems at Nuclear Power Plants," IAEA, Vienna, pp. 575-595 (1975)

Duckworth, H.E., Duckworth, H.W., Porter, A., and Rogers, J.S., Environmental Aspects of Nuclear Power Development in Canada, Canadian Environmental Advisory Council, Occasional Paper No., 2, Environment Canada, Ottawa, 53 p. (1977)

Economic Commission for Europe, January 12, Environmental Aspects of Energy Production and Use, With Particular Reference to New Technologies, Report ENV/r. 43/Add. 2, 15 p. (1978)

Economic Commission for Europe, December 15, Environmental Aspects of Energy Production and Use, With Particular Reference to New Technologies, Report ENV/r. ECE, Geneva, 42 p. (1975)

Electric Power Research Institute, Transmission Line Reference Book 345 kV and Above, Fred Weidner and Son, New York (1975)

Esch, G.W., and McFarlane R.W., Editors, Thermal Ecology II Proceedings of a Symposium, Augusta, Georgia, April 2-5, 1975, ERDA Symposium Series CONF-750425, Energy Research and Development Administration, Washington, D.C., 404 p. (1976)

Fern, W.J., and Brabets, R.I., Field Investigations of Ozone Adjacent to High Voltage Transmission Lines, IEEE Paper T 74 057-6, Institute of Electrical and Electronics Engineers, New York (1974)

Fry, F.E.J., The Effect on Environmental Factors on the Physiology of Fish, in "Fish Physiology, Edited by W.S. Hoar and D.J. Randall, Vol 6, Academic Press, New York, pp. 79-134 (1971)

Gibbons, J.W., Thermal Alteration and the Enhancement of Species Populations, in "Thermal Ecology II," Edited by G.W. Esch and R.W. McFarlane, ERDA, Washington, D.C., pp. 27-31 (1976)

Hammond, A., Metz, W., and Maugh, T., Energy and the Future, American Association for the Advancement of Science, Washington, D.C., p. 102 (1973)

Harleman, D.R.F., and Stolzenbach, K.D., Fluid Mechanics of Heat Disposal From Power Plant Generation, Annual Review of Fluid Mechanics, Vol. 4, pp. 7-32 (1972)

Heath, W.G., Ecological Significance of Temperature Tolerance in Gulf of California Shore Fishes, Journal, Arizona Academy of Sciences, Vol. 4, pp. 172-178 (1967)

Hendrickson, P.L., Bahl, R.W., Gray, B.A., and Maynard, W.S., Measuring the Social Attitudes and Aesthetic and Economic Considerations Which Influence Transmission Line Routing, Report BNWL-1837, Battelle Pacific Northwest Laboratories, Richland, Washington, pp. 1-7 (1974)

Hochachka, P.W., and Somero, G.N., Strategies of Biochemical Adoption, W.B. Saunders Co., Philadelphia (1973)

Hutchison, V.H., Factors Influencing Thermal Tolerances of Individual Organisms, in "Thermal Ecology II," Edited by G.W. Esch and R.W. McFarlane, ERDA, Washington, D.C., pp. 10-26 (1976)

International Atomic Energy Agency, Urban District Heating Using Nuclear Heat, Proceedings of a Symposium, Vienna, March 15-19, 1976, Doc. STI/PUB/461, IAEA, Vienna, 209 p. (1977)

International Atomic Energy Agency, Directory of Nuclear Reactors, Vol. X Doc. STI/PUB/397, 402 p. (1976)

International Atomic Energy Agency, Environmental Effects of Cooling Systems at Nuclear Power Plants, Proceedings of a Symposium, Oslo, August 26-30, 1974, Doc. STI/PUB/378, IAEA, Vienna, 831 p. (1975a)

International Atomic Energy Agency, Combined Effects of Radioactive, Chemical and Thermal Releases to the Environment, Proceedings of a Symposium, Stockholm, June, 2-5, 1975, Doc. STI/PUB/404, IAEA, Vienna, 360 p. (1975b)

International Atomic Energy Agency, Thermal Discharges at Nuclear Power Stations: Their Management and Environmental Impacts, Doc. STI/DOC/10/155, IAEA, Vienna, 155 p. (1974)

International Atomic Energy Agency, Nuclear Power and the Environment, Doc. STI/PUB/321 (Rev.), IAEA, Vienna, 85 p. (1973)

International Atomic Energy Agency, Environmental Aspects of Nuclear Power Stations, Proceedings of a Symposium, New York, August 10-14, 1970, Doc. STI/PUB/261, IAEA, Vienna, 970 p. (1971)

Jackson, R.H., Hudman, L.E., and England, J.L., Assessment of the Environmental Impact of High Voltage Transmission Line, Journal of Environment Management, Vol.6, No. 2, pp. 153-170 (1978)

Janes, D.E., October, Background Information on High Voltage Fields, Environmental Health Perspectives, Vol. 20, pp. 141-147 (1977)

Kaufman, G.E., and Michaelson, S.M., Critical Review of the Biological Effects of Electric and Magnetic Fields, in "Biologic and Clinical Effects of Low-Frequency Magnetic and Electric Fields," Edited by J.G. Llaurado et al., Charles C. Thomas, Springfield, Illinois (1974)

Kitchings, J.T., Shugart, H.H., and Story, J.D., Environmental Impacts Associated with Electric Transmission Lines, Report ORNL-TM-4498, Oak Ridge National Laboratory, Oak Ridge, Tennessee, pp. 1-41 (1974)

Lee, S.S., and Sengupta, S., Waste Heat Management and Utilization, Proceedings of a Conference, Miami Beach, May 9-11, 1976, 3 Vols., Mechanical Engineering Department, University of Miami, Coral Gables, Florida (1977)

Nuclear Energy Policy Study Group, Nuclear Power Issues and Choices, Ballinger Publishing Co., Cambridge, Mass., 418 p. (1977)

Solandt Commission, Transportation: A Public Inquiry into the Transmission of Power Between Lennox and Oshawa, Report to the Government of Ontario, Toronto, 213 p. (1975)

Strefer, C., Interaction Mechanisms of Radioactive, Chemical and Thermal Releases from the Nuclear Industry, in "Combined Effects of Radioactive, Chemical and Thermal Releases to the Environment, "IAEA, Vienna, pp. 3-11 (1975)

Sylvester, J.R., Possible Effects of Thermal Effluents on Fish: A Review, Environmental Pollution, Vol. 3, No. 3 pp. 205-216 (1972)

United Nations Environment Programme, Review of the Impact of Production and Use of Energy on the Environment, Report of the Executive Director, Document GC/61/Add. 1, UNEP, Nairobi, Kenya, 107 p. (1976)

U.S. Atomic Energy Commission, Office of Planning and Analysis, Nuclear Power Growth 1974-2000, Report WASH-1139, AEC, Washington, D.C. (1974)

U.S. Department of the Interior, and Department of Agriculture, Environment Criteria for Electric Transmission Systems, U.S. Government Printing Office, Washington, D.C., p. iii (1970)

U.S. Nuclear Regulatory Commission, Office of Special Studies, Nuclear Energy Centre Site Survey - 1975, Part III, Technical Considerations, U.S. Nuclear Regulatory Commission, Washington, D.C. (1976)

Workshop on Alternative Energy Strategies, Energy: Global Prospects 1985-2000, McGraw-Hill, New York, 291 p. (1977)

5

THE ENVIRONMENTAL IMPACT OF RADIOACTIVE RELEASES FROM ACCIDENTS IN NUCLEAR POWER REACTORS

J. R. Beattie, R. F. Griffiths, G. D. Kaiser and G. H. Kinchin

Safety and Reliability Directorate
United Kingdom Atomic Energy Authority

INTRODUCTION

Nuclear power as a means of producing electrical energy is well established in the United Kingdom, and the thermal neutron power reactors now in operation and under construction there will provide nearly 10 GW(e) of the nation's electrical generating capacity. Nuclear power reactors that are considered in this paper and examples of which exist in the United Kingdom include thermal neutron reactors (gas-cooled) and fast neutron reactors (sodium-cooled). No serious accident having any significant environmental impact has ever occurred to any of these reactors in the United Kingdom, and because of the extraordinarily thorough measures taken throughout all stages of design, construction and operation (as well as safety supervision and licensing procedures) it is thought that none is likely to occur in the near or not-so-near future. Nevertheless, 'scenarios' for such accidents have been envisaged and have been the subject of intensive and prolonged scientific predictive studies, it being acknowledged that accidents causing significant radioactive environmental impact could indeed be realised, though only with a very low probability (as a result of a highly unlikely combination of what are themselves individually highly unlikely circumstances and events). The fundamental reason for designating the power reactors as potentially, at least, a main source of accidental radioactive environmental impact, is that any such reactor combines in one installation (namely the fuel elements which together form the power-producing core of the reactor) a very large inventory of radioactive substances (approximately 10^9 - 10^{10} Curies) and a variable energy source (up to about 3×10^9 Watts thermal power) from which heat must be removed at an adequate rate (partly as electrical power and partly as waste heat according to circumstances and requirements) and with adequate reliability.

By comparison with the possible consequences of accidents to nuclear power reactors, the consequences and estimated risk levels from accidents thought possible for fuel reprocessing plants are small. For this reason, fewer studies of this kind have been carried out for such plants, and the methods of probability - consequence analysis have been developed to a lesser extent than has been the case for reactors. It was therefore considered inappropriate to include a discussion of reprocessing plant accidents in this paper, which is devoted exclusively to accidents in nuclear power reactors. It should also be stated that it is accidental releases of radioactive material to the atmosphere that this paper will consider. Accidental releases to the aquatic environment, and in particular to the sea, are unlikely to have serious effects on the environment and are not considered.

Hazards of Radiation and of Radioactive Substances Accidentally released from Nuclear Installations

X-Rays and radioactivity were discovered in 1895 and 1896, and the practical measures required for protection were first formalised internationally in 1928; later the International Commission on Radiological Protection (ICRP) was established. The Recommendations of ICRP regularly reviewed and published at intervals over the years embody today more than fifty years of experience. The most recent example of the Commission's Recommendations is ICRP Publication 26 (1977), which as ever proclaims a philosophy of protection derived from the most up-to-date information and the best sources of knowledge and wisdom in the field. The observance of ICRP Recommendations over the years has resulted in an almost total absence of proved radiation-indiced injuries among the many thousands of people who have worked in the rapidly expanded nuclear industry since the 1940's, and in the remarkably good health record of this industry despite the very large quantities of radioactive materials it has created and used. There is no evidence of any injury or deleterious effect on the environment, or on the members of the general public who live in that environment, from the small and carefully controlled quantities of gaseous and liquid effluents routinely discharged in the course of normal operations from power reactors in the United Kingdom. Routine radioactive effluent discharge rates, and any direct radiation from reactors, are so limited by design provisions and by operational control that it is in the highest degree unlikely that there is any measurable or significant effect on the lives of flora, fauna, domestic animals or man; that this is so is regularly checked by measurements in the environment.

There have never been any accidents to large electricity-producing power reactors in operation (or in the shut-down state) in the United Kingdom, or for that matter, elsewhere in the world, that have caused to be released to the atmosphere as much radioactivity as would make any significant impact on man or his environment. A rational man might well ask why one troubles oneself by studying the subject of the environmental impact of nuclear power reactor accidents, for if there are no accidents, there is presumably no impact. However, as was said above, a large power reactor contains large quantities of radioactive substances, many of them dangerous if liberated. Some are radioactive fission products such as xenon, krypton, iodine, caesium, ruthenium and strontium (named here as important examples); others are the actinides, mainly plutonium, americium and curium, created by successive neutron capture in uranium and other heavy elements. Some are readily liberated by heat or chemical action, but many others such as the actinides are much less easily set free. The problem is that despite the highest engineering skills being deployed and all reasonable precautions being taken there will always be a small but non-zero probability that an accident could happen, and that therefore an uncontrolled release of radioactivity to the atmosphere might occur; for this reason one studies the possibilities in order to be prepared. The environmental impact could, if circumstances combined to produce the worst or nearly the worst, be large, with land and dwellings contaminated with radioactivity for some distance downwind and people killed or injured in the area. However, for the gas-cooled graphite-moderated reactors presently operating in the United Kingdom, such an accident (and its outcome) is not considered credible: the basic design of these reactors, as is explained later, implies that the probability of an accident occurring which could lead to such consequences is believed to be negligibly small. Later in the paper one attempts to define a low level of probability per annum which may be deemed consonant with "incredible". In this introduction we merely note that the design accident, referred to as the "Maximum credible accident" for such reactors is a loss of carbon dioxide coolant pressure leading to a release of around $10^2 - 10^3$ Curies of iodine 131, which would invoke the pre-arranged emergency arrangements to protect local people from the effects of inhaling radio-iodine and a temporary ban on local milk being consumed.

Pressurised Water Reactors (PWR's) may be introduced into the United Kingdom. These will be an addition and alternative to the Advanced Gas-cooled Reactors (AGR's) now in service, alongside the "Magnox" gas-cooled reactors that have formed the first stage, about 4 GW(e), of the United Kingdom nuclear power programme. The environmental impact of accidents to PWR's has been examined by Rasmussen et al in the Reactor Safety Study published by the United States Nuclear Regulatory Commission (1975). Analogous studies relating to a PWR accident on a possible site in the United Kingdom are being carried out by staff of the Safety and Reliability Directorate (SRD) of the United Kingdom Atomic Energy Authority (UKAEA) using SRD's computer code TIRION (Kaiser, 1976; Fryer, 1978).

The UKAEA for many years has studied the possible environmental consequences of hypothetical accidents to fast reactors. This interest, beginning in the 1950's with the first UK fast reactor at Dounreay, continued in the 1970's with the much larger 250 MW(e) Prototype Fast Reactor (also at Dounreay), and is now concentrated on future designs for a Commercial Demonstration Fast Reactor (CDFR) of about 1000 MW(e). This is being undertaken in consultation with the Nuclear Power Company (NPC) and the Electricity Generating Boards (CEGB and SSEB). Studies have been carried out by SRD staff of UKAEA using SRD's computer code TIRION, by research staff of the Berkeley Laboratories of the Central Electricity Generating Board (BNL and CEGB) using BNL's code WEERIE (Clarke, 1976), and also by staff of the National Radiological Protection Board (NRPB) using the SRD code TIRION (Kelly et al, 1977).

For any reactor system the problem is - "What is a credible accident?" (in simple terms how severe and how unlikely is the worst accident we must consider?) Definitive answers do not appear to be possible. This paper responds to the difficulty by describing and enumerating the environmental impact of typical and representative accidents, and offers only general answers to questions on their probability of occurrence.

Mention of probability leads inevitably to discussions of risk, since risk is most commonly defined as the product of consequence (however measured) and probability of occurrence per unit time. Two aspects of risk emerge as being of special interest, namely <u>Individual Risk of Death</u>, expressed as the number of deaths per year from a particular cause divided by the total population, and <u>Community Risk of Death</u>, which may be expressed as the frequency of occurrence of an accident in which N or more people are killed. Community Risk may also be expressed as the frequency of occurrence of an accident in which a number of people in the range N_1 to N_2 are killed. The two forms convey the same information in different ways, and the first can be converted to the second by differentiation or subtraction (and <u>vice versa</u> by integration or summation). Definition of the whole risk cannot be achieved unless information on the whole spectrum of values of N is available. In this paper Community Risk is expressed exclusively in terms of the frequency of occurrence of an accident in which N or more people are killed. From discussion of risk one progresses to the question of "what is an acceptable risk?", and this is taken up in a later section of this paper. Of course, to state the obvious, acceptable risk is that degree of risk a person will accept. The scientist's answer must be to collect and examine evidence on the magnitude of other risks that people accept at and away from work, voluntarily and involuntarily, and from such comparisons deduce what may be generally acceptable to most rational people.

After a survey of facts in the field of risk studies, the paper reaches conclusions about the acceptability or otherwise of the environmental impact of accidents to nuclear reactors. It is concluded, from the overall study made in this paper, that this environmental impact is of a nature, magnitude and probability that should be acceptable to the people of the United Kingdom.

NATURE OF THE ENVIRONMENTAL IMPACT OF RADIOACTIVE RELEASES

Natural and Artificial Sources of Background Ionising Radiation in the Environment

The environment is permeated by naturally occurring radioactive materials and by ionising radiations from other natural sources. There is a small amount of uranium and associated radioactive substances in almost all rocks and soils, and consequently in all building materials. Potassium, one naturally occurring isotope of which is radioactive, is an essential component of our bodies and contributes about 20% of the natural background radiation we experience. The planet Earth is also exposed to streams of radiation from outer space and from the sun. A substantial part of this cosmic radiation is absorbed in the upper atmosphere but some penetrates to ground level. The radiation doses from the various components of the natural background are summarised in Table 1 - more detailed information can be found in a report by Webb. The cosmic ray component depends on latitude as well as height, and the dose of radiation from the earth depends on the character of the local rocks and soil. Consequently in some localities the total natural doses can be nearly twice that indicated in Table 1. That this background radiation is due to natural causes does not necessarily mean that it is harmless. However, although the background radiation varies from place to place, no correlation has ever been demonstrated with regional variations in the incidence of disease, including cancer.

Table 1 Annual Radiation Doses Making Up the Natural Background

Cosmic rays at sea level	30 millirads
Radiation from the earth	50 millirads
Radiation from within the human body	20 millirads
TOTAL	100 millirads

Some radioactivity to be found in the environment is the result of world-wide fall-out from distant nuclear explosions past and present, and the quantities which are generally small at present are regularly reported on by NRPB (Hunt et al, 1976), UKAEA, Government Departments and Ministries, and UNSCEAR (1977) as appropriate. Probably the most widely distributed sources of man-made radiation in the community are the X-ray sets used for medical diagnosis and the radioactive materials used to make luminous watch and clock dials. There is also a very small amount of radiation from television sets and from some luminous devices other than clocks and watches. None of these sources makes a contribution to the total dose to the population approaching that of natural background.

Routine and generally continuous discharges of radioactive gases and liquids are made from nuclear installations to the atmosphere and to the aquatic environment respectively, under close governmental supervision. These discharges need not concern us in this paper, since this source of radiation and radioactivity makes only a minute contribution to the population dose compared to natural background radiation.

The natural radiation background has been briefly reviewed above since that is the normal radioactive impact of the environment. One might wish to view the environmental impact of nuclear accidents against this normal background. However this would be largely academic, since the potential radiation dose to individual members of the general public from some nuclear accidents would in some circumstances be much larger than the annual dose to the same individual from natural background radiation.

The Effects on Human Health of Various Levels of Radiation and Radioactivity

A population may be exposed to radiation in various ways following the accidental release of radioactivity from a nuclear installation. These include external exposure of the whole or parts of the body to different types of radiation from the ground contaminated with activity deposited from the cloud. In addition, airborne radioactive particles could be inhaled, either during the passage of the cloud or as a result of resuspension of activity, and contaminated foodstuffs or water may be consumed. All this would result in irradiation of either the lungs or gut (or both), and a fraction of the incorporated radionuclides that are soluble in body fluids would translocate from the site of entry to be deposited in and irradiate other tissues of the body. Depending upon the radiation dose, both 'early' and 'delayed' somatic damage could occur to individual members of the exposed population. Also, hereditary or genetic effects could occur among their descendants.

These effects were quantitatively assessed by Rasmussen and his team of co-workers and advisers in Appendix VI of the Reactor Safety Study (USNRC, 1975), and more recently by Smith and Stather of NRPB (1976). Some qualitative aspects are worth mentioning first, since they will assist understanding. The early effects of radiation are dose dependent; in the human there appears to exist a dose, equivalent to about 100 rads for whole body radiation below which clinical effects are unlikely. Above this dose, pathological changes are due mainly to damage to cell membranes allowing leakage of fluids and electrolytes, and mainly also to loss of reproductive capacity of stem cells leading to a diminution in numbers of mature cells (an early effect in vital tissues such as bone marrow and gut because in these there is normally a rapid turnover of cells). Cancer is the main late somatic effect of radiation. The risk of occurrence is assumed to increase linearly with dose and there is believed to be no threshold dose. Radiation-induced cancer differs in no known way from cancer caused by other factors. Regarding genetic effects, although radiation dosage of the gonads may express itself through an increase in the incidence of abortion, the 'late' or 'delayed' effect, if any, will be hereditary, giving rise to abnormalities appearing in the immediate descendants (children or grand-children) or possibly in later generations.

Knowledge of the early effects of radiation in man is derived from studies of individuals exposed to the atom bombs dropped on Hiroshima and Nagasaki, the Marshall Islanders who were contaminated with radioactive fallout, radiation accident cases that have occurred, and patients treated by radiotherapy. Data from these studies are supplemented by other data from animal experiments where necessary. The dose mortality relationship shown in Figure 1 has been proposed in respect of irradiation of the bone marrow resulting from the accidental release of radioactivity in a reactor accident. It will be noted that a dose of

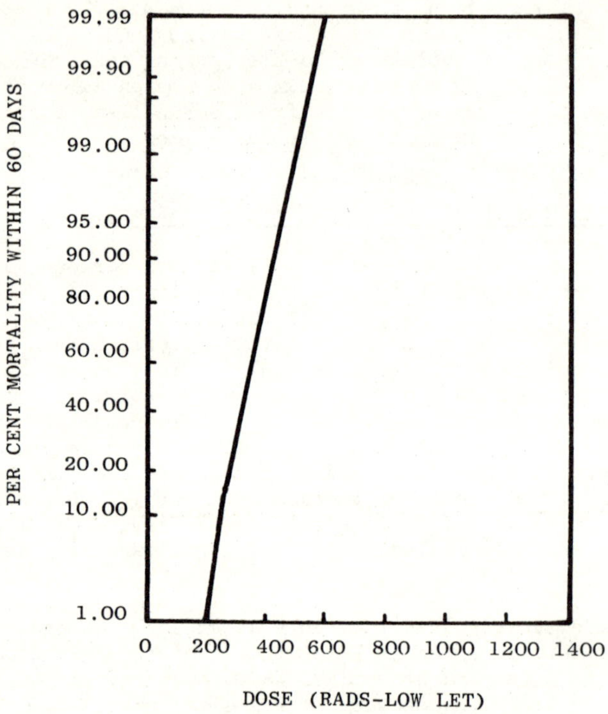

Fig. 1 Estimated dose-mortality curve for whole body irradiation (due to bone marrow damage).

about 200 rads (low LET*) is suggested below which virtually no deaths would occur. For the radionuclides expected to be released from a reactor accident, the external and internal doses to bone marrow should be added. The effective radiation dose to the bone marrow is taken to be that dose received in the first 7 days plus half the dose received between 8 and 30 days (Smith and Stather, 1976). In many of the hypothetical cases of accidental releases from nuclear installations which we shall examine, much or most of the bone marrow irradiation is in fact external gamma radiation from, for example, radioactivity deposited on the ground. Early radiation injury following external radiation represents an overlap of both bone marrow and gastro-intestinal tract injuries. The critical dose to gut required to produce death in man is not known but it has been shown that the gut can recover from doses (of external radiation delivered to the stem cells of the small intes-

* low LET = low Linear Energy Transfer from radiation to tissue, i.e. essentially beta and gamma radiation only.

tine) in excess of 1000 rads (low LET*). The gut dose-mortality relationship used in calculations for this paper is shown in Figure 2. Following a release, irradiation of the lungs may occur from external sources or from internally deposited radionuclides (or both). Adverse reaction of the lungs to radiation may be seen at an early stage as inflammation of the air sacs and the conducting airways (pneumonitis). In severe reactions, death occurs within days due to oedema of the lungs followed by cardiovascular collapse. However, a person may survive the acute phase of pneumonitis only to develop progressive lung changes, which include fibrosis or reduction in elasticity of the air sacs and loss of cilia from the conducting airways. These changes result in difficulty in breathing, a loss of efficiency in gaseous exchange and increased susceptibility to lung infections. Such damage may eventually lead to early death. The lack of numerical data on

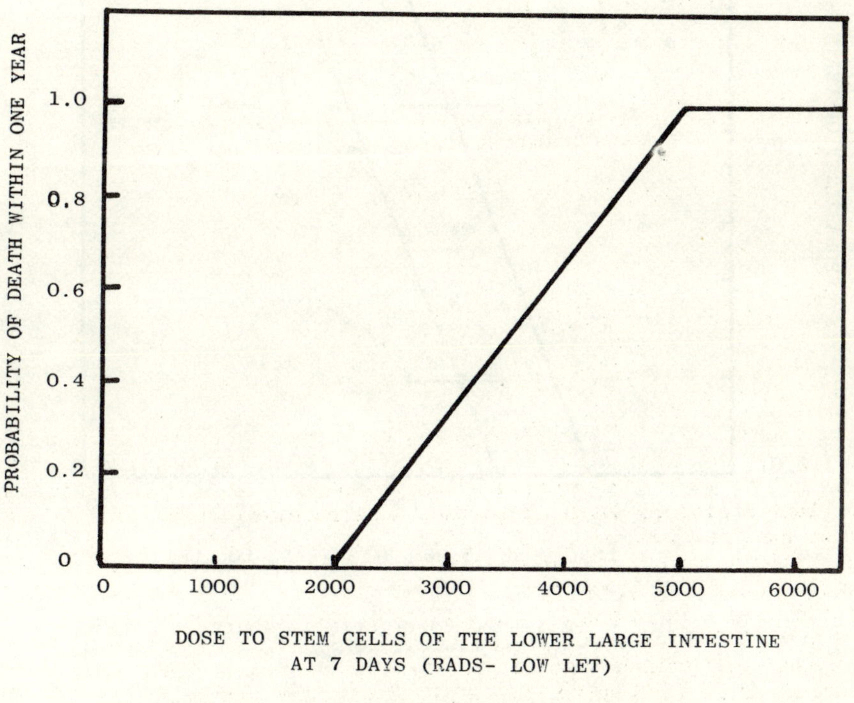

Fig. 2 Estimated dose-mortality curve for irradiation of the lower large intestine.

the effects on humans makes us consider animal data and extrapolate these to man. For example, beagle dogs have been exposed to aerosols of insoluble particles 'tagged' by fusing in yttrium 90, yttrium 91, strontium 90 and cerium 144. The data on yttrium 90 and 91 have been used to derive lung dose-mortality curves for the low LET radiation based on cumulative lung dose up to 1 year, and these are shown in Figure 3. Other data on dogs that had inhaled plutonium 239 suggest that an RBE* of 10 be used for estimating the early effects of alpha-emitters in

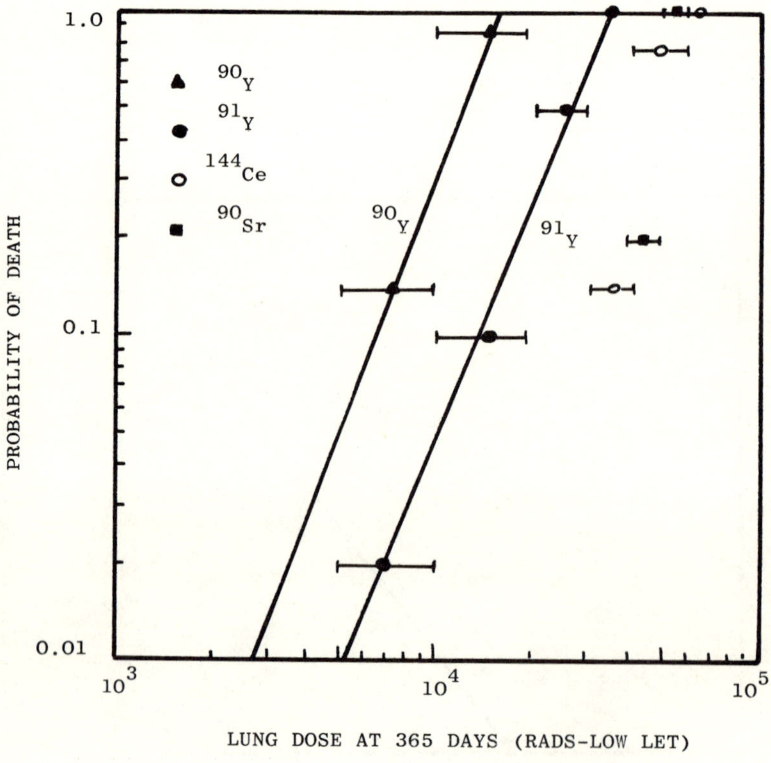

Fig. 3 Estimated dose-mortality curve for irradiation of the lung

---

* RBE - Relative Biological Effectiveness. If the RBE = 10, 1 rad from alpha emission would cause the same damage as 10 rads from beta or gamma emission.

the lung. Data points for cerium 144 and strontium 90 have also been included in the figure. Estimates of lung morbidity (sickness) caused by radiation have also been made, of which the essential features are a threshold at 2500 rads (low LET) and 100% incidence at 6000 rads (low LET), both doses assumed cumulative over 1 year; this estimate is summarised by Figure 4. In addition to the major effects described above, other forms of radiation damage may occur which are not lethal, such as skin damage and hypothyroidism, but no attempt is made to formulate dose-response relationships because reliable quantitative data are not available and it is unlikely that these effects would present problems as serious as those already discussed.

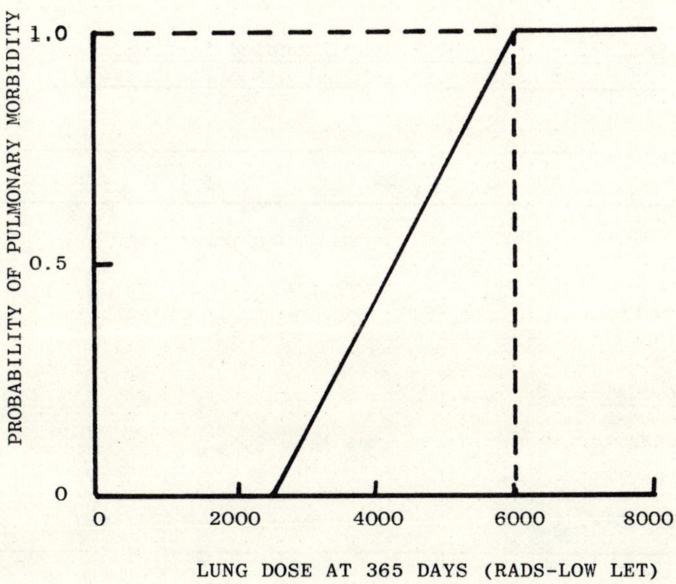

Fig. 4 Estimated dose-morbidity curve for irradiation of the lung.

The major late (or 'delayed') somatic effect of radiation on man is cancer. The risks of radiation-induced cancer have been summarised in various reviews (USNRC, 1975; Smith and Stather, 1976; ICRP, 1966; Dolphin and Marley, 1969; BEIR, 1972; Medical Research Council, 1975; Thorne and Vennart, 1976). Data available on cancer incidence usually relate to small groups of people exposed to high doses of radiation at high dose rates, and it is, therefore, necessary in most cases to extrapolate the data to obtain the assessment of cancer risk we require. In this paper, as is usual, a linear no-threshold model is used, i.e., the probability of cancer death is considered to be directly proportional to the total dose. It should be noted that linear extrapolation from effects at high doses and dose rates may in some cases overestimate possible risks. Using this model, the absolute cancer risk is obtained expressed as the number of excess cancers expected to develop (e.g., cancers per 10^6 man-rads). Only human data have been used for calculating these risk coefficients for the late (delayed) effects of radiation. Estimates of risk coefficients have been made for radiation-induced deaths from leukaemia and cancers of the lung, bone, liver, grastrointestinal tract, breast and all other tissues taken together; these values are shown in Table 2. An estimate is also available of the risk of benign thyroid nodules and this is shown in Table 3. The risk coefficients, which are derived mainly from Smith and Stather (1976), and which are not dissimilar to those in WASH-1400, have been rounded so as not to imply greater accuracy than the facts justify. It is also assumed that there is a period after irradiation during which there is a negligible increase in cancer incidence, and that this is followed by a period of increased but constant cancer incidence that lasts for a number of years. For leukaemia the period of incidence is assumed to be from 5 to 20 years after irradiation, and for other cancers the incidence period is taken as from 10 to 40 years after irradiation. It is recognised, of course, that this must be an oversimplification of the real situation.

Table 2 Number of Deaths Expected from Leukaemia and Other Cancers (after Smith and Stather, 1976)

	Cancer deaths per 10^6 man rads (low LET)
Leukaemia	20
Lung Cancer	20
Bone Cancer	10
Liver Cancer	10
Gastrointestinal tract Cancer	20
Breast Cancer	20
Thyroid Cancer	5
All other cancers	20
TOTAL	125

Regarding hereditary effects of irradiation the reader is referred to Smith and Stather (1976) for detailed information. As stated there, damage to the germinal cells (i.e. gonads) can result in spontaneous abortion or hereditary disease. However, the risk of abortion is difficult to quantify because many occur so early in pregnancy as to be undetectable and, for that reason, abortions are not considered. On hereditary disease, Smith and Stather (1976), conclude that "in a standard population exposed to radioactivity following a reactor accident, a total of 57 cases of serious hereditary disease per 10^6 man rads (low LET) is predicted over many generations, of which 15 and 9 cases will appear in the first and second generations respectively". (However, it is stated that these values derive from studies carried out mainly on the mouse.)

Table 3 Risk Estimates for Incidence of Benign Nodules and Thyroid Cancer (after Smith and Stather, 1976)

	Nodules of cancers per 10^6 man rads (low LET)	
	External Radiation*	I-131 and longer-lived isotopes**
Benign nodules***	100	10
Thyroid cancer	100	10
Deaths from thyroid cancer	5	0.5

* These values apply to doses up to 1500 rads (low LET); for 1500 to 5000 rads take half the values; for > 5000 rads assume no risk. For internal emitters these dose ranges should be increased by a factor of 10.

** Isotopes with half-lives shorter than I-131 are assumed to be equivalent to external radiation.

*** For children, double this value.

Maximum Permissible Levels and Emergency Reference Levels of Radiation and Radioactivity

The ICRP recommended in 1965 that the "cautious assumption" should be made "that any exposure to radiation may carry some risk" (ICRP, 1966). In July 1977 a new version of ICRP recommendations was published as ICRP Publication 26, embodying a system of dose limitation, one main feature of which, carried over from the previous recommendations, was that "all exposures should be kept as low as is reasonably achievable, economic and social factors being taken into account". ICRP Publication 26 distinguishes between "stochastic" effects ("those for which the probability of an effect occurring, rather than its severity, is regarded as a function of dose, without threshold") and "non-stochastic" effects ("for which the severity of the effect varies with the dose, and for which a threshold may therefore occur"). Some somatic effects are "stochastic", and "of these carcinogenesis is considered to be the chief somatic risk of irradiation at low doses and therefore the main problem in radiation protection". These latest ICRP Recommendations have not yet been incorporated in UK Regulations, although they have been interpreted by NRPB in a short paper, ASP 2, published through Her Majesty's Stationery Office in February 1978. The recommended dose limits for individual members of the general public, based on ICRP (1966), are listed in Table 4. These dose limits are intended to minimise the risk of somatic effects of radiation occurring among individual members of the general public, and in order to conform to the ICRP's enjoinment quoted above, the operators of nuclear installations, and supervising and licensing agencies, take pains to ensure that any dose received by members of the public is as low as may reasonably be achieved in ordinary circumstances, the limits being approached, if at all, only for short transitory periods.

Table 4 ICRP Recommended Dose Limits for Individual Members of the General Public (ICRP, 1966)

Whole body, gonads, red bone marrow	0.5 rem/yr
Skin, bone, thyroid (but thyroid for children up to 16 yrs. of age, 1.5 rems/yr).	3.0 rem/yr
Other single organs	1.5 rems/yr

The UK Medical Research Council has from time to time published recommendations concerning levels of radiation and radioactivity in the environment and population in the aftermath of a future accident to a reactor or other nuclear facility. The term "Emergency Reference Level" (ERL) has been coined in an attempt to describe more accurately the purpose of recommended levels of this type. According to the Medical Research Council (1975), an Emergency Reference Level (ERL) of dose is briefly defined as the radiation dose below which countermeasures are unlikely to be justified. There are also derived ERL's either of exposure or of activity in environmental materials, which correspond to the ERL's of dose. The ERL's are put forward not as firm action levels but as dose levels at which the responsible authorities should judge whether countermeasures should be introduced, full account being taken of any disadvantages and risks from these countermeasures. ERL's are given in Table 5 for iodine 131, caesium 137, ruthenium 106 and strontium 90. (The ERL of external dose from gamma radiation would be 10 rads average in body tissues, approximately 15 Roentgen measured in free air.) The ERL carries with it only a very low risk that a person might be hurt by the dose - of the order of 1 chance in 1000 that he will die in the next 40 years. Nevertheless, it should not be thought of as a maximum permissible dose for consideration before an accident has happened, but rather as a guide for action in the aftermath of the accident. The maximum permissible doses must remain those tabulated at the beginning of this sub-section.

Nature of Radiological Hazards from the Passing Cloud of Airborne Radioactive Material

At the first International Conference on the Peaceful Uses of Atomic Energy, held in Geneva in 1955, Marley and Fry of AERE Harwell presented a paper entitled "Radiological Hazards from an Escape of Fission Products and the Implications in Power Reactor Location", and most of the general ideas they introduced are still valid today (Marley and Fry, 1956). They pointed out that a cloud of radioactivity released to the atmosphere may be effective biologically in two ways: first by external beta and gamma radiation from the cloud to the person and also by inhalation of the contaminated air of the cloud; and second by the effects due to activity deposited on the ground, buildings, and vegetation that will remain after the cloud has passed downwind. Marley and Fry pointed out that it is unlikely that evasive action could be organised in time to avoid the direct effects of the radioactivity in the cloud on persons in the open during the passage of the cloud.

External gamma radiation emitted by the cloud as it passes over, around or even a few hundred metres to the side of a member of the public located downwind of the point of release, will deliver a whole body dose. If the dose is small there will be no 'early' effect recognisable, but some risk of 'delayed' effects such as leukaemia or other cancer will have been imparted. If the dose is large (the release being relatively large and the person quite near the installation) then radiation sickness may develop within a short period and death

Table 5 Emergency Reference Levels for I-131, Cs-137, Ru-106 and Sr-90

Isotope	Critical organ	ERL of dose (rems)	ERL of cloud-dosage (Ci-sec/m^3)	ERL in milk (uCi/litre)
Iodine-131[a]	Thyroid	30	0.020	0.25
Caesium-137[b]	Whole body	10	1.5	5.5
Ruthenium-106[c]	Lung	30	0.014	-
Strontium-90	Bone marrow	10	0.05	0.15

[a] ERL of cloud-dosage makes allowance for iodine and tellurium in equilibrium proportions.

[b] External gamma radiation from caesium-137 deposition is more limiting than the ERL in milk.

[c] Ruthenium is released as insoluble ruthenium oxide, and does not appear in cows' milk.

may result within days or weeks for those most heavily irradiated. External beta radiation exposure of those immersed in the cloud must also be considered, some of this coming from radioactive material in the cloud, and some from material deposited on the person from the cloud (though much of the latter could be removed by washing). The dose to skin so delivered is less serious as a hazard than the whole body dose from gamma rays.

For those immersed in and, therefore, inhaling the cloud as it passes downwind, a significant fraction of the airborne radioactive vapours and particles will be deposited in the lungs. If the deposited material is biologically insoluble it may rest in the lung for many weeks or months, and if the radioactive half-life of the material is long, the time-integrated dose to lung tissues may be large. For example, consider the fission product ruthenium which may be released from a thermal reactor accident if the gas surrounding overheated fuel is sufficiently oxidising; the biological half-life of ruthenium in the lung is firmly established at around 230 days and the radioactive half-life of ruthenium 106 is 1.0 year. A considerable dose to lung may, therefore, result from ruthenium released in some types of thermal reactor accident. In the hypothetical core disruptive accident postulated for fast reactors, the potential dose to lung from inhaled plutonium and other alpha-emitting actinides is of interest, because these nuclides have long radioactive half-lives and long retention times in the lung, and they emit alpha particles which are radio-biologically most harmful.

Many nuclides are released in a biologically soluble state, which allows them to pass rapidly from the lung into the bloodstream, from which they may be taken up by organs of the body. Iodine produces a dose to the thyroid gland in this way, and the inhalation of radioiodine following a reactor accident is generally regarded as potentially a serious hazard particularly to young children. However, tests have shown that the dose to the thyroid may be reduced significantly by giving stable potassium iodide or iodate tablets within an hour or so of the radioiodine being inhaled. Inhalation of caesium, a radionuclide which is readily released by reactor accidents and which is biologically soluble, produces a dose to the whole body. However, this dose is relatively small, because the biological half-life of caesium is quite short (the radioactive half-life of caesium 137 is long - 30 years). Against this, the caesium dose is to the whole body, and this must be considered at least one order of magnitude more important than a dose to a single organ.

Nature of Radiological Hazards from Radioactive Material deposited on the Ground

As will be explained in a later section of this paper, passage of the cloud of radioactive material at ground level downwind of the source would contaminate the ground, vegetation and buildings over which it passed. The presence of this contamination will require emergency remedial measures to be taken if possible.

The dose of external gamma radiation from fission products deposited on the ground (and to a lesser extent the beta radiation) which would be received by persons if they remained in the area, must be considered, and evacuation of the local population from the area may be required for a time. If a release of volatile fission products had occurred from a thermal reactor, deposited isotopes of iodine and tellurium, which are abundant and have short half-lives, would deliver a significant gamma dose in the first 2-3 weeks; if evacuation of the local population had been required they could then return home. If ruthenium were also released in quantity, a significant gamma dose would be delivered over a period of 1-2 years, determined mainly by the radioactive half-life of ruthenium 106 (1.0 year) (although this period might be reduced considerably if the ruthenium were washed away by rain, which on available evidence appears quite likely (Scott Russell, 1966). Caesium 137 is less abundant but is readily released and has a radioactive half-life of 30 years. Through the gamma emission of its short-lived daughter barium 137m, deposited caesium could deliver persistent levels of whole body radiation exposure embarrassing at low levels and dangerous to life at higher levels if no way could be found of removing the contamination. In many soils a large part of deposited caesium 137 is firmly trapped in soil minerals after a period of about a year. Measurements of gamma radiation were made over a period of years above various soils contaminated with caesium 137 and the results are shown in Figure 5, which is based on data of Gale et al (1963).

If the release we consider is from a hypothetical core disruptive accident to a fast reactor, the above applies to some degree where external gamma radiation from deposited radioactivity is concerned. But in addition to the activity of ruthenium spread over several months we should have to consider the beta-gamma emitting isotopes of several fission products such as zirconium, niobium, cerium etc., which have radioactive half-lives measured in months.

When external radiation from ground deposition had subsided sufficiently for re-occupation of land and houses to be permissible, inhalation of activity resuspended in the air by the wind and by the activities of man and animals might nevertheless prevent re-occupation. There is a general formula relating resuspended concentrations in air to levels of ground contamination which expresses the fact (unfortunately not as yet very accurately established) that over the years contamination becomes worked or weathered into the soil and consequently less available for resuspension. Resuspension is probably of most importance as a potential hazard following a release from a core disruptive accident to a fast reactor, when finely divided plutonium 239 and other very long-lived alpha-emitting actinides could be deposited on the ground.

Significant skin contamination would probably be picked up from a contaminated environment by children playing outdoors and by men working in the open

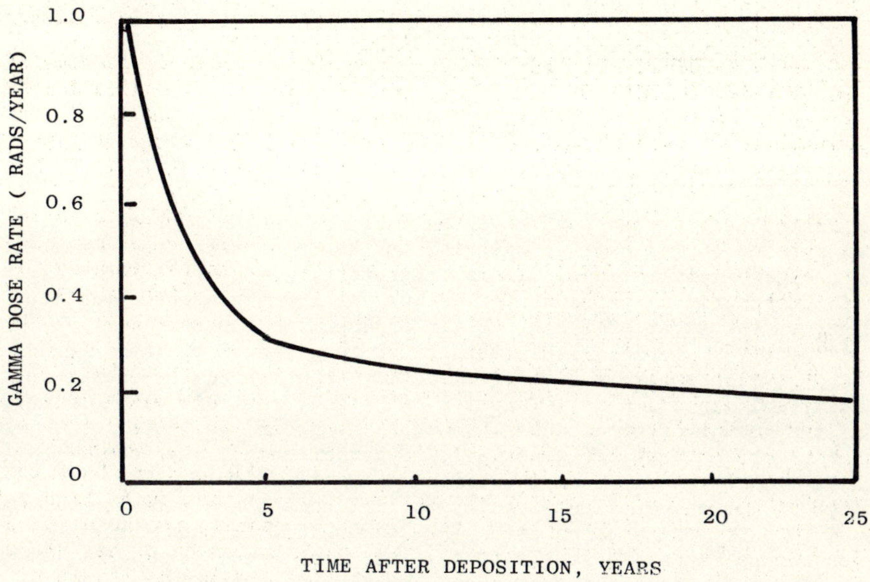

Fig. 5 Temporal variation of gamma dose-rate from Cs-137 deposition.

air, but in less contaminated areas (from which evacuation would be unlikely to be required) reasonable care and regular washing could be expected to provide adequate control of this type of hazard. However, there are other ways in which radioactive material originally deposited on ground and vegetation can cause hazard. For example, activity could be ingested accidentally from contaminated hands, particularly in the case of young children; food contaminated by radioactivity in houses and shops could be consumed, and there would be radioactivity deposited on green vegetables in gardens and on other growing crops. All such routes of entry into the human body can be covered by a general statement about ingestion from a contaminated environment, based on studies made at AERE Harwell of uptake of fallout from distant weapons tests; the simple concept emerges that, irrespective of age, one will accidentally ingest the activity from an equivalent area of ground, of about 10^{-3} m^2/day.

One route, probably the most important, for ingestion of deposited activity is excluded from consideration above, and must now be considered. This is ingestion by humans of iodine 131 in the milk produced by cows grazing contaminated pastures. ERL's of iodine 131, caesium 137 and strontium 90 were given in Table 5, but only that of iodine 131 need concern us here in the context of reactor accident releases, since if milk consumption is banned because of iodine 131, the other isotopes mentioned will be automatically taken into account (Beattie and Bryant, 1970).

As outlined above, environmental impact from the viewpoint of man is the impact on man himself. This conclusion is one often implied but seldom stated clearly. If man at the top of the ecological 'tree' is safe from the radiological impact of accidental releases from nuclear installations, then so are the flora and fauna that compose that 'tree'. This may be a proposition that requires further examination and detailed proof, but there this paper must leave it. This paper states environmental impact in terms of injuries and deaths caused to man, and in terms of land areas that might be made temporarily uninhabitable or unusable by man.

ATMOSPHERIC DISPERSION

The Nature of the Problem

In the next section, there is an analysis of the consequences of the accidental release of radioactive material to the atmosphere from a number of differing nuclear installations. It is, therefore, necessary to consider the subject of atmospheric dispersion particularly within the atmospheric boundary layer, that is, the layer of turbulent air adjacent to the surface. The movement of radioactive material within this layer is governed by the interaction between many physical phenomena among which may be numbered the change in the wind velocity and direction as a function of height, the rate of incoming solar radiation, the amount of cloud cover, whether the ground is wet or dry, the nature of the surface (whether grassland, forest or city), the topography (whether flat or hilly) and the source of the effluent itself which may, for example, be large enough or hot enough significantly to influence the dispersion process. Any attempt to carry out a calculation that takes into account these and other relevant phenomena leads inevitably to a solution of great complexity, requiring the use of powerful computers. Moreover, experimental verification of the behaviour of effluent in the atmosphere is certainly not available for many cases of interest (e.g., dispersion in regions of complex topography). It is, therefore, necessary and desirable to simplify the dispersion model used in order to reduce it to manageable proportions.

The Gaussian Model

The most widely used method for the calculation of the dispersion of radioactive material in the atmosphere employs the conventional Gaussian formula (Slade, 1968; Gifford, 1976). A typical example of the way in which the airborne concentration of effluent, released at ground level, falls off as a function of the distance above the ground or the distance across the wind, is displayed in Figure 6.

The use of the Gaussian model can be justified on the basis of the random properties of atmospheric turbulence. A small radioactive particle, while being carried downwind at the mean windspeed, is also thrown about by the turbulent forces acting upon it so that it may be regarded as taking a 'random walk'. A typical accidental release of radioactive material is made up of an extremely large number of such small particles and there is a standard mathematical theorem (the central limit theorem) which may be used in order to show that the distribution of the particles may be described by the Gaussian formula. This derivation is a simplification of the extremely complicated processes that occur in the atmospheric boundary layer; nonetheless, the Gaussian model is widely accepted as an adequate description of dispersion in the atmosphere and is often said to be accurate to within a factor of two or three (Scriven, 1974).

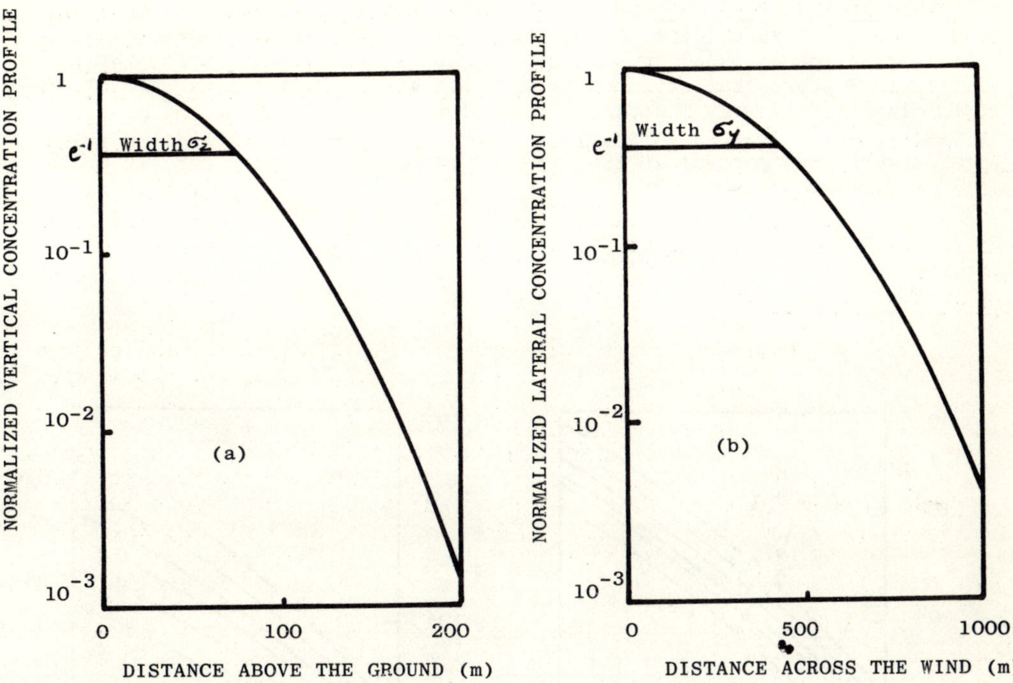

Fig. 6 (a) Vertical concentration profile
(b) Lateral concentration profile

Parametrization of Standard Deviations

The quantities σ_y and σ_z, which are the lateral and vertical widths or standard deviations of the Gaussian distribution as shown in Fig. 6, are monotonically increasing functions of travel distance or travel time. They also depend on the weather category, which is defined by reference to several variables such as the windspeed measured at a height of 10 m, the rate at which the sun is heating the earth's surface, the atmospheric temperature gradient and the amount of cloud cover. At one extreme of the range of weather categories there is a hot, sunny summer's day when the atmosphere is well mixed by convectively generated turbulence and any effluent released into it disperses rapidly. At the other extreme there is a still, cold winter's night with a temperature inversion from the ground upward, when effluent disperses very slowly. Between these categories is a cloudy day with a moderate to brisk wind, which is typical of conditions in which dispersion occurs at an 'average' rate.

In the widely used Pasquill-Gifford scheme the weather is conventionally divided into six categories (A to F) ranging from conditions of rapid to slow dispersion (Gifford, 1976). An up-to-date method* for describing the behaviour of σ_y and σ_z over flat grassland up to a distance not exceeding 100 km from the source is shown in Figure 7 for each of the weather categories (Hosker, 1974). This parametrization is based on work carried out at the UK Met Office (Smith, 1974). Figure 7 shows that effluent will remain in a relatively narrow and concentrated plume in category F weather. The plume becomes broader and more dilute as the weather moves towards category A. For a typical UK coastal site, category F weather occurs five percent of the time and category D seventy percent of the time.

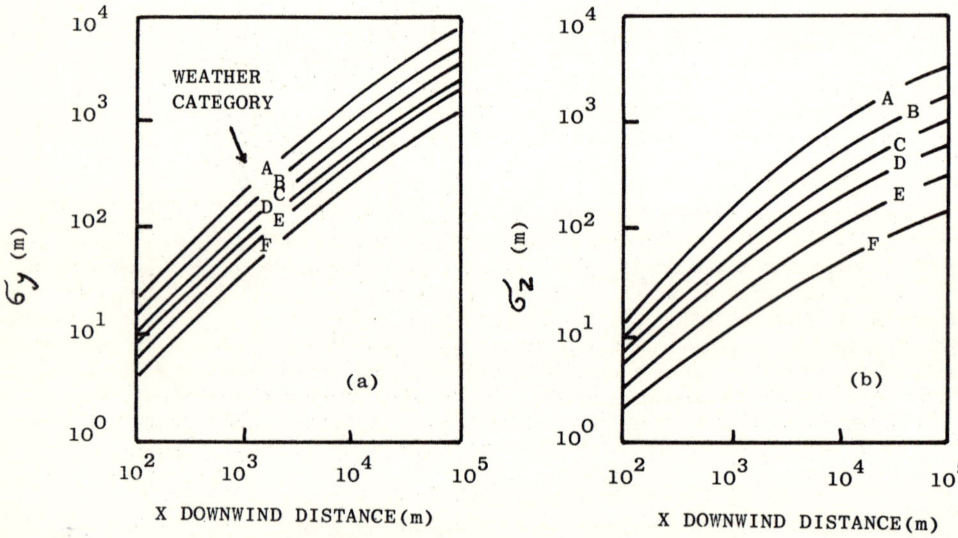

Fig. 7 Behaviour of σ_y and σ_z over flat grassland.

* This is only one of a great number of schemes, comprehensively reviewed by Gifford (1976).

For departures from the ideal flat countryside specified, it may be possible to identify modifications to the σ_y's and σ_z's which take account of topographical and other features specific to the site. There is no substitute, however, for detailed numerical calculations or the accumulation of atmospheric dispersion data in experiments conducted at and around the installation in question.

Modifications of the Basic Gaussian Equation

Allowance must be made for depletion of the plume by radioactive decay, which is easily accomplished using standard methods. The cloud of radioactive material can also be depleted by dry deposition as it travels downwind. An approximate way of treating this problem is to assume that, if the concentration of the effluent immediately above the ground is χ, the rate of deposition of material onto the surface is χ Vg in suitable units; Vg is the deposition velocity, a quantity which is introduced to cover a number of mechanisms which can cause deposition, for example, gravitational settling, uptake by plants and geometrical trapping by objects on the ground. A survey of published data on the value of Vg produces figures varying between 0.0001 cm/sec. and 20 cm/sec. (Hosker, 1974). A conventional representative value for Vg, however, is 0.3 cm/sec. The cloud may also be depleted by washout (rain falling from above it) or rainout (rain formed within the plume). These are complex phenomena best dealt with within a framework of time varying meteorology as described by Ritchie et al (1976).

Another important effect is the dependence of the quantity σ_y on the duration of the release (Griffith, 1977). Strictly, the values of σ_y that appear on Figure 7 are applicable to the calculation of concentration profiles averaged over about half-an-hour. If the release in question persists for a greater time, fluctuations in wind direction effectively spread the time averaged plume across a wider arc; that is, σ_y is an increasing function of the duration of release T. If the release rate remains constant, the average concentration on the plume axis, is, therefore, lower for increasing T. Figure 8 shows some examples of how downwind concentrations vary as the deposition velocity Vg and the duration of release T change in category D ('average') and category F ('worst case') weather.

Plume Rise

Yet another modification to the basic Gaussian formula must be introduced if plume rise is taken into account. A plume of radioactive material emerging from a damaged reactor containment may rise for a number of reasons. (i) It may be very hot since it is emerging from a region in which there is much sensible heat and in which there may also have been explosions, nuclear excursions or fires. (ii) It may possess high upward momentum if it is emerging from an orifice. (iii) The fission products within the plume are highly radioactive and thus add continuously to the heat content of the plume. (iv) If moisture is present, latent heat can be released by condensation.

In general, one of the most important contributors to plume rise is the sensible heat content of the plume, which can be very large. For example, the Rasmussen report identifies an accident sequence in PWR's, PWR1b, in which a rate of sensible heat emission of 100 MW is sustained for about ten minutes. In the case of a commercial fast reactor, a hypothetical core dispersive accident such as the notional case considered by NRPB (Kelly et al, 1977), supposed to involve the release of several percent of the core to the atmosphere, would be expected to be accompanied by the dispersal and burning of several tonnes of sodium in the secondary containment. This would liberate several hundred Gigajoules; a rate of

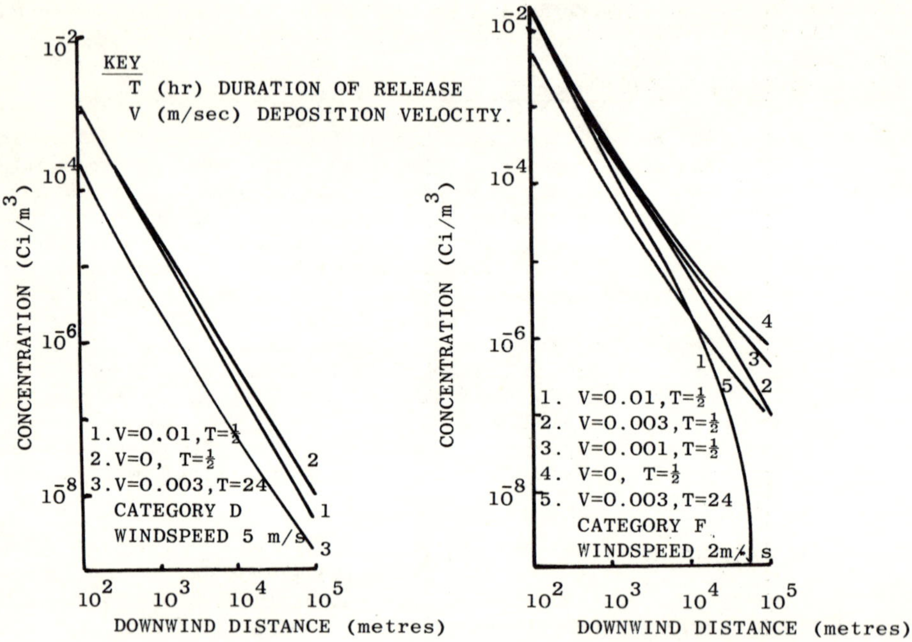

Fig. 8 Variation of downwind concentrations.

release of heat of 100 MW could be sustained for up to half-an-hour, so causing the plume to rise to great heights and thereby reducing airborne dosages at ground level considerably.

The foundations of any plume rise model rest on a few simple conservation laws such as mass, enthalpy and momentum (Russo, 1976). The implementation of these laws in the case where the plume rises into a turbulent atmosphere is described by Briggs (1975), who also reviews the many plume rise models available in the literature. The adaptation of the Briggs model to the reactor safety field appears in Kaiser (1977). Figure 9 gives examples of how airborne dosages can be affected by the incorporation of plume rise due to an initial rate of heat emission of 100 MW into the basic Gaussian model. Note that the effect of the reactor building wake is taken into account. This causes some initial dilution of the plume (Gifford, 1976).

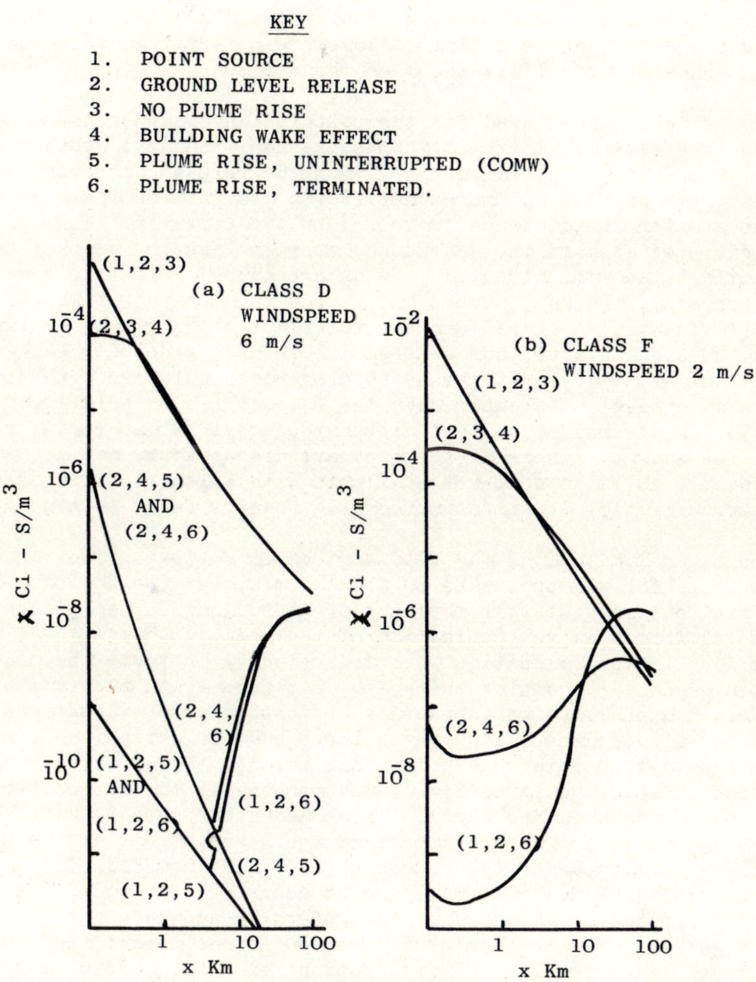

Fig. 9 Airborne concentrations as a function of distance downwind for various assumptions about plume rise release of 1 Ci of a long-lived fission product.

Limitations of the Model

In order to provide SRD with a flexible tool with which nuclear safety studies can be carried out, most of the work described in the foregoing has been incorporated into a computer code, TIRION, which has been used to derive some of the results which are presented in the following section of this report. TIRION is now being used by several groups and organizations as is exemplified by the fact that it was used at the Windscale Inquiry by both BNFL (1978) and an anti-nuclear protest organization, the Political Ecology Research Group.

Codes, developed within other organizations and embodying similar methods - for example, the CEGB code WEERIE (Clarke, 1976) - are also in widespread use. Notwithstanding this widespread acceptance it is important to remember that the models described in this section are idealised and are not absolute predictions of the consequences of releasing radioactive material to the atmosphere. It is pertinent to consider the limitations on the model.

Many of the methods used for the calculation of the consequences of releasing radioactive material to the atmosphere assume that the weather category, wind speed and wind direction remain fixed once the release has started. Clearly, this assumption becomes more and more unrealistic the greater the duration of the release or the greater the distance travelled by the effluent. The model used in the calculations presented in the Rasmussen Report allows the weather category and wind speed, but not the wind direction, to change hourly. If time varying meteorology is incorporated, the most severe early consequences follow when rain brings relatively large quantities of radioactive material to the ground in large centres of population, so causing a serious problem due to high levels of background radiation (McGrath et al, 1977). If the meteorology does not vary with time, the most severe early consequences occur when the radioactive material travels large distances during an inversion, so that airborne concentrations remain relatively high far from the source. These differences are discussed by McGrath et al (1977). There are, however, no methods as yet available with which the effects on consequences of assuming time dependent meteorology can be fully examined.

Some other limitations may be described as follows. The Pasquill-Gifford scheme is not valid for windspeeds below about 2 m/s, but the few experiments that have been carried out suggest that the rate of diffusion is then greater than would be expected from a naive application of the model, because there usually are rapid fluctuations in wind direction when the velocity is low. The model is also inapplicable in regions of complex topography but here again the rate of spreading of the plume often appears to be enhanced over forests or a mountainous region. By contrast, the rate of dispersion over a large body of water is often inhibited by the presence of a temperature inversion and because the effective surface roughness, which affects the intensity of the mechanical turbulence, is exceedingly small even when the waves are large. These and other phenomena are discussed by Gifford (1976) and illustrate an important truth which must always be borne in mind when using generalised safety codes such as TIRION or CRAC (that used in the Rasmussen report): these codes do not purport accurately to model the detailed meteorology of a particular site. They are generally adequate for 'rapid assessments of the magnitude of a problem' but when conditions depart from ideal 'there is no alternative to the services of a diffusion meteorology specialist' (Slade, 1968) and site specific experiments may be required.

Interactions with Populations: Probabilities and Results of Consequence Calculations

If a release of radioactive material of given magnitude is assumed to have occurred and the weather conditions are also assumed to be known, it is possible to calculate the areas within which the ground is contaminated above any postulated level by fission products and/or actinides. If the wind direction is also fixed, and the distribution of the population around the reactor is known, it is also possible to estimate the number of people who may contract cancer, or the number who may suffer an early demise, or how many genetic defects may be transmitted to succeeding generations, using the techniques described in the preceding section of this paper.

It follows that there is no unique answer for the consequences of a given accidental release of radioactive material, since the weather conditions and wind direction cannot be fixed in advance. For example, if the nuclear installation in question is on the coast, the probability that there will be no consequences equals the probability that the wind will blow out to sea. If there is a large city nearby, there will be no consequences in that city unless the wind is blowing towards it. Even if the wind does blow in that direction, certain consequences will be avoided if the dose received by individuals is below a corresponding threshold - there are thresholds for early effects. It may, therefore, be the case that, as well as it being necessary for the wind to blow towards the city, weather conditions would have to be unfavourable before there would be any chance of there being a large number of early effects. If the weather conditions are allowed to vary through the whole spectrum of possibilities, the results emerge as probability distributions of early effects, late effects and/or contaminated areas.

In practice, a reactor system is so complicated that a spectrum of releases of quantities of fission products and actinides that vary over many orders of magnitude is possible (USNRC, 1975). For each of these releases, the probability distribution of consequences may be calculated and then weighted by the frequency with which it is estimated that the given release may occur. The sum of these distributions, taken over the set of all possible releases, gives an overall frequency-number curve known as an f - N line, or in the precise mathematical terminology employed by the Rasmussen report, a complementary cumulative distribution function, examples of which are given in the following section. The presentation of the results of consequence calculations in this probabilistic fashion is unavoidable partly because of the statistical nature of the processes occurring in the atmosphere and partly because it is not possible to present the results of calculations of the possible releases from a nuclear installation other than in the form of a frequency-magnitude distribution. It is on the basis of these f - N lines that judgements of the acceptability or otherwise of nuclear installations can be made (Kinchin, 1978).

ENVIRONMENTAL CONSEQUENCES OF NOTIONAL ACCIDENTAL RELEASES FROM NUCLEAR POWER REACTORS

Inventories of Fission Products and Actinides in Reactors

As stated in the Introduction, the reason for considering power reactors as potentially main sources of accidental radioactive impact on the environment is the coexistence in irradiated fuel of a large inventory of radioactive substances and of a variable energy source from which heat must be removed at an adequate rate and with adequate reliability. These radioactive substances are of two main types - the fission products, beta-gamma emitters mostly of short to medium-long half-life, and the actinides, mostly alpha emitters of very long half-life. The fractions of the various fission product or actinide elements composing the reactor inventory, which are likely to be released in a hypothetical reactor accident depend, of course, on the type of reactor and the nature of the particular accident imagined. These aspects are discussed in the following subsections The inventories themselves are, however, the origin of the hazards. Table 6 lists inventories of activity of isotopes which are of importance in relation to environmental hazards from accidental releases. These inventories were taken from CEGB sources (the RICE computer code (Nais, 1977); the FISP computer code (Kelly et al, 1977), and USNRC sources (1975) as indicated in Table 6. The reactors considered are Magnox Gas-cooled Reactors, Advanced Gas-cooled Reactors, Pressurised Water Reactors and the Commercial Demonstration Fast Reactor. Generally, fission product inventories do not vary much from one reactor type to another, for a given level of thermal power; however, the actinide inventory does so vary, the inventory of higher actinides in particular increasing considerably with burn-up. Each of the four reactor types mentioned above and listed in Table 6 is considered in turn in the following sub-sections concerned with environmental impact of accidents.

<u>Environmental Impact of Accidents to Magnox Gas-Cooled Reactors</u>

In gas-cooled reactors such as Magnox the most serious kind of accident that has to be considered is a failure of the pressure circuit causing a rapid loss of coolant (Matthews, 1975). As in all reactor types, very high reliability of shut-down is ensured by providing a number of automatic shut-down devices (more control rods than needed are provided to allow for random failure of a few) of different independent kinds (to avoid some unforeseen fault being common to all). The nuclear reaction having been quickly shut down, enough gas flow can be maintained even at atmospheric pressure to cool the fuel and prevent damage to it, provided that conditions in the channel are otherwise normal. In all UK gas-cooled reactors, ample emergency supplies of carbon dioxide are available to be fed to the reactor circuit and prevent ingress of air. In assessing the safety of magnox reactors in steel pressure vessels with separate heat exchanger shells, the arbitrary assumption is made that a bottom (inlet) duct fractures, and that some air enters before the damaged circuit can be isolated. Thereafter carbon dioxide is fed in, gas flow is begun by the blowers driven by auxiliary motors and finally main motor driven gas flow is restored. It is an important and remarkable feature of Magnox (and AGR) reactors that they are able comfortably to survive sudden complete loss of coolant pressure. This is illustrated by a typical temperature transient taken from Beattie et al (1964), displayed in Figure 10. This is for a Magnox reactor with well-irradiated fuel (high fission product after-heat) and long-irradiated graphite (appropriate levels of Wigner energy and the effects of irradiation and impurity-enhanced graphite oxidation rate have been taken into account in the calculations). It will be seen from this example that the design ensures safe clad temperatures are not normally exceeded in the first few minutes when fuel internal temperatures equalise radially across the fuel rod, gas cooling being temporarily diminished. Temperatures would remain under control for several hours with only auxiliary motor driven gas flow, and when main motor driven flow is restored temperatures fall very rapidly to entirely safe levels. The later Magnox reactors have prestressed concrete pressure vessels which contain the reactor core and all the heat exchangers: any accidental breach in the gas circuit, if it occurred, would probably be very small and the consequent variations of reactor temperatures would be still easier to control. The conclusion is drawn from studies of this kind that if there should ever occur an accidental depressurisation of a Magnox reactor, by far the most probable outcome will be that there is no release of fission products (or actinides) at all, and therefore no environmental impact.

Safety studies show that for the earlier steel pressure vessel magnox reactors, should a sudden loss of coolant pressure occur there is a small chance,

Table 6 Fission Product and Actinide Inventories (Total Curies)

NUCLIDE	Magnox notional 3000 MW(t)*: (Nair, 1977)	AGR notional 3000 MW(t)*: (Nair, 1977)	PWR 1320 MW(e): USNRC (1975)	CDFR 1300 MW(e) Kelly et al (1977)
Kr^{85}	1.3×10^{6}	1.3×10^{6}	5.6×10^{5}	-
Kr^{85m}	2.2×10^{7}	1.7×10^{7}	2.4×10^{7}	-
Kr^{87}	4.3×10^{7}	3.2×10^{7}	4.7×10^{7}	2.4×10^{7}
Kr^{88}	6.0×10^{7}	4.5×10^{7}	6.8×10^{7}	3.4×10^{7}
Rb^{86}	3.9×10^{4}	2.0×10^{5}	2.6×10^{4}	2.8×10^{5}
Sr^{89}	7.9×10^{7}	5.9×10^{7}	9.4×10^{7}	4.2×10^{7}
Sr^{90}	1.1×10^{7}	1.1×10^{7}	3.7×10^{6}	1.1×10^{6}
Sr^{91}	1.0×10^{8}	7.8×10^{7}	1.1×10^{8}	6.2×10^{7}
Y^{90}	1.1×10^{7}	1.2×10^{7}	3.9×10^{6}	-
Y^{91}	4.4×10^{7}	3.3×10^{7}	1.2×10^{8}	5.8×10^{7}
Zr^{95}	1.4×10^{8}	1.2×10^{8}	1.5×10^{8}	1.1×10^{8}
Zr^{97}	1.4×10^{8}	1.2×10^{8}	1.5×10^{8}	1.3×10^{8}
Nb^{95}	1.4×10^{8}	1.2×10^{8}	1.5×10^{8}	1.0×10^{8}
Mo^{99}	1.5×10^{8}	1.4×10^{8}	1.6×10^{8}	1.4×10^{8}
Tc^{99m}	1.3×10^{8}	1.2×10^{8}	1.4×10^{8}	1.2×10^{8}
Ru^{103}	1.3×10^{8}	1.3×10^{8}	1.1×10^{8}	1.3×10^{8}
Ru^{105}	8.2×10^{7}	9.4×10^{7}	7.2×10^{7}	1.2×10^{8}
Ru^{106}	5.0×10^{7}	6.3×10^{7}	2.5×10^{7}	4.2×10^{7}
Rh^{105}	8.0×10^{7}	8.1×10^{7}	4.9×10^{7}	1.2×10^{8}
Te^{127}	7.2×10^{6}	7.0×10^{6}	5.9×10^{6}	8.4×10^{6}
Te^{127m}	1.2×10^{6}	1.1×10^{6}	1.1×10^{6}	1.2×10^{6}
Te^{129}	2.4×10^{7}	2.2×10^{7}	3.1×10^{7}	3.0×10^{7}
Te^{129m}	6.5×10^{6}	6.1×10^{6}	5.3×10^{6}	5.6×10^{6}
Te^{131m}	1.2×10^{7}	1.1×10^{7}	1.3×10^{7}	1.1×10^{7}
Te^{132}	1.2×10^{8}	1.1×10^{8}	1.2×10^{8}	1.1×10^{8}
Sb^{127}	7.2×10^{6}	7.1×10^{6}	6.1×10^{6}	8.6×10^{6}
Sb^{129}	2.5×10^{7}	2.4×10^{7}	3.3×10^{7}	3.2×10^{7}
I^{131}	7.9×10^{7}	7.3×10^{7}	8.5×10^{7}	8.2×10^{7}
I^{132}	1.2×10^{8}	1.1×10^{8}	1.2×10^{8}	1.2×10^{8}
I^{133}	1.7×10^{8}	1.5×10^{8}	1.7×10^{8}	1.5×10^{8}
I^{134}	1.8×10^{8}	1.7×10^{8}	1.9×10^{8}	1.6×10^{8}
I^{135}	1.6×10^{8}	1.4×10^{8}	1.5×10^{8}	1.5×10^{8}
Xe^{133}	1.7×10^{8}	1.5×10^{8}	1.7×10^{8}	1.5×10^{8}
Xe^{135}	8.4×10^{7}	2.4×10^{7}	3.4×10^{7}	1.6×10^{8}
Cs^{134}	7.3×10^{6}	3.3×10^{7}	7.5×10^{6}	1.5×10^{6}
Cs^{136}	2.9×10^{6}	6.3×10^{6}	3.0×10^{6}	5.2×10^{6}
Cs^{137}	1.4×10^{7}	1.7×10^{7}	4.7×10^{6}	3.0×10^{6}
Ba^{140}	1.5×10^{8}	1.3×10^{8}	1.6×10^{8}	1.3×10^{8}
La^{140}	1.5×10^{8}	1.4×10^{8}	1.6×10^{8}	1.3×10^{8}
Ce^{141}	1.4×10^{8}	1.2×10^{8}	1.5×10^{8}	1.3×10^{8}
Ce^{143}	1.3×10^{8}	1.1×10^{8}	1.3×10^{8}	1.1×10^{8}
Ce^{144}	1.2×10^{8}	1.0×10^{8}	8.5×10^{7}	4.6×10^{7}
Pr^{143}	1.3×10^{8}	1.1×10^{8}	1.3×10^{8}	1.1×10^{8}
Nd^{147}	5.4×10^{7}	5.0×10^{7}	6.0×10^{7}	4.8×10^{7}
Np^{239}	2.0×10^{9}	1.7×10^{9}	1.6×10^{9}	1.9×10^{9}
Pu^{238}	4.4×10^{4}	2.7×10^{5}	5.7×10^{4}	3.6×10^{5}
Pu^{239}	1.1×10^{5}	3.6×10^{4}	2.1×10^{4}	1.2×10^{5}
Pu^{240}	1.2×10^{5}	4.8×10^{4}	2.1×10^{4}	2.6×10^{5}

Table 6 Fission Product and Actinide Inventories (Total Curies) cont'd.

NUCLIDE	Magnox notional 3000 MW(t)*: (Nair, 1977)	AGR notional 3000 MW(t)*: (Nair, 1977)	PWR 1320 MW(e): USNRC (1975)	CDFR 1300 MW(e) Kelly et al (1977)
Pu^{241}	1.6×10^7	2.1×10^7	3.4×10^6	3.2×10^7
Am^{241}	2.5×10^4	4.7×10^4	1.7×10^3	8.0×10^4
Cm^{242}	2.6×10^6	8.8×10^6	5.0×10^8	9.2×10^6
Cm^{244}	4.5×10^3	4.9×10^5	2.3×10^4	2.2×10^5

* The largest UK Magnox reactors (Wylfa) generate about 1500 MW(t) each. AGR reactors also generate about 1500 MW(t) each. The inventories listed for Magnox and AGR are for a notional thermal power of 3000 MW to facilitate comparison with PWR and CDFR.

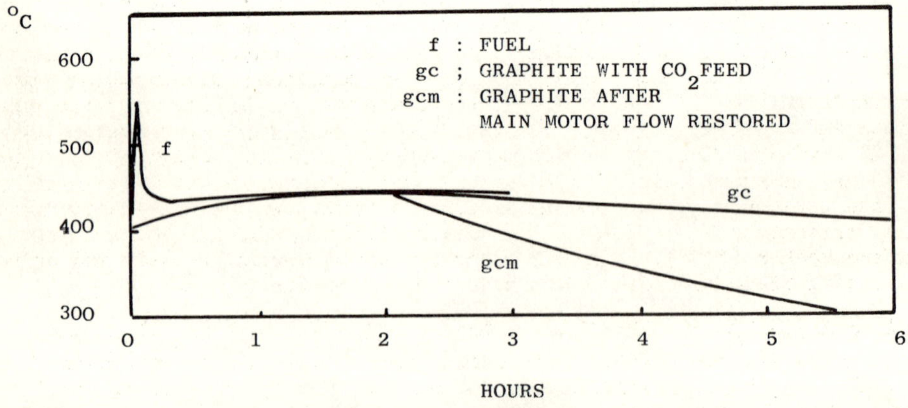

Fig. 10 Magnox Reactor Depressurisation Accident: Calculated Temporal variation of Fuel and Graphite Temperatures.

less than 1 in 100, of a single channel melt-out occurring in the most highly rated fuel channel. This and other aspects of gas-cooled reactor accidents have been discussed by Macdonald et al. (1977), where experimental evidence is reviewed and theoretical aspects are studied using the CEGB reactor safety analysis code WEERIE (Clarke, 1976). Macdonald and his co-workers group the fission products into broad classes of volatility, in which for example iodine and tellurium are considered "volatile", caesium and ruthenium are "mid-volatile" unless there is air present when they are classified as "volatile", and strontium and the rare earths are considered "non-volatile". The release fractions from fuel to coolant are assumed to be 10%, 1% and 0.1% respectively for a channel melt; for a channel fire they are increased 5 times. Macdonald et al. (1977) state that for the assumed releases the ERL of inhalation dose (by implication a thyroid dose mainly from iodine and/or a lung dose mainly from ruthenium and other beta emitters) is estimated to occur at 2.4 km for the steel vessel Magnox reactor and 1.1 km for the concrete vessel Magnox reactor. It may be inferred approximately that releases broadly in the range 10-100 Ci iodine-131 are under consideration. The deposition of released activity within the reactor coolant circuit and containment is an important factor determining the discharge to atmosphere, and in the WEERIE calculations this is taken into account using elemental plate-out factors which for the non-gaseous fission products are assumed to lie in the range 0.01-0.15 min^{-1}; these values are typical of those observed in operating gas-cooled reactors under channel melt-out conditions (Weir, 1970; Knockaert et al., 1969). Other factors which have been investigated include the reduction in inhalation doses at short downwind distances due to the entrainment of effluent in the wake of the reactor building. This is illustrated in Figure 11, which shows representative dose versus distance curves calculated using the WEERIE code for a Magnox depressurisation accident accompanied by a single channel melt-out. The dashed curves show the effect of building entrainment as predicted by a simple virtual source model, which wind tunnel experiments have demonstrated may be appropriate in many circumstances (Macdonald et al., 1977). Figure 11 shows that, for the accident and dispersion conditions considered, the inhalation dose to the thyroid from radioactive iodine could be 1 ERL out to 450 m downwind at most. Other environmental impacts to be considered include contributions to thyroid dosage from ingestion of cow's milk (in practice this would mean the distance downwind to which prohibition of milk supplies should extend) and from ingestion via routes other than milk e.g. via green vegetables, hand contamination. The distances at which these effects might independently give rise to 1 ERL of thyroid dose can be calculated. The picture emerges, supplementing the information contained in Figure 11, of a prohibition for a few weeks on the sale and consumption of milk from farms within about 3 km downwind of the reactor, and possibly of a temporary warning about general iodine contamination and against eating locally grown green vegetables for members of the public, if any, who elect to remain within perhaps 150 m downwind of the reactor. These and any other aspects of the 'environmental emergency' would be confidently handled and contained by the elaborate, carefully planned and frequently practised emergency arrangements at CEGB nuclear power stations, which have been described by Orchard and Walker (1977). Similar emergency arrangements are made and exercises carried out at SSEB and UKAEA/BNFL nuclear power reactor sites.

Environmental Impacts of Accidents to Advanced Gas-Cooled Reactors

The Advanced Gas-Cooled Reactor (AGR) is a development of the Magnox reactor, allowing higher fuel ratings and burn-up through the use of uranium oxide fuel in assemblies of stainless steel clad pins and of re-entrant coolant flow through the core. All the AGR's operated by the UK Generating Boards have prestressed concrete pressure vessels. A sudden loss of coolant pressure due to an accident would be most unlikely to lead to fuel melt-out and the maximum transient fuel temperature would be several hundreds of degrees Celsius below the melting poing of stainless steel. There is a possibility, however, of a few pins of highest rating puncturing under internal fission gas pressure at the clad tempera-

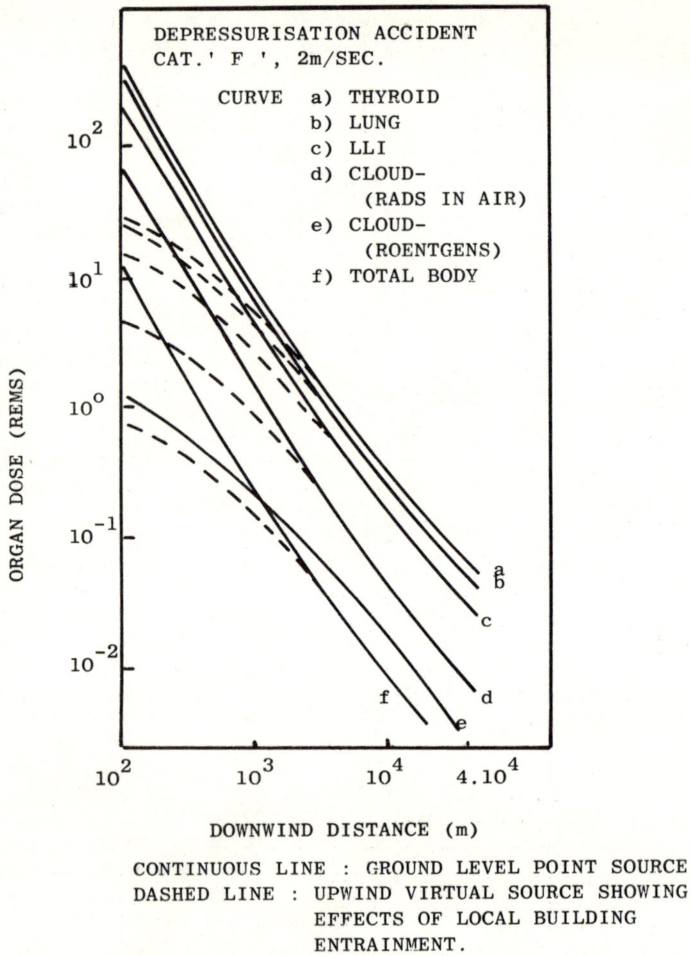

Fig. 11 Magnox Reactor Depressurisation Accident with Single Channel Melt-Out; Integrated Inhalation and Cloud Doses.

tures reached transiently, external gas pressure being reduced. A small percentage of the inventory of gaseous and volatile fission products in the pins will have diffused to the fuel-clad interspace during normal operation and will be available for release. The manner and magnitude of fission product release from fuel to coolant in such conditions has been much studied (e.g., Findlay, 1973), and their deposition in the reactor gas circuit must also be taken into account (see also Macdonald et al, 1977). The estimated size of release to the atmosphere after a loss-of-pressure accident is generally less than 10 Ci iodine-131. The only environmental impact would be a temporary ban on consumption of cows' milk up to about a mile downwind. Partly as a result of these favourable environmental features some AGR's are located quite close to urban areas in the United Kingdom. However, there is, as yet, an absence of published studies of AGR (and Magnox reactor) accidents similar to the USNRC studies of water reactor safety.

Environmental Impact of Accidents to Pressurised Water Reactors

The development and consequences of hypothetical accidents to PWR's were analysed on a probabilistic basis by the USNRC and published in the Reactor Safety Study or "Rasmussen Report" (USNRC, 1975). Of these accidents, that designated PWR 9 approximates to the "design basis accident" (d.b.a.) or "maximum credible accident" (m.c.a.), and is perhaps the analogue of the depressurisation accident for the Magnox and AGR reactors, described before. If, as in PWR 9, there were an accidental pipe break in the primary pressurised coolant circuit of a PWR, the greater part of the water would flash off as steam carrying some water along with it. Steam is a very poor coolant compared to liquid water, and if nothing were done fuel would melt. In the PWR 9 accident description, the emergency core cooling system comes into action automatically as designed and cooling water is quickly conveyed to the fuel. However, a small percentage of the inventory of gaseous and volatile fission products which will have collected in the fuel clad interspace is assumed to escape from fuel pins punctured by internal fission gas pressure. The environmental impact is minimised by the containment which is designed to contain the steam pressure. This would be high initially and some leakage of fission products would occur to atmosphere. However, the containment water sprays would come into action to reduce the steam pressure and terminate the leakage after about half an hour. According to the Reactor Safety Study the release to atmosphere would consist of about 10 Ci iodine-131 with similar small quantities of other gaseous and volatile fission products. The release foreseen for this accident is thus similar to that envisaged for the steel pressure vessel Magnox reactor. It is perhaps partly for this reason that it is believed in some quarters that if PWR's are adopted in the United Kingdom they should at first be constructed on sites similar to those occupied by Magnox reactors.

The environmental impact of more severe PWR accidents has also been analysed by USNRC in the Reactor Safety Study. In these accidents, at low levels of probability, it is envisaged that engineered safeguards such as the emergency core cooling system and/or the containment spray system fail to come into operation as required following primary coolant circuit depressurisation. Environmental consequences are thus larger. As an example of this class of severe but highly improbable accidents, accident type PWR 2 is discussed briefly below. Reading the Reactor Safety Study one learns that the frequency of occurrence of this accident is just less than 1 in 10^6 years per reactor, although one surmises that this figure would depend somewhat on design details of the particular PWR in question. Some experts in the UK consider that the true frequency of this class of event is less than 1 in 10^7 years per reactor. The main features of the sequence of events imagined to culminate in this accidental release, which is very large, are as follows. The category 2 releases are associated with the failure of core-cooling systems, leading to core melting, concurrent with the failure of containment spray and heat-removal systems. Containment rupture would be caused by overpressure due to hydrogen burning and steam pressure. A substantial fraction of the contents of the breached containment would be released in a puff over a period of about 30 minutes, with the remaining fraction continuing to leak out at a relatively low rate thereafter. The total release would include 6×10^7 Curies of Iodine-131, along with about 2.5×10^6 Curies of Caesium-137. Because of the presence of hot pressurised gases in the containment at the time of failure, this release would be characterised by a relatively high rate of release of sensible heat.

The Rasmussen study made use of nine categories in all to represent the spectrum of possible releases from PWR's. An expected frequency of occurrence was calculated for each of the associated release inventories, and the consequences were estimated in each case. Figures 12, 13 and 14 are reproduced from the report

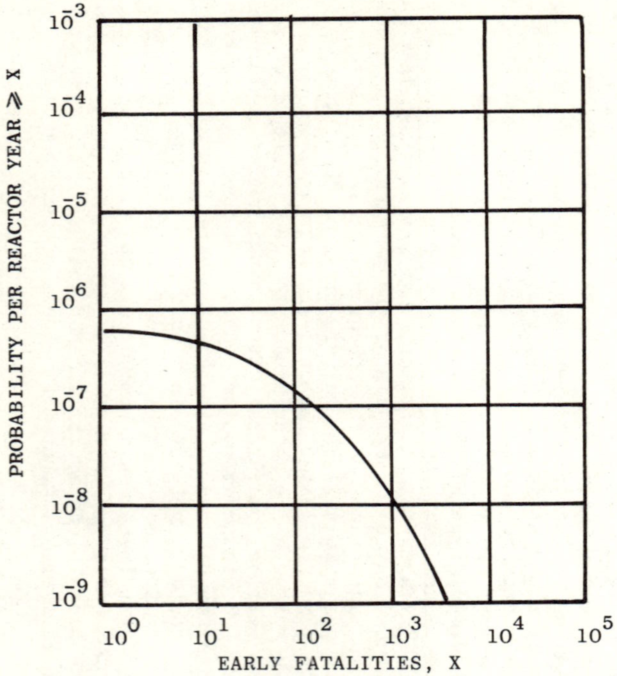

NOTE: APPROXIMATE UNCERTAINTIES ARE ESTIMATED TO BE REPRESENTED BY FACTORS OF 1/4 AND 4 ON CONSEQUENCE MAGNITUDES AND BY FACTORS OF 1/5 AND 5 ON PROBABILITIES

Fig. 12 Probability Distribution for Early Fatalities per Reactor Year (Pressurised Water Reactor).

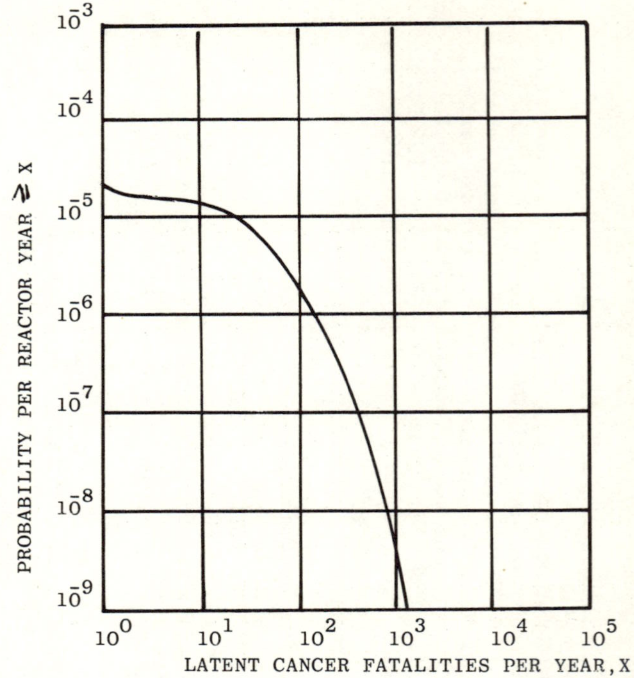

NOTE: APPROXIMATE UNCERTAINTIES ARE ESTIMATED TO BE REPRESENTED BY FACTORS OF 1/6 AND 3 ON CONSEQUENCE MAGNITUDES AND BY FACTORS OF 1/5 AND 5 ON PROBABILITIES

Fig. 13 Probability Distribution for Latent Cancer Fatality Incidence per Reactor Year (Pressurised Water Reactor).

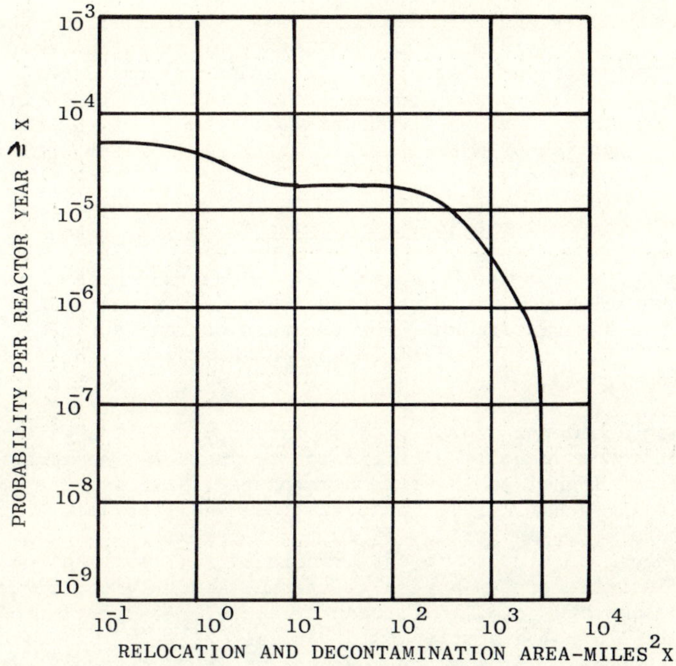

NOTE: APPROXIMATE UNCERTAINTIES ARE ESTIMATED TO BE REPRESENTED BY FACTORS OF 1/5 AND 2 ON CONSEQUENCE MAGNITUDES AND BY FACTORS OF 1/5 AND 5 ON PROBABILITIES

Fig. 14 Probability Distribution for Decontamination Areas per Reactor Year. (Pressurised Water Reactor).

as representative examples of the results of the study. These figures show the probability per reactor year versus the magnitude of the consequence for three specific effects, namely early fatalities, latent fatalities, and size of the area of land requiring decontamination (principally due to deposited Caesium). As can be seen from Figure 12 the frequency of an accident that results in more than 10 early fatalities is about 5×10^{-7} per reactor year; accidents involving 100 or more such fatalities are predicted to have a frequency of about 10^{-7} per reactor year.

The frequency distributions displayed in these graphs reflect the particular characteristics of the population distributions and weather patterns appropriate to the United States. In view of the recent decision to consider the option of building PWR's on UK sites, it is necessary to examine PWR accident consequences in the context of UK conditions; studies are currently under way at SRD to investigate the topic.

The numbers of people that could be affected by a given release would be rather larger in the UK compared with the US figure, because of the higher population density. However, this would be counterbalanced by a decreased probability of occurrence of class F weather (the least favourable in the context of dispersion of hazardous material) in the UK. The range of engineering options available in future PWR designs permits sufficient flexibility to ensure that the plant can be built to meet future UK nuclear safety criteria and proposed guidelines for the limitation of potential risk. According to the Generic Safety Study published in July 1977 by the Nuclear Installations Inspectorate, "The Inspectorate consider that there is no fundamental reason for regarding safety as an obstacle to the selection of a Pressurised Water Reactor for commercial electricity generation in Britain".

Environmental Impact of Accidents to Fast Reactors*

The safety case for the UK commercial demonstration fast reactor (CDFR) is based on the inherently advantageous characteristics of the sodium cooled pool type system. The use of an efficient, low pressure coolant acting as a large heat sink, together with a highly reliable shut-down system, reduce the risk of accident to very low levels.

Re-fueling accidents probably offer the most easily envisaged route for the escape of significant quantities of fission products to the atmosphere of the secondary containment (if not to the free atmosphere), although present designs now make this exceedingly unlikely. The situation usually examined is one in which an error occurs while an irradiated fuel sub-assembly is being transferred from the coolant vessel or circuit to the irradiated fuel examination booths or to the irradiated fuel store. As a result, the sub-assembly could overheat and melt, releasing gaseous and volatile fission products. At the time of reactor shut-down a typical CDFR sub-assembly would contain $\sim 2 \times 10^5$ Ci of iodine-131 and ~ 5000 Ci of caesium-137, but since movement of the sub-assembly out of the coolant would be preceeded by a stay of about 100 days in the coolant clear of the reactor core, only some 50 Ci of iodine-131 would remain. If this and the 5000 Ci of long-lived caesium-137 should escape into the secondary containment, its pathway to the atmosphere would lie through a battery of sprays, HEPA filters (High Efficiency Particulate Air Filters) and charcoal filters (any actinides and non-volatile fission products becoming airborne as a dust would be trapped by the HEPA filters). The radioactive material released to the atmosphere, assuming all these safety features operate correctly, would contain less than 1 Ci of iodine-131 and 1-10 Ci of caesium-137: this would lead to very small environmental effects, and it can be shown that it would be expected that significant levels of dose would not be reached beyond the boundary fence.

Another possible accident could involve localized fuel pin damage. In this case volatile fission products such as iodine-131 would be released from the fuel but would be efficiently absorbed by the sodium coolant - the affinity of sodium for certain fission products, leading to their retention in the primary circuit, is an additional safety feature of the fast reactor. Minor damage to a fuel pin would thus lead to a bubble, containing gaseous fission products only, escaping to the blanket gas. Any release to the atmosphere would be trivial.

Considerations such as those in the foregoing show that it should be possible to operate fast reactors to at least the standard of safety already achieved in thermal reactor operations. The only safety feature that is more difficult to establish for fast reactors than it is for thermal reactors is the effect of a highly improbable nuclear excursion that might take place, for example, as a result of a geometrical compaction of the core giving a reactivity increase. The whole design concept of the reactor seeks to avoid 'whole core accidents' but

---

* In this section, liquid metal cooled fast breeder reactors (LMFBR's) only are considered since, in the UK and elsewhere, R and D programmes are almost exclusively devoted to these and not to a possible alternative, the gas-cooled FBR.

these still attract much attention in the safety study of the reactor.

In these hypothetical accidents, complete loss of flow or some arbitrarily large increase of reactivity is assumed to occur and in addition the highly reliable automatic protection system is assumed to fail and cause the core to melt down. The fuel could then be heated to its boiling point and the boiling of the fuel and the heating of the gases within it will cause core expansion which will finally terminate the excursion. Even in such a violent excursion as this the total energy released will still be relatively small and will be unlikely to breach the containment, although it will never be possible to calculate the exact course of such speculative events. There is, however, sufficient evidence to show that the probability of whole core accidents is acceptably low, less than one such accident in a million years of reactor operation. Nonetheless, it is the intention that CDFR will have a strong containment.

This containment will be provided by a thick prestressed concrete pressure vessel with a strong roof supporting the reactor components, and an outer secondary containment building to prevent the escape of any radioactivity which finds its way into the space above the operating floor. This structure will have a very high probability of completely containing any accident, though in scientific and engineering terms it cannot be claimed that the containment of all whole core accidents can be absolutely guaranteed, and it is reasonable to look at the consequences that could arise should a whole core accident result in a breach of the containment and a release of activity.

Recently, the UK Nuclear Installations Inspectorate has commissioned the National Radiological Protection Board to undertake a 'theoretical study of the possible outcome of a range of events extending to extremes in which all protective measures have failed' (Kelly et al, 1977). The preamble to this work, written by the NII, speaks of a 'sudden and very serious release' which can only take place 'if the containments failed catastrophically'. NRPB consider releases to the atmosphere containing up to 10% of the core plus the volatiles from the remainder of the core, though there is no suggestion that a whole core accident would actually result in such a level of release.

The NRPB report concentrates on early deaths, early morbidities, cancers and hereditary effects. Taking the assumption that 5% of the core of a 3,000 MW(t) fast reactor is vaporized and escapes to the atmosphere, together with volatile fission products from the remainder of the core, typical results from Kelly et al, 1977, are as follows: on a remote site, there could be 300 early deaths, 200 morbidities, 3,900 deaths from cancers and 60 or so hereditary effects within 30 years (for a definition of the various forms of hereditary effects see Smith and Stather (1976).

The figures quoted above refer only to category C-D, 'average' weather conditions, which alone are considered by NRPB. It has been shown, however, that for a given release, the results should properly be presented as a probability distribution of consequences. In Figure 15 the probability distribution of early deaths shown as curve 1 is obtained assuming that the 5% release described above occurs from a reactor on a remote site. This curve was calculated using the SRD computer code TIRION assuming that there is no plume rise and that there is early evacuation of people from areas heavily contaminated by deposited γ-emitters.

It is possible to demonstrate that the calculated early consequences of a large release from a fast reactor can probably be considerably reduced if proper account is taken of plume rise, an aspect to which only limited attention is given in Kelly et al (1977). The burning of several tonnes of sodium could liberate 100 plus GJ of heat so that rates of heat release of 100 or 1000 MW can in principle

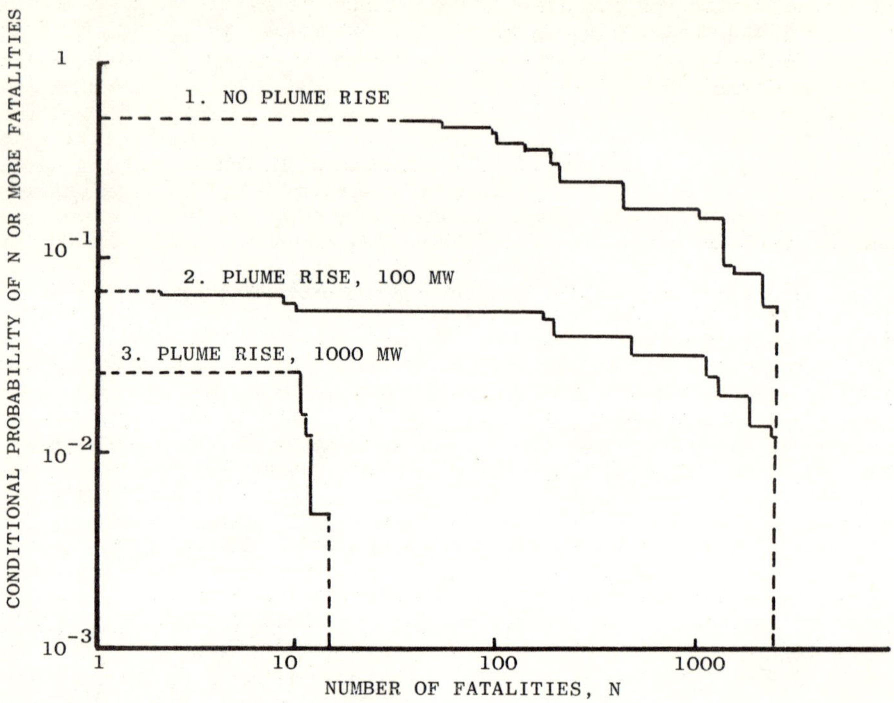

Fig. 15 Conditional Probability of Early Death Assuming the Escape to the Atmosphere of 5% of the Core plus Volatiles from a Fast Reactor, 1000 MW(e), on a 'Remote' Site in the UK.

be sustained for some 30 or 3 minutes respectively. Assuming that the plume lifts cleanly off the ground, Figure 15 shows that the calculated numbers of early deaths are very sensitive to assumptions about plume rise. The effect of plume rise on the probability distributions of cancers and of unacceptably contaminated areas of land is much less marked, however.

The areas likely to be contaminated in the long term by deposited γ-emitters are comparable to those predicted for the most severe of Rasmussen's postulated PWR accident sequences since, in both cases, of the order of 10^6 Ci of caesium-137 are assumed to escape from the reactor. For the fast reactor case, the areas significantly contaminated by deposited actinides, which are hazardous when resuspended by the wind, are larger than for the biggest of the PWR sequences, partly because the fast reactor core contains more actinides than does the PWR - see Table 6 - and partly because, in the most severe of these hypothetical whole core incidents, a relatively large percentage of the inventory of non-volatiles is assumed to escape to the atmosphere.

To conclude, dispersion of a substantial fraction of a fast reactor core (or of a thermal core) would cause a large number of casualties. Continuing developments of theoretical studies of the course of the whole core accidents and their aftermath will enable the designers of the UK's CDFR to ensure that the frequency of occurrence of accidents severe enough to release significant quantities of radionuclides to the atmosphere will be low enough to satisfy the high standards practised by the nuclear industry. Examples of these standards are the guidelines recently suggested by Kinchin (1978).

THE CONCEPT AND COMPARISON OF RISK

The Nature of Risk

It is a common experience that certain human activities are attended by a high degree of risk that something will go wrong, resulting in damage, injury or death as a consequence. Perhaps the most obvious example is the use of motor transport vehicles, which constitutes the largest single cause of accidental deaths in the UK. Other pursuits involve lesser risks, and there are some activities that people do not think of as risky at all, and therefore have no hesitation in participating in them. However, it is misleading to think in such black and white terms, and to characterise things as "safe" or "unsafe"; no human activity can be free from the risk of injurious consequences arising from error or malfunction of some kind. Discussion of the topic of risk is notoriously fraught with difficulties arising from differences of opinion over what degree of risk is acceptable, and how different risks are to be compared. In order to engage in a sensible dialogue on this subject it is necessary to distinguish quite clearly at the outset between two aspects, namely the measurement of risk, and the acceptability of risk. Each of these aspects displays its own particular sort of difficulty, and each raises problems arising from the necessity to acknowledge these difficulties.

Of the two, it is the acceptability of risk that poses the more intractable problem, since in the end this is an issue which must be decided through the democratic process at Government level, whereas the measurement of risk is susceptible to a satisfactory degree of quantitative description. Decisions on the acceptability of risk can be more rationally achieved if there is a strong basis of quantitative measurement of risk. In view of these characteristics the following sections are mainly devoted to the problem of the measurement of risk, but they also contain some indicators of the nature of the difficulties associated with the acceptability of risk, in order to make explicit the distinction between the two aspects.

The Measurement of Risk

Various ways of expressing risk have been devised, each suited to the particular activity with which the risk is associated. Although these have their merits in their particular applications, there is considerable difficulty in comparing different measures of risk, since the choice of a common unit is a matter of some debate. Rather than attempt to unravel this contentious issue, and perhaps thereby confuse the matter further by adding yet another scheme of interpretation, it is simpler to reduce the problem to the consideration of two concepts of risk, namely Individual Risk, which is the probability of death per person per year (calculated as the number of deaths per year from a particular cause divided by the total population), and Community Risk, which can be expressed as a graph showing the frequency of occurrence (f, yr^{-1}) of incidents resulting in N or more fatalities. Such a graph is known as an f-N line, and risk is then defined as the product f times N.

In order to gain a perspective of the risks to which people are exposed it is instructive to look at Individual and Community Risk data for the United

Kingdom. Readers of the Rasmussen Report will be familiar with the tables and graphs of these risks presented in that study, based on data and predictions for the United States. These figures have been used by many authors in discussions of risk measurements as though they applied to any comparable industrialised nation. It is perhaps worth noting that there are substantial differences in the risk data for the UK compared with the USA, as evidenced in two recent UKAEA reports (Grist, 1978; Griffiths and Fryer, 1978); the overall individual risk of accidental death in the USA is about twice the UK value, with some causes displaying risks that are as much as three times higher in the USA compared with the UK.

Table 7 and Figure 16 are reproduced from Grist (1978) and Griffiths and Fryer (1978) respectively. To give some perspective to the individual risk data in Table 7 it is instructive to note that accidental death constitutes only

Table 7 Individual Death Risks in Great Britain

ICD* No's	Cause of death	Risk per person per year
000-999	All Causes	1.18×10^{-2}
000-799	Natural Causes	1.14×10^{-2}
390-458	Disease of Circulatory System	6.17×10^{-3}
410-414	Ischaemic Heart Disease	3.12×10^{-3}
140-239	Neoplasms	2.47×10^{-3}
800-999	Other than Natural Causes	4.42×10^{-4}
800-949	All Accidents	3.25×10^{-4}
810-823	Road Accidents	1.22×10^{-4}
880-887	Falls	1.11×10^{-4}
890-899	Fires	1.6×10^{-5}
910	Drowning	1.2×10^{-5}
800-807	Rail Transport	3.3×10^{-6}
925	Electrocution	2.5×10^{-6}

* International Classification System for causes of death.

2.75% of the overall risk from all causes, and that about two-thirds of the overall accidental death risk arises from fatalities due to road accidents and falls. Figure 16 showing frequency of occurrence of incidents resulting in N or more fatalities for 5 different causes in the UK, displays some features that require comment. It will be noted that there is a relatively high risk of death through fires in the N = 1 to 10 range. This may indicate that a greater degree of care and regulation is applied to fire safety in premises where large numbers of people assemble, as distinct from domestic premises where less control may be exercised. The distinctly higher risk in the aircraft passengers line in the range N = 10 to 100 doubtless reflects the preponderance of aircraft carrying this number of passengers. For ships' crews, rail passengers and Public Service Vehicle occupants the data correspond approximately to lines of equal risk with the risk product fN being typically about 10 yr^{-1} over much of the range (i.e. the product fN does not vary much with N), whereas the fire and aircraft lines both depart substantially from this behaviour. This is referred to again in the discussion of the acceptability of risk.

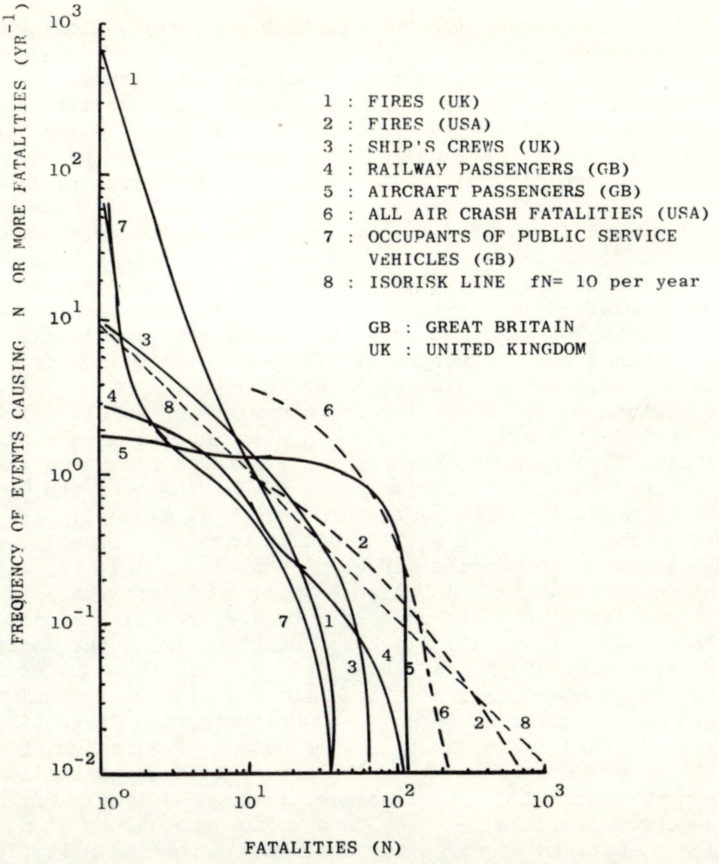

Fig. 16 The Incidence of Multiple Fatality Accidents.

Difficulties in the Measurement of Risk

As noted above the use of Individual Risk and Community Risk as indicators involves certain assumptions and qualifications. It is only fair to acknowledge these, and mention is now made of some of the facets that should be borne in mind in interpreting these data. Concerning Individual Risk, the assumption in the definition is that the population at hazard is the total population. This is a simplifying assumption that has to be examined in some detail. Obviously some occupations are attended with a higher risk than others (e.g. divers on North Sea oil rigs have a very high risk associated with their work), and these departures from average values can be very large indeed, so that the risk associated with a particular cause of death can be quite different for different sections of the community. Again, it is clear that some classes of risk are significantly underestimated by assuming that it is the total population that is exposed to the hazard; for example not everyone travels by road, or by rail, and any individual may choose to forego these forms of travel, thereby eliminating that contribution to his total individual risk. Further, people that do engage in a particular pursuit do not generally share the risk equally among their number. In activities of this sort, where the risk can reasonably be argued to increase in proportion to the extent to which the individual engages in the pursuit, other things being equal, it is appropriate to express the risk in terms of, for example, deaths per person per mile travelled, or deaths per person per hour of activity.

This serves to illustrate the difficulties encountered when trying to compare risks. Although the definition of Individual Risk adopted here does not overcome these problems, it does have the virtue of simplicity, and its assumptions are unambiguous. With these qualifications in mind the reader may gain a useful insight into the spectrum of individual risk catalogued in Table 7.

In considering the Community Risk, as represented by the f-N lines of Figure 16, one needs to be aware of the limitations imposed on their significance by the nature of the data. It is important to bear in mind that there are two conflicting characteristics inherent in the construction of such a graph. First, it is clear that one cannot base such a graph on only one year's data, since the statistical significance will be very low. It is necessary therefore to use many years' data to increase the significance. However, because the total number of incidents in a given period involving large numbers of fatalities will usually in practice be considerably smaller than the number involving only a few fatalities, the statistical significance decreases as N increases. This means that even for an activity in which standards of safety are not changing over the years, the significance of the product fN for large values of N will usually be quite small. Again, the accident rates in different activities vary considerably, so that it will usually be necessary to compare different periods of data for different activities. This necessity may also arise by virtue of different practices of record keeping in different fields, or simply because records may not have been kept for very long. The second aspect, which conflicts with the desirability from the statistical point of view of using data periods that are as long as possible, is that safety standards and practices may have changed considerably over a period of years. For example, it is noted by Griffiths and Fryer (1978) that over the last twenty years there has been a reduction by a factor of about 3 in the annual total number of accidental deaths incurred in the operation of the railways, a trend which could not be displayed in the f-N line. Given these limitations, the concept of the f-N line is a useful one, since it does provide the basis for the comparison of risks for accidents involving multiple fatalities, which undoubtedly have a special significance in the perception of risk.

The Acceptability of Risk

The foregoing survey of Individual and Community Risk provides information on the magnitudes of the contributions made by various activities to the overall burden of risk. This does not reveal anything about people's willingness to accept these risks, except insofar as the fact that they are allowed to continue to exist implies some degree of acceptance.

However, something useful can be deduced from the fact that where there is a risk of death approaching 10^{-3} per year, steps are usually taken to reduce this if possible - hence the current concern over the death toll due to traffic accidents. As the risk diminishes, concern and countermeasures are less in evidence, and for risks less than about 10^{-6} per year it seems that the individual does not think it necessary to take steps to reduce the risk further. This suggests that society in practice regards individual risks of about 10^{-6} or less per year as acceptable. It is often argued that a distinction needs to be made between voluntary and involuntary risks, and it is clearly reasonable that this should be the case, difficult though it is in some instances to draw such a clear distinction. However, this argument becomes less forceful for risks that are of such a small magnitude that the public generally are unaware of them at any significantly conscious level.

Many authors, in discussing the concept of the f-N line as a measure of community risk, have pointed out that an industry that gives rise to, say, 100 deaths one at a time over a period of 10 years is more readily tolerated than one causing 100 deaths all at once every 10 years, even though the two are equal in terms of the definition of community risk. The reason for this decrease in the acceptability of equal risks as N increases is probably that an incident causing 100 deaths in a single community is much more disruptive than 100 deaths spread over a longer time, and over a wider community. Does this mean that in general the acceptability of community risk falls as N increases? Referring to the aircraft passenger f-N line in Figure 16, there is a very much higher risk in the N = 10 to 100 range than in the N = 1 to 10 range, and yet practice suggests that this is in fact acceptable. Although one would hesitate to draw any conclusions from this, it does demonstrate the powerful influence of economics and perceived benefit as factors that affect our willingness to be exposed to a given risk.

Risk Criteria for Nuclear Reactors

In view of the desirability of comparing nuclear power plant risks with others to which we are exposed it is logical to express design criteria in terms of individual and community risk criteria. Accordingly the UKAEA Safety and Reliability Directorate has been developing such criteria for use as design targets for power reactors. Because of the potential for both early and delayed fatalities as a result of radiation exposure, two criteria for both individual and community risk have been proposed. The rationale behind these proposals is discussed by Kinchin (1978). For Individual Risk it is suggested that 10^{-6} per year should be the limit for early deaths, and 3×10^{-5} per year for delayed deaths. For Community Risk it is proposed that the risk product fN should take the value 10^{-4} per year for early deaths, and 3×10^{-3} per year for delayed deaths.

The intended interpretation of the Community Risk Criteria, as discussed by Kinchin (1978) is that the value of the ordinate p(N) represents the limiting probability per year (or frequency) of producing a number of fatalities roughly in the range N/3 to 3N, or approximately a factor of 10 spanning the number N. These criteria are therefore non-cumulative forms of fN lines; they are displayed in Figure 17 together with their equivalent cumulative forms, which show the limiting frequency of incidents producing N or more fatalities. It will be seen by comparing these cumulative forms of Kinchin's criteria with the cumulative fN lines for actual accidents in Figure 16 that they fall many orders of magnitude below the fN product for multiple fatality accidents in the UK, for which values of the product fN of about 10 per year are representative.

To illustrate the level of risk implied by the individual early death criterion of 10^{-6} per year, one may consider the following: suppose that an individual decides to reduce his risk of early death from a nuclear power plant accident by moving house away from the plant, and that in consequence he has to drive a somewhat greater distance to get to work. From Table 7 the individual risk of death from road accidents is 1.22×10^{-4} per year in the UK. The average annual mileage covered by the British motorist is about 10,000 miles (16,000 km), so the risk may be expressed as $(1.22 \times 10^{-4})/10^4 = 1.22 \times 10^{-8}$ per mile per year. It follows that the risk of 10^{-6} is equivalent to driving an extra 82 miles (122 km) per year, which corresponds to an extra 300 yards (270 m) per journey to or from work for a man working a 5 day week for 48 weeks a year. Thus, if moving away from the nuclear power plant increases the distance to work by more than 300 yards (270 m) it is safer to live next to the plant. Put in these terms, most people would find this risk acceptable.

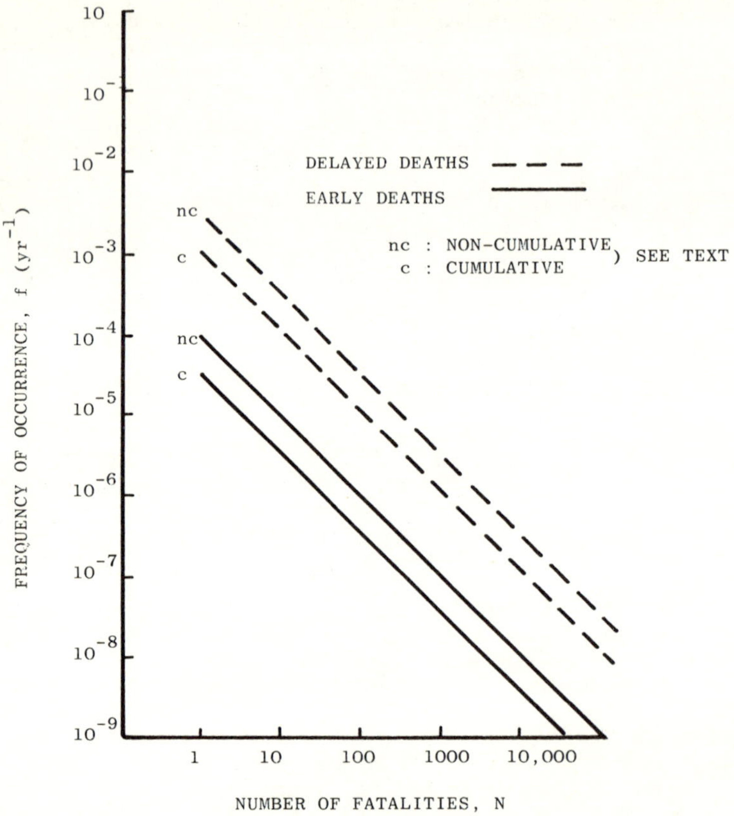

Fig. 17 Proposed Community Risk Criteria

CONCLUSIONS

The impact on the environment of notional accidents to nuclear power reactors is principally, this paper concludes, the impact on man of the radioactivity released. Man, at the top of the environmental 'tree', is apparently the prime target of environmental impact. This impact consists of the early and late effects of radiation on man, and the paper reviews these, from 'early' deaths caused principally by high radiation dose to bone marrow, to 'late' deaths caused by radiation-induced cancer occurring perhaps some decades after the accidental release. In the event of a reactor accident release, doses of radiation could reach man more or less directly from the cloud of airborne radioactivity, or indirectly by radiation from activity deposited on the ground, some coming via various ingestion routes (food chains). Physical and other mechanisms whereby radiation and radioactive substances would reach man following such an event are described in this paper.

From the description of notional accidents to various reactor systems and their environmental consequences, it emerges that reactor systems and their engineered safeguard systems are so designed that the environmental impact of an accident, should one occur, will in all probability be very small indeed. In this context the 'design basis accident' is the familiar model of the most probable outcome. Its only environmental impacts worthy of note will be, in the first instance, a dose to the thyroid and/or lung by inhalation, amounting to up to about 30 rads to individuals who are about 10^2 to 10^3 metres downwind. Such a dose implies an individual risk of 'late' death in the range 10^{-4} to 10^{-3}, but the risk would diminish rapidly with distance for persons further away downwind. In the second instance there would be a need to ban the sale and consumption of cows' milk from farms out to about 3 kilometres downwind of the reactor, for a period of only a few weeks.

Considering notional accidents of much greater severity but very much lower probability of occurrence, further possible aspects of environmental impact are revealed. The impact of more severe PWR accidents as discussed on the basis of the analysis published by USNRC, and a discussion of recent UK analyses of the impact of more severe fast reactor accidents is also given. The "impact" is displayed in the form of graphs of numbers of 'early' and 'late' deaths against probability per reactor-year. Thousands of fatalities can be calculated if one neglects all favourable factors, but most postulated severe accident releases would be accompanied by the simultaneous release of great quantities of heat. Plume rise would therefore be a feature of most large releases (although downwash in the lee of reactor buildings would be a contrary factor to take into account). With plume rise taken into account, numbers of 'early deaths could be reduced to 10-100 at a conditional probability level of 1-10% as is shown in Figure 15. However, it has been pointed out that the effect of plume rise on the probability distributions of cancers and of unacceptably contaminated areas of land would be much less marked. For both PWR and fast reactor accidents, of this class of severity and low probability, land areas likely to be contaminated in the long-term by deposited gamma-emitting nuclides would be a problem in the aftermath of the accident; for the fast reactor case, land areas significantly contaminated by deposited actinides which are hazardous when resuspended by the wind would be larger than for the PWR.

The paper concludes by reviewing the risks to which man is exposed at all times because of his environment - his risk of death as an individual, for instance by chance accident at home, at work or at play, and also by the apparently chance incidence of fatal illness. These risks are by no means insignificant, though the average man may often view them with equanimity. Finally, community or collective risk is examined. It appears that from fires, railway accidents and aircraft crashes the community accepts without too much demur a group risk product f-N which is typically about 10 years^{-1}, but sometimes even higher.

Viewed against this, criteria that the nuclear industry has set itself seem very safe indeed. The criteria proposed by Kinchin (1978) are, for Individual Risk, 10^{-6} per year as the limit for 'early' death and 3×10^{-5} per year for 'late' death; for Community Risk the Kinchin criteria propose that the risk product fN should be 10^{-4} per year for early deaths and 3×10^{-3} per year for 'late' deaths. These criteria for community risk fall well below the risk products already indicated for actual multiple fatality accidents in the United Kingdom. Expressed in terms of the product fN, as defined in this paper, the risk associated with accidents to nuclear power reactors is estimated to be very small indeed.

ACKNOWLEDGEMENTS

The co-operation of certain members of the staff of several United Kingdom organisations who commented at short notice on a preliminary draft of this paper is gratefully acknowledged; comments by Mr. G. C. Dale and Dr. B. M. Wheatley of CEGB, Dr. R. H. Clarke of NRPB and Mr. P. S. Bland of SSEB were extensive, detailed and of particular value to the authors. British Nuclear Fuels Ltd, the Nuclear Installations Inspectorate, the Department of Energy and the Department of the Environment were also consulted during the short time available for preparation of the paper. However, the material presented (except where referenced to other authors), the opinions expressed and the conclusions reached are the responsibility of the authors of the paper.

REFERENCES

Atom, Generic Safety Issues of Pressurised Water Reactors, 251, 218 (1977).

Beattie, J.R. and P.M. Bryant, Assessment of Environmental Hazards from Reactor Fission Product Releases, UKAEA Report AHSB(S)R135 (1970).

Beattie, J.R., J.B. Lewis and R. Lind, Graphite Oxidation and Reactor Safety. Proc. 3rd Int. Conf. Peaceful Uses At. Energy, 13, paper 185 (1964) Geneva.

BEIR Report, The Effects on Populations of Exposure to Low Levels of Ionizing Radiation, Advisory Committee on the Biological Effects of Ionizing Radiations, US National Academy of Sciences/National Research Council, Washington D.C. (1972).

Briggs, G.A., Plume Rise Predictions, USDOC, National Oceanic and Atmospheric Administration, Air Resources, Atmospheric Turbulence and Diffusion Laboratory, Oak Ridge, ATDL Contribution File No. 75/15 (1975).

Clarke, R.H., The WEERIE Program for Assessing the Radiological Consequences of Airborne Effluents from Nuclear Installations, Health Physics, 25, 267 (1976).

Dolphin, G.W. and W.G. Marley, Risk Evaluation in Relation to the Protection of the Public in the Event of Accidents at Nuclear Installations. UKAEA Report AHSB (RP)R93 (1969).

Findlay, J.R., The Birth, Abundance and Movement of Fission Products through Fuel, Paper 1, Conference on Fission Products in Reactor Circuits, University of Birmingham, BNES (1973).

Gale, H.J. et al, The Weathering of Caesium 137 in Soil, UKAEA Report AERE R4241 (1963).

Gifford, F.A., A Review of Turbulent Diffusion Typing Schemes, Nucl. Safety, 17, 68 (1976).

Griffiths, R.F., Effect of Duration of Release on the Dispersion of Effluent Released to the Atmosphere, UKAEA Report SRD R85 (1977).

Griffiths, R.F. and L.S. Fryer, The Incidence of Multiple Fatality Accidents in the UK, UKAEA Report SRD R110 (1978).

Grist, D.R., Individual Risk - A Compilation of Recent British Data, UKAEA Report SRD R125 (1978).

Hosker, R.P., Estimates of Dry Deposition and Plume Depletion over Forests and Grassland, in: Physical Behaviour of Radioactive Contaminants in the Atmosphere, IAEA STI/PUB/354, 291 (1974).

Hunt, G.J. et al, Fallout in Rainwater and Airborne Dust - Levels in the UK during 1975, NRPB Report NRPB R49 (1976).

ICRP (International Commission on Radiological Protection). Publication 26 (1977). Pergamon Press, Oxford, UK.

ICRP (International Commission on Radiological Protection). The Evaluation of Risks from Radiation, ICRP Publication 8 (1966). Pergamon Press, Oxford, UK.

ICRP (International Commission on Radiological Protection). Recommendations of the International Commission on Radiological Protection, ICRP Publication No. 9 (1966). Pergamon Press, Oxford, UK.

Kaiser, G.D., A Description of the Mathematical and Physical Models incorporated in TIRION 2. UKAEA Report SRD R63 (1976). Fryer, L.S., A Guide to TIRION 4, UKAEA Report SRD R120 (1978).

Kaiser, G.D., Radioactive Plumes, Paper 60, IVth Int. Cong. of IRPA, Paris, April 1977.

Kelly, G.N. et al, An Estimate of the Radiological Consequences of Notional Accidental Releases of Radioactivity from a Fast Breeder Reactor, NRPB Report NRPB-R53 (1977).

Kinchin, G.H., The Assessment of Hazards in Engineering Work, Proc. Inst. Civil Eng., 1, 64, 431 (1978).

Knockaert, A. et al, St. Laurent-des-Eaux Reactor 1 Fuel Melt-down Accident October 1969, Paper 4, Inst. Mech. Eng./BNES London (1970).

Macdonald, H.F. et al, Development and Current Status of Emergency Monitoring Procedures at CEGB Nuclear Power Stations, J. British Nucl. Energy Soc., 16(2), 177 (1977).

Marley, G.W. and T.M. Fry, Radiological Hazards from an Escape of Fission Products and the Implications in Power Reactor Location, Proc. Int. Conf. Peaceful Uses At. Energy, Geneva, Paper A/Conf 8 (1956).

Matthews, R.R., Nuclear Power and the Environment, International Symposium on Man and his Environment, 19th September 1975, Birmingham University, England.

McGrath, P.E., D.M. Ericson and I.B. Wall, The Reactor Safety Study (WASH 1400) and its Implications for Radiological Emergency Response Planning, Proc. Int. Symp. on the Handling of Radiation Accidents, Vienna, IAEA-SM-215/23 (1977).

Medical Research Council, Criteria for Controlling Radiation Doses to the Public after Accidental Escape of Radioactive Material. HMSO London (1975).

Medical Research Council, The Toxicity of Plutonium, HMSO London (1975).

Nair, S., A Reactor Invent ory Code for Calculating Actinide and Fission Product Arisings using a Point Source Model, CEGB Berkeley Nuclear Labs. Report RD/B/N4138 (1977).

Orchard, H.C. and C.W. Walker, Experience in the Provision and Exercising of Emergency Arrangements at CEGB Nuclear Power Stations, IAEA SM 215/30, Symposium on the Handling of Radiation Accidents, Vienna (1977).

Ritchie, L.T., W.D. Brown and J.R. Wayland, Effects of Rainstorms and Run-Off on the Consequences of Nuclear Reactor Accidents, Sandia Laboratories Report SAND 76-0429 (1976).

Russo, A.J., Reactor Accident Plume Rise Calculations, Sandia Laboratory Report, SAND 76-0340 (1976).

Scott Russell, R., Radioactivity and Human Diet (1966). Pergamon Press, Oxford, UK.

Scriven, R.A., Physical Behaviour of Radioactive Contaminants in the Atmosphere, IAEA, Vienna STI/PUB/354, 511 (1974).

Slade, D.H., Meteorology and Atomic Energy, US Atomic Energy Commission/Division of Technical Information, Report TID-24190 (1968).

Smith, H. and J.W. Stather, Human Exposure to Radiation following the Release of Radioactivity from a Reactor Accident: a Quantitative Assessment of the Biological Consequences. NRPB Report NRPB-R52 (1976).

Smith, F.B., A Scheme for Estimating the Vertical Dispersion of a Plume from a Source near Ground Level, Unpublished Met. Office Note (1974).

Thorne, M.C. and J. Vennart, The Toxicity of Strontium 90, Radium 226 and Plutonium 239, Nature, 263, 553 (1976).

United Nations Scientific Committee on the Effects of Atomic Radiation (UNSCEAR), Sources and Effects of Ionizing Radiation, United Nations E.77.IX.1 (1977).

United States Nuclear Regulatory Commission (1975), Reactor Safety Study: WASH-1400.

Webb, G.A.M., Radiation Exposure of the Public - the Current Levels in the United Kingdom. NRPB Report NRPB-R24.

Weir, C.G., Chapelcross Melt-out and Recovery Operations, Paper 3, Inst. Mech. Eng/BNES Discussion on Active Working on Reactors, London (1970).

Windscale Inquiry, Report by the Hon. Mr. Justice Parker, HMSO (1978).

6

THE ENVIRONMENTAL IMPACT OF REPROCESSING

P.W. Mummery, H. Howells and A. Scriven
British Nuclear Fuels Ltd.
Risley, Warrington,
Cheshire, WA3 6AS, United Kingdom

INTRODUCTION

Experience of reprocessing and its environmental impact extends over many plants in different countries and over a time span of about 35 years; UK experience represents a significant part of this total. Furthermore, British Nuclear Fuels Limited hold a unique position since its proposals for a new oxide fuel reprocessing plant (THORP) for fuel from thermal nuclear reactor power plants was the subject of a wide ranging public inquiry under the Town and Country Planning Acts. This inquiry lasted a hundred days and was conducted by a High Court Judge, Mr. Justice Parker, assisted by two eminent technical assessors, Sir Edward Pochin and Sir Frederick Warner. The scope of the inquiry included environmental impact particularly in relation to: (1) implications to the safety of the public both in respect to the disposal of solid, gaseous and liquid wastes and in respect of the safety of operations; (ii) infra-structure generally including traffic, amenities, employment, housing and public services. Furthermore, the current and immediate future exploitation of nuclear power worldwide is based on the use of oxide fuel of the kind appropriate to THORP. Accordingly, it is considered appropriate that this paper discusses the environmental impact of the proposed THORP plant; it also includes a discussion of the significance of accidents as well as routine operation. At the Windscale Inquiry and in other circumstances social/political issues have been raised, in particular, the potential activity of terrorists, the most appropriate ways of achieving non-proliferation of nuclear weapons and the security precautions considered necessary and their political impact on society. It is not proposed to discuss these issues in the paper. It was concluded in Mr. Justice Parkers report (1978) on the Windscale Public Inquiry that planning approval should be given for the proposed reprocessing plant. A Special Development Order giving approval was subsequently approved in Parliament.

The activities to be considered could include in principle all the various aspects associated with a programme for the reprocessing of spent nuclear fuel, the wastes arising from these activities and the environmental impact of the operations. Such activities include: transport of the irradiated fuel from the power reactors to the reprocessing site; the operations on the reprocessing site;

immediate disposals of wastes from the site; conditioning and storage of wastes; and transport and disposal of wastes which have been stored.

BASIS OF HAZARD ASSESSEMENT

Principles of Assessment of Risk from Exposure to Ionising Radiation:

Throughout the development of the nuclear power industry close attention has been paid to the question of risk to the workforce and the general public arising from exposure to ionising radiation. The methodology of assessment of risk to any specified group of people can be divided into four sequential steps, as follows:

i. Source of the radioactive material
Identification of each point of emission of radioactive effluent to the environment; and determination of the rate of emission of each specific radionuclide or group of nuclides; and consideration of means and effectiveness of effluent control.

ii. Environmental routes for transport of radionuclides to man.
Determination of the pathways by which the radionuclides can reach man. Studies of the dispersal of the radionuclides in the environmental media and of ways in which the activity can reach man in air, in water and in food.

iii. Assessment of radiation dose
Consideration of concentrations of radionuclides in the media and food, rates of intake into the body, distribution of the material in the body, biological half-lifes of the radionuclides, absorbed dose in individual organs and biological effectiveness of the particular radiation.

iv. Derivation of risk
Estimation of risk in terms of biological effect resulting from exposure to the assessed dose of radiation. In the case of low levels of radiation exposure resulting from the operation of reprocessing plant the relationship between dose and a specified health effect, is statistical. The effects on population of exposure to low levels of ionising radiations were discussed in a report (2) of the Advisory Committee on the Biological Effects of Ionising Radiation (BEIR).

System of Dose Limitation Recommended by the International Commission on Radiological Protection (ICRP): The ICRP, in ICRP Publication 26 (1977), recommends a system of dose limitation, the main features of which are as follows: (i) no practice shall be adopted unless its introduction produces a positive net benefit; (ii) all exposures shall be kept as low as reasonably achievable, economic and social factors being taken into account; (iii) the dose equivalent to individuals shall not exceed the limits recommended for the appropriate circumstances by the Commission. These are three fundamental principles to be followed in the design and operation of reprocessing plant. The emphasis is clearly on justification of the plant, in that 'benefit' should outweigh 'detriment', and on keeping exposures as low as reasonably achievable within the recommended limits.

However the qualification that economic and social factors be taken into account is important. If a plant is designed, constructed and operated to achieve radiation doses which are low in comparison with the ICRP limits, then little benefit may be gained by reducing the dose levels still further, when balanced against the cost, in terms of human resources and capital expenditure. The cost/benefit ratio for each further decrement of dose will increase as the dose decreases. Beyond a given point, and where this point lies is a matter of judgement in each case, any further expenditure could be more usefully deployed in reducing risk associated with other activities of society. The considerations in determining what is 'reasonably achievable' in reduction of dose below the recommended limits are discussed in ICRP Publication 26. A cost-benefit approach is presented in ICRP Publication 22. This provides a technique of optimising the collective dose equivalent due to radioactive waste management. It includes estimates of the cost to effect increasing reductions in the radiation exposure. It is apparent that as this process proceeds the cost of each decrement achieved increases and a judgement has to be made as to how far this process should be continued. Such collective dose equivalent studies must include both occupational and public exposures. Additionally the radiation exposure of individual workers and members of the public must also be considered in this optimisation process. In those cases where there is a justification for making such detailed assessments the analysis must be made over the complete waste management cycle, this will include the generation of the waste, immediate disposal or short term storage to permit radioactive decay and long term storage, conditioning and disposal.

Acceptance of ICRP Recommendations: There has, in general, been worldwide acceptance of ICRP recommendations which have become the basis of radiological protection legislation in many countries. An EEC Council Directive (1976) laying down basic radiological protection standards is based on ICRP recommendations, and members of the Community are required to make arrangements to conform with the Directive. Prior to the issue of that Directive relevant legislation in the UK and other European countries was already based on ICRP recommendations. In the UK the National Radiological Protection Board, in accordance with a Direction of the Health Ministers (1977), has a duty to advise government departments and statutory bodies on the acceptability of the recommendations of the ICRP for application in the UK. The Board also advises on recommendations of the UN, OECD/NEA and EEC proposals. The Board has recently advised (1977) that the system of dose limitation recommended by ICRP (in Publication 26) provides a satisfactory basis for controlling the exposure of persons to ionising radiations in workplaces and in the general environment.

Dose Limits - Recommended by the ICRP: The Commission's recommendations are intended to prevent "non-stochastic" effects (i.e. those for which the severity of effect varies with dose, and for which a threshold may therefore occur) and to limit to an acceptable level the occurrence of "stochastic" effects (i.e. those for which the probability of an effect occurring, rather than its severity, is regarded as a function of dose, without threshold). Exposure from natural background radiation and from medical procedures is not taken into account. The Commission believes that non-stochastic effects will be prevented by application of a dose-equivalent limit of 50 rem in a year to all tissues except the lens of the eye for which a limit of 30 rem in a year is recommended. In the case of stochastic effects the recommended system of dose limitation is based on the principle that the risk should be equal whether the whole body is irradiated uniformly or not. The recommended dose-equivalent limits for the whole body are 5 rems in a year for occupationally exposed persons and 500 millirems in a year for individual members of the general public.

Weighting factors are attributed to different tissues/organs representing the proportion of the risk resulting from specific tissues/organs to the total risk when the whole body is irradiated uniformly. Weighting factors recommended by the Commission are given in Table 1.

Table 1

Organ/Tissue		Weighting Factor (compared to whole body irradiation)
Gonads		0.25
Breast		0.15
Red Bone Marrow		0.12
Lung		0.12
Thyroid		0.03
Bone Surfaces		0.03
Remainder*		0.30
	Total	1.00

The ICRP no longer ascribes dose limits as such to individual organs but dose limits for a specific organ can be derived if it is assumed that the individual organ only is exposed. The dose limit applicable to any individual organ is the lower of the relevant stochastic and non-stochastic limits. The stochastic limit is derived from Table 1.

For example if the thyroid is the only organ exposed the stochastic limit for an occupationally exposed person is 5 rems/y (whole body) x $\frac{1}{0.03}$ (weighting Factor) = 167 rems/y. However the non-stochastic limit is 50 rems/y so the limit applicable is 50 rems/y. Some annual dose limits derived in this manner are shown in Table 2. The limits derived in this way for the individual organs/tissues apply only on the assumption that the single organ only is exposed. If more than one organ is exposed the weighting factors must be used in order to attribute the proportion of the risk from each organ to the total risk if the whole body were irradiated uniformly.

---

* A weighting factor of 0.06 is ascribed to each of the five organs or tissues of the remainder receiving the highest dose equivalents. In the special case of the "skin" a weighting factor of 0.01 is assigned to relate the stochastic risk of the radiation dose received by the skin to that resulting from whole body irradiation.

Table 2

Organ or Tissue	Dose-equivalent Limit (rem)	
	Worker	Member of the Public
Whole Body	5	0.5
Gonads	20	2.0
Breast	33	3.3
Red Bone Marrow	42	4.2
Lung	42	4.2
Thyroid	50	5.0
Bone Surfaces	50	5.0
Lens of the Eye	30	3.0

The ICRP no longer recommends a limit for application to populations as it did in earlier recommendations in 1965. The former dose limit of 5 rem in 30 years was a genetic dose limit based on hereditary effects. In ICRP Publication 26 the Commission does not propose dose limits for populations because such limits could suggest the acceptability of higher exposure than is necessary or probable, and a higher risk than is justified by any present or easily envisaged future development. Furthermore, the Commission states that knowledge gained over the past 2 decades indicates that genetic defects, while important, are unlikely to be of overriding importance. The Commission now takes the view that observance of the general principles presented in ICRP Publication 26, together with the system of dose limitation established there, is likely to ensure that the average dose equivalent to the population will not exceed 50 millirems per year.

The former limit, based on genetic considerations was 5 rems per generation, i.e. 5 rems in 30 years equivalent to 170 millirems per year. In the UK a view was taken in a 1959 White Paper on the Control of Radioactive Wastes that the gonad dose to the population from waste disposal could be kept to 100 millirem per person per 30 years (i.e. 3.3 millirems per year) at most and that it would never be necessary to exceed 1 rem per person per 30 years (i.e. 33 millirem per year). The White Paper is currently under review in the UK.

The National Radiological Protection Board (1977) has recently issued advice to the expert group engaged in the review. In its advice the Board endorses the ICRP Publication 26 recommendations relating the annual dose equivalent limits for individual members of the public; and it is noted that the use of a limit of 500 millirems in a year, combined with the technique of optimisation discussed in ICRP Publication 22 will in most cases result in an average dose rate equivalent to a critical group of less than 100 millirem per year of life-long whole body exposure from all sources of radiation (this does not include exposure from natural background radiation or medical procedures). Hence the life-time whole body dose equivalent of an individual will not normally exceed 7 rem.

On the subject of population exposure the Board refers to the ICRP's view that the system of dose limitation established in ICRP Publication 26 is likely to ensure that the annual dose equivalent average over the population will not exceed 50 millirems. The Board considers that the maintenance of the annual dose equivalent below 50 millirems when averaged over the whole UK population is a reasonable objective; this objective being related to exposure from all sources of radiation excluding natural background and medical procedures. The Board also considers it most unlikely that exposure of the population of the UK will approach this value but is keeping the matter under review. The contribution from all UK waste management practices is not likely, in the Board's view, to exceed one-tenth of this value i.e. an annual average dose equivalent of 5 millirems.

Risk Arising from Exposure to Radiation: In ICRP Publication 26 the Commission considered the risk to body tissues arising from exposure to radiation. The risk factors for the purposes of radiation protection are given as follows:-

Red Bone Marrow	2×10^{-5} per rem	(leukaemia)
Bone	5×10^{-6} per rem	(cancer)
Lung	2×10^{-5} per rem	(cancer)
Thyroid	5×10^{-6} per rem	(Mortality from cancer)
Breast	2.5×10^{-5} per rem	(cancer)
All other tissues combined	5×10^{-5} per rem	(cancer)
Whole Body	10^{-4} per rem	(Mortality risk for radiation induced cancer, as an average for both sexes and all ages)

COMPARISON OF STANDARDS AND PRACTICE IN THE UK AND THE USA IN THE LIMITATION OF EXPOSURE OF THE PUBLIC AS A CONSEQUENCE OF DISCHARGES OF RADIOACTIVE EFFLUENT TO THE ENVIRONMENT

The UK Approach: UK policy with regard to exposure arising from waste disposal is set out as follows in the White Paper referred to earlier:

(a) To ensure, irrespective of cost, that no member of the public shall receive more than one tenth of ICRP level for occupational workers.

(b) To ensure, irrespective of cost, that the whole population of the country shall not receive an average dose of more than 1 rem per person in thirty years;

(c) To do what is reasonably practicable, having regard to cost, convenience and the national importance of this subject, to reduce the doses far below these levels.

With regard to (b) it was expected that the average dose could be kept to 0.1 rem per person per thirty years at most (3.3 millirem per year). This has indeed been proved to be so. The need for discharges, and the nature and quantity and means of control of such discharges, is considered on a site by site basis. Discharges are permitted only under the terms of authorisations granted by the Department of the Environment and the Ministry of Agriculture, Fisheries and Food and also the Scottish Development Department and the Welsh Office, in the case of discharges occurring in Scotland or Wales, in accordance with the Radioactive Substances Act 1960. The UK policy on control of radioactive wastes is under review and the NRPB has given advice (on the application of ICRP recommendations) to the expert group engaged in the review. The basis of the present policy is a case-by-case pragmatic approach to the question of discharge authorisations, in the light of the overall developing situation. This philosophy provides for firm control with flexibility. It has stood the test of time and has served well but is being re-examined as part of the review.

Standards Set by the US Environmental Protection Agency (EPA): The EPA has established environmental standards (1973, 1977) applicable to Uranium fuel cycle operations in the USA (excluding mining and waste disposal sites). The standards require that normal operations shall be conducted in such a manner as to provide reasonable assurance that:-

(a) Annual dose equivalent to any member of the public as the result of exposures to planned discharges to the general environment from uranium fuel cycle operations, and to radiation from these operations does not exceed -

25 millirems to the whole body
and 75 millirems to the thyroid
and 25 millirems to any other organ

(b) The total quantity of radioactive materials entering the general environment from the entire uranium fuel cycle contains less than -

Radionuclide	curies per gigawatt-year of electrical energy produced
Krypton 85	50,000
Iodine 129	0.005
Plutonium 239 and other alpha emitting trans-uranic radionuclides with half-lives greater than one year	0.0005

The standards provide for variances to be granted by the regulatory agency for unusual operations. The effective date of the standards in (a) above is December 1979. The effective date of the standards in (b) above is December 1979, except that the standards for Krypton 85 and Iodine 129 shall be effective from January 1983, for any such radioactive materials generated by the fission process after these dates. The "general environment" means "the total terrestrial, atmospheric and aquatic environments outside sites upon which any operation which is part of the nuclear fuel cycle is conducted".

It should be noted that the US Environmental Protection Agency "Standards" are related to the particular conditions which apply in that country: the "standards" were to apply to fuel reprocessing plants operated by privately owned utilities and to many more such plants than envisaged for the UK. Furthermore the limits were derived assuming only airborne releases of radioactivity. The basic requirement recommended by ICRP, that the total radiological impact of waste discharges should be 'as low as reasonably achievable', can be satisfied by a number of different approaches. At the end point where there are differences in the numerical limits, described as "standards" in some countries, these do not mean that there will necessarily be significant differences in actual doses received from given practices.

NATURE OF REPROCESSING OPERATIONS

The Thermal Oxide Fuel Reprocessing Plant (THORP):

The plant will be constructed on the Windscale site of British Nuclear Fuels Limited (BNFL) in the North West of England. It has a designed maximum throughput of 1200 tonnes of irradiated uranium per year and the fuel for reprocessing has a maximum irradiation rating of 37,000 MWD/tonne. In the course of this study these data will be assumed to apply and the further, conservative, assumption will be made that the spent fuel will have a storage time of only one year prior to commencing reprocessing. The different steps in fuel reprocessing are briefly considered below. Each stage will generate radioactive wastes; the sources and nature of these are also indicated.

Storage Ponds: On arrival at the reprocessing plant the spent nuclear fuel is initially stored in specially constructed 'ponds' for periods up to a few years. This permits the radioactive decay of the shorter lived fission products. The fuel is received from the nuclear power stations in robust, heavy shielded flasks, from which the fuel is removed either directly into the ponds or transferred to water filled storage bottles provided with vents to release accumulated gases. The pond water becomes contaminated primarily with radiocaesium and radiostrontium isotopes. Decontamination of this water is to be effected by filtration to remove particulate and by ion exchange units to absorb activity in solution. The fuel storage ponds give rise to low level liquid wastes, contaminated filtering media and solid ion-exchange wastes.

Fuel Shearing Operations: After removal of the fuel from the storage ponds it is transferred within fuel flasks to the shearing facility. Here the fuel is mechanically chopped into small pieces. Dust from the shearing cave extract is directed into the dissolver prior to stack discharge.

Leaching: The chopped fuel is dropped into a perforated basket placed within the dissolver unit which contains hot nitric acid. The fuel is leached from the chopped cladding by boiling the nitric acid.

Clarification: After leaching the fuel hulls (chopped cladding) are washed and transferred to storage under water. The liquid in the dissolver will contain some insoluble material including fission product alloys from the fuel and both coarse and fine particulate from the cladding material together with debris from the structural components of the fuel elements. Fine particulates will be separated by centrifuging while the coarse materials will be collected by settling. Both these wastes will need to be stored prior to disposal.

Solvent Extraction Process: The liquor from the dissolver, after the removal of solids, is passed into the first extraction system where the process separating the fission product wastes and the products, uranium and plutonium, starts. The solvent used is tri-n-butyl phosphate in a paraffin diluent, the separation and purification involves a number of extraction cycles using this solvent. After each cycle the solvent undergoes a washing process before being re-used. This wash liquor is a major component of the liquid effluent arisings. The products from the solvent extraction cycles are uranium and plutonium nitrates, while the bulk of the fission products are removed as high level wastes for storage prior to ultimate disposal.

Finishing Processes: The two products of the solvent extraction process are first concentrated by evaporation. The uranyl nitrate solution is then decomposed to its oxide by heating, subsequently this product can be sent to another site for conversion to uranium hexafluoride and used in a re-enrichment cycle to be ultimately part of further material for fuel fabrication; alternatively, it may be held in storage awaiting introduction into the fuel cycle. The plutonium nitrate is converted first to plutonium oxalate by a precipitation process and subsequently to the oxide by thermal decomposition and calcination.

WASTE MANAGEMENT

The reprocessing operations outlined above give rise to liquid and gaseous effluents together with radioactive solid wastes. The source and nature of these wastes are considered below.

Liquid Effluents:

The majority of radioactivity arising from reprocessing is associated with the high level liquid waste. It contains more than 99% of the fission product activity. This waste arising from the first extraction cycle of the reprocessing plant, is reduced in volume by evaporation and stored in stainless steel tanks. The tanks are fitted with cooling coils to remove the fission product heat. The system is designed to very high standards of integrity, with multiple containment of the liquor and adequate redundancy provision for the cooling systems. It is intended to convert this high level waste into a solid form by vitrification. The resultant conditioned waste will be stored, finally to be disposed of into suitable deep geological formations or alternatively on, or buried in, the deep ocean floor. Effluents which may be categorised as low activity wastes arise, after treatment, from the fuel storage ponds and from the chemical plant, where steam-stripping, steam ejector condensates and cell decontamination washings give rise to such effluents. These wastes can, with suitable control of the rate of activity discharge, be disposed of by pumping into the sea within limits specified by those central government departments who control and authorise such disposals. Finally, there are intermediate level wastes which require conditioning and storage prior to disposal. These are all plant liquid wastes, including the aqueous raffinate from all stages of extraction in the plant other than the first cycle extractor, scrubber liquors, wastes from the highly active liquor evaporators and decontamination and plant wash liquors. For some cases most of the activity is associated with fission products such as zirconium, niobium and ruthenium, with relatively short half lives and where delayed storage can significantly reduce the activity. These wastes are concentrated by evaporation and then held in tank storage till such time that they can safely be discharged to sea. Other waste streams contain sufficient actinide activity to require treatment prior to disposal and this is effected by flocculation treatment to precipitate actinides (plutonium and americium principally) for storage on site. The floc formed in this way may later be incorporated in the vitrified solid waste stream or in a solid matrix such as bitumen for disposal by the same route as that for the high level waste.

Gaseous Effluents:

The gaseous and airborne effluents may arise from three possible sources, ventilation from the working or operating areas, from within the cells containing plant vessels and finally by the extract on the plant vessels themselves. Since the plant cells are sealed the operating area has a ventilation and extract system separate from that of the cells and vessels. Cell air and more especially vessel extract air are liable to be contaminated. It is for this reason that the vessel extract system is designed to operate so that the cell air is at a higher pressure than that in the vessels, any leakage will permit air to pass only from the cell into the vessel. Since dust in the air will pick up activity, the plenum air to the ventilation system is filtered prior to entry into the plant while the exhaust air is heated to vapourise entrained liquor and filtered prior to discharge to atmosphere through a chimney stack. Additional to this general treatment, appropriate local plant treatment is given in some cases. For instance the off-gas from the dissolver vessel is treated by alkali scrubbing which will remove most of the radio-iodine together with tritium and some of the carbon-14 present. Further clean up of the effluent for any residual radio-iodine can be achieved by a specially developed solid absorber containing silver. Most of the radioactivity left in the gas steam is krypton-85 and processes to remove this are being developed. The evaporator systems associated with the Highly Active Liquor Storage tanks can have high entrained activity in the extract gas; to deal with this problem electrostatic precipitators and a high efficiency scrubber unit are used to remove such activity.

Solid Wastes:

The bulk of the solid wastes is of low activity. This is of a heterogeneous nature and comes from plant operations and general routine maintenance work and can be contaminated with 'alpha' and/or beta/gamma emitting nuclides. Most of this waste can be immediately disposed of by shallow land burial. Solid wastes of activity too high for such disposal can be held in retrievable stores, in some cases to permit radioactive decay, and to be further conditioned prior to disposal by the same route as that to be adopted for the high level waste. Wastes in this category will include fuel hulls, air filters and the ion exchanges and sand bed filters used to decontaminate liquid effluent together with sludges from flocculation treatment.

Radioactive Waste Management

Low-level liquid effluents are discharged immediately to sea under controlled conditions. Intermediate level liquid effluents are concentrated by evaporation, stored to permit natural radioactive decay and conditioned prior to sea discharge. High level liquid waste, on the other hand, is concentrated by evaporation, conditioned and stored and will be vitrified, prior to further storage, and ultimately disposed of to either secure geological formations or, alternatively, on or under the ocean bed. Gaseous effluents are conditioned prior to aerial discharge. Low level solid wastes are disposed of by shallow trench burial. Intermediate level wastes are, after storage and conditioning, disposed of either by deep sea dumping or, when available, as for the high level waste. Plutonium and other transuranic contaminated wastes are, after storage and conditioning, disposed of either by deep sea dumping or, when available, as for the high level waste.

WASTE DISCHARGES

The wastes discharged into the environment immediately, as for the low level liquid effluents and the gaseous wastes, and the intermediate level liquid wastes which are discharged after conditioning and storage are presented in Table 3.

Table 3*

Radionuclide present in waste	Estimated Discharge Level, curies/year	
	A	B
Atmospheric Discharges		
Krypton 85	1.4×10^7	1.4×10^7
Tritium	5×10^4	10^4
Carbon 14	600	600
Iodine 129	0.6	0.015
Ruthenium	4	0.4
Strontium 90	1	0.1
Alpha emitters	0.1	0.01
Liquid Discharges (to sea)		
Caesium 134/137	8,000	1,600
Strontium 90	1,600	320
Tritium	10^6	10^6
Iodine 129	48	48
Ruthenium 106	7,500	1,500
Zirconium/Niobium 95	1,400	280
Alpha emitters	1,800	360

* The data are presented in two columns A and B. The waste discharge levels quoted in column A refer to the estimated maximum release rates which might arise from the operation of the THORP plant. In the case of Krypton 85 and Carbon 14, atmospheric discharges, no retention has been assumed; this also applies to liquid discharges of tritium and iodine 129 although in practice some will occur in the plant.

- Arrangements are being made to provide for a facility for the removal of Krypton 85 subject to satisfactory development and to the expenditure being justified on a cost-effectiveness basis, as recommended by ICRP Publication 26. This might lead to a reduction in this discharge by approximately a factor of 10.
- For the other isotopes where the discharges are being specifically limited the figures in column B represent the current flow sheet estimates. The figures in column A allow for a substantial margin which was considered prudent at this stage of design.

ENVIRONMENTAL IMPACT OF NORMAL OPERATIONS

Identification of the Critical Groups:

The fundamental approach to assessing the risk from exposure to ionising radiation has been presented. Essentially the risk relates to four factors:

i. the source of the radioactive material introduced into the environment.
ii. the environmental routes for the transport of radio-nuclides to man.
iii. assessment of the consequential radiation dose and
iv. evaluation of the associated risk.

ICRP Publication 26 suggests that the best method of effecting such an evaluation and of controlling the dose to individual members of the public is to identify and study those groups who receive the highest doses from each particular source of radiation. Such groups have been termed 'critical groups'; in this terminology the use of the word 'critical' is used as in critical path analysis. The actual doses received by an individual will depend on many factors. If the group is small enough it may be possible to study such factors for each person; if not, a sample must be chosen.

In the particular case of waste discharges from Windscale the following factors are taken into account in establishing the 'critical groups'.

i. Continuous residence near Windscale so as to be exposed by immersion and inhalation to the highest ground level airborne concentrations of radio-nuclides discharged from chimney stacks.

ii. Consumption of a quantity of vegetable produce grown in the vicinity of Windscale.

iii. All milk consumed is from dairy farms in the close vicinity of Windscale.

iv. Fish (including shell fish) caught in the Irish Sea in the close vicinity of the Windscale Sea discharge pipeline is consumed at a rate of 265 g per day.

v. A total of 300 hours each year is spent in the Ravenglass Estuary on the silt banks between high and low water marks.

Exposure of Critical Groups:

It is on the basis of this form of analysis that Table 4 has been compiled with respect to the reprocessing of thermal oxide fuel at Windscale.

Table 4. Radiation exposure of Critical Groups arising from THORP

Radionuclide	Critical Pathway	Critical Organ	Organ Dose (millirems per year)	Dose Limit (ICRP 26) (millirems per year)	Organ Dose as percentage of ICRP(26) Limit (%)	Equivalent Whole Body Dose: Weighting Factor (ICRP 26) (2)	Equivalent Whole Body Dose: Exposure x Weighting Factor (millirems per year)	
ATMOSPHERIC DISCHARGES								
Krypton 85	Immersion	Skin	27	4000 (1)	0.7	0.01	0.27	
Tritium	Inhalation and food	Whole Body	1	500	0.2	1	1.0	
Carbon 14	Inhalation and food	Whole Body	0.75	500	0.15	1	0.75	
Iodine 129	Milk	Thyroid	18	5000	0.4	0.03	0.54	
Ruthenium 106	Inhalation	Lung	7.5	4200	0.2	0.12	0.90	
Strontium 90	Milk	Bone	108	5000	2.2	0.03	3.24	
Alpha emitters	Inhalation	Lung	37.5	4200	0.9	0.12	4.5	
						Sub total		11.20
AQUEOUS DISCHARGES (TO SEA)								
Caesium 134/ Caesium 137	Fish	Whole Body	5.5	500	1.1	1	5.5	
Strontium 90	Fish	Bone	7.5	5000	0.2	0.03	0.23	
Tritium	Fish	Whole Body	0.5	500	0.1	1	0.5	
Iodine 129	Fish	Thyroid	15	5000	0.3	0.03	0.45	
Ruthenium 106	Silt (direct radiation)	Whole Body	3.5	500	0.7	1	3.5	
Zirconium 95/ Niobium 95	Silt (direct radiation)	Whole Body	1.25	500	0.3	1	1.25	
Alpha emitters	Resuspension from silt (inhalation)	Lung	1.5	4200	0.04	0.12	0.18	
						Sub total		11.6
						Total (Rounded)		23

Notes:

(1) Derived from ICRP 26 occupational lifetime limit of 2000 rem to prevent cosmetically unacceptable changes.

(2) The Weighting Factors applicable when considering stochastic effects which may arise from exposure to low level.

Table 4 indicates that the summated dose, provided in the final column, of 23 millirem per year can be regarded as the dose equivalent to the uniformly irradiated whole body which carries the same risk to the individual as the sum of all the risks arising from exposure of all the body organs specified in the table. It should be noted that the summated dose of 23 millirem per year applies to a hypothetical person who receives radiation from all the possible radiation pathways also it is based on assumed waste discharge rates which contain subtantial safety margins. It is considered that provided most of the design targets are achieved that the dose equivalent would be more realistically about a factor of three lower than this figure. ICRP have recommended that a whole body dose-equivalent limit of 500 millirem in a year be applied to individual members of the critical groups. Consequently the dose equivalent conservatively deduced in the table is 5% of this limit and a more realistic value might be about a third of this value.

Environmental Risk:

ICRP Publication 26 indicates that for the purpose of radiation protection involving individuals the mortality risk for radiation induced cancer is about 10^{-4} per rem of radiation. The estimated maximum exposure in any individual member of the general public due to all operations on the Windscale site, including both magnox and Thorp reprocessing, is 50 millirem per year. This is equivalent to a risk level of 5×10^{-6} per year. This is commented on in the report of the Windscale Inquiry as follows:

"A member of the public receiving the intended maximum dose (i.e. 50 millirems per year) will therefore be 100 times more likely to die in an accident of some sort in the year in which he receives it than to die years later from cancer as a result of it. He will be about 10 times more likely to die of leukaemia from natural causes and, if he regularly smokes 10 cigarettes a day, he will be 500 times more likely to die from that cause than from emissions from THORP and Magnox combined. The annual risk from this source will be about the same as that involved in travelling 250 miles by car or being a male aged 60 for 1 hour 40 minutes".

Occupational Exposure and Associated Risk: An estimate, based on design objectives, of the exposures of those to be engaged in the operation and maintenance of THORP was given in evidence at the Windscale Inquiry as an average of 0.8 rem per year, which is 16% of the ICRP dose limit of 5 rem per year applicable to occupational exposure of the whole body. If account is taken of the Magnox reprocessing operations, which will continue after THORP is commissioned, the average dose is 0.9 rem per year. Taking the ICRP 26 dose-risk relationship of 10^{-4} per rem quoted above, the mortality risk for radiation induced cancers from exposure at 0.9 rem per year is approximately 10^{-4} per year. The risk from exposure at 1 rem per year is related, in the Windscale Inquiry report, to other risks as follows:

"The dose to workers of 1 rem per annum is the equivalent of regularly smoking three cigarettes a week and one-fifth of the risk of accidental death in the year in question".

Collective Dose Commitments: Estimates were given in evidence at the Windscale Inquiry of collective dose commitments to local, national (UK) and world populations arising from routine discharges from THORP of specified radionuclides. The estimates relating to the world population were presented in the report of the Windscale Inquiry, and the collective dose commitments relating to specified organs were converted to equivalent whole body dose commitments using the weighting factors from ICRP 26 as shown in Table 5.

Table 5

Radionuclide	Collective dose* commitment per year of reprocessing (man rem)	Tissues or organ irradiated	Equivalent whole body dose commitment per year of reprocessing (man rem)	
ATMOSPHERIC DISCHARGES				
Carbon 14	5,400	whole body	5,400	
Strontium 90	300	bone	12	
Plutonium **	76	lung	20	
Krypton 85	1,750	whole body	1,750	
Krypton 85	370,000	skin	3,700	
		Sub total		10,882
AQUEOUS DISCHARGES (TO SEA)				
Caesium 134/137	2,141	whole body	2,141	
Tritium	450	whole body	450	
Iodine 129	2,800	thyroid	112	
Strontium 90	1,600	bone	64	
Plutonium (soluble)	1	bone	0	
		Sub total		2,767
		TOTAL		13,649
		ROUNDED TO man rems per year of reprocessing		14,000

* Collective dose commitments are given per year of reprocessing (assuming 10 years of operation) and integrated over a period of 100 years.

** Assumed here to be insoluble.

If the estimated collective dose to the THORP work-force (800 man rem per year) is added to the above total of collective dose commitment for the world population, a total of 14,800 man rem per year is obtained. Using the risk factor, from ICRP 26, of 10^{-4} per rem the mortality risk from radiation induced cancer, relating to a collective dose commitment of 14,800 man rem per year, is about 1.5 per year, i.e. less than 2 in the whole world population per year of reprocessing.

Genetic Risk Estimates: In commenting on exposure of populations, in ICRP Publication 26, the ICRP say that knowledge gained over the past two decades indicates that genetic effects, while important, are unlikely to be of overriding importance, and need to be related to the sum of all other effects. This general view appears to be borne out in the specific case of the THORP plant. Of the total collective dose commitment of 14,800 man rem per year quoted above only a relatively small proportion is genetically significant. The dose to the gonads is the sum of all the whole body doses contributed by those radionuclides which irradiate the whole body each of these being adjusted to account for the relationship between whole body dose and gonad dose. The gonad dose is then adjusted to take account of age structure and mean age of conception. The genetically significant fraction of the gonad dose is estimated to be about 0.4 in the case of the general public and about 0.3 for those occupationally exposed. The genetically significant collective dose commitment integrated over a period of 100 years is estimated as about 3,200 man rem to the world population and about 240 man rem to the THORP work-force for each year of operation of THORP. Authoritative estimates (NAS, 1972; ICRP-26; EPA, 1977) of genetic risk factors lie between 2×10^{-4} and 3×10^{-4} as the risk of serious genetic abnormality per genetically significant man rem. This results in an estimate of about 1 serious abnormality in the total of all subsequent generations, in the whole world population, from each year of operation of THORP.

Gonad Dose to the UK Population: The collective gonad dose commitments to the population of the UK can be derived from evidence submitted at the Windscale Inquiry. Collective whole-body dose commitments were given as in Table 6. These are adjusted to give collective gonad dose commitments. The dose commitments integrated over 100 years refer to one year of THORP operation.

Table 6

Radionuclide	Collective whole body dose commit-ment (man rems per year)	Gonad dose / Whole body dose	Collective gonad dose commitment (man rems per year)
Krypton 85	89	1.2	107
Tritium	210	1.0	210
Carbon 14	350	0.6	210
Caesium 134/137	1,200	1.0	1,200
		TOTAL	1,727

Assuming that the population of the UK is 55 million the collective gonad dose commitment of 1,727 man rems per year of THORP operation results in an average gonad dose of 0.03 millirem per year. This is seen to be very low in comparison with the average gonad dose from natural background radiation which is estimated to be about 87 millirem per year in the UK. (NRPB, 1974). This average dose of 0.03 millirem per year may also be considered by reference to the advice given by the National Radiological Protection Board (NRPB) on this topic. NRPB (1977) has stated that the contribution to the annual average dose equivalent resulting from all UK waste management practices is not likely to exceed 5 millirem. It can be seen from this that operation of the THORP plant would contribute only about 0.6% to this figure.

Summary of Risks:

Risks to the THORP workforce, to members of the general public near Windscale and to wider populations have been quantified above, with some comparison of risks to individuals with other risks in life. The risk estimates are summarised in Tables 7 and 8.

Table 7

Risks to Individuals in THORP Workforce and in the Critical Groups of the General Population

Group	Nature of effect	Risk (Per person per year)*	Equivalent Risks (per year of exposure to the risk)**
Workforce	Death from radiation induced cancer	10^{-4}	1/5 of risk of accidental death. or 1/25 of risk from regular smoking of 10 cigarettes per day. or 2 times risk of leukaemia from natural causes.
Individual member of critical groups	Death from radiation induced cancer	2.5×10^{-6}	1/200 of risk of accidental death. or 1/1200 of risk from regular smoking of 10 cigarettes/day or 1/20 of risk of leukaemia from natural causes.

* Assuming operation of THORP at design capacity, 1,200 t/y, 37,000 MWD/t, cooled fuel
Includes risk from Magnox reprocessing

** From the Sixth Report of the Royal Commission on Environmental Pollution (1976).

Table 8

Risk to World Population

Group at risk	Nature of effect	Risk (for each year of operation)*	Comment
World population (including THORP workforce)	Death from radiation induced cancer	2 (in entire world population)	
World population (including THORP workforce)	Serious genetic defect	1 (in entire world population)	applies to total of all subsequent generations

The Royal Commission on Environmental Pollution, (1976) quotes a range of risks arising from a variety of specific activites. The range varies from a risk of 1/400 from smoking ten cigarettes a day to 1/2,000,000 from being struck by lightning, the risk being the probability of death for an individual per year of exposure. The Royal Commission comment that their intention is to give an appreciation of the scale of risks associated with a range of activities, some of which are accepted voluntarily by individuals, others being concomitants of everyday life, and some being regarded as acts of God. The Commission suggests that risks greater than about 1 in 1000 per year are considered unacceptable, those of the order of 1 in 10,000 per year justify the spending of public money to try to eliminate the causes and mitigate their effects, those below 1 in 100,000 per year are considered as individual risks for which warnings may be given, while those below 1 in 1,000,000 per year are generally accepted without concern.

There is probably no activity of man, no matter how innocuous it may seem for which it is possible to demonstrate unequivocally that the risk associated with that activity is acceptable to individuals or groups or to general society. The activity of reprocessing of nuclear fuel is certainly no exception to this general rule. It is a question which cannot be dealt with automatically by critical examination and weighing of evidence. A balance of judgement is required which may have to take into account political and social factors. It was evident from the proceedings of the Windscale Inquiry that many individuals and groups are prepared to acknowledge that risk is, and will be, associated with any means of production of energy and that the risk of each means should be compared one with the other and with other activities in industry and in life generally. However there are others who are prepared to tolerate very little, if any, risk from nuclear power production even though risk, and indeed

* Risk based on integration of dose over a total period of 100 years arising from one year of operation at design capacity.

a greater risk, arises from production of power by other means. In these circumstances it is necessary that the nuclear industry should present its safety case clearly to the public so that the question of risk can be considered by an informed public with the facts at their disposal.

The view of the ICRP as expressed in ICRP 26 are as follows:

i. On dose equivalent limits for workers -
"The Commission believes that for the foreseeable future a valid method of judging the acceptability of the level of risk in radiation work is by comparing this risk with that for other occupations recognised as having high standards of safety, which are generally considered to be those in which the average annual mortality due to occupational hazards does not exceed 10^{-4}".

ii. On dose equivalent limits for individual members of the public -
"From a review of available information related to risks regularly accepted in everyday life, it can be concluded that the level of acceptability for fatal risks to the general public is an order of magnitude lower than for occupational risks. On this basis, a risk in the range 10^{-6} to 10^{-5} per year would be likely to be acceptable to any individual member of the public".

The risk estimates relating to the operation of THORP meet these criteria.

Comparison with UK, US and ICRP Criteria:

Table 9 compares the exposure predictions for the general public with UK, US and ICRP data.

Table 9. Environmental Impact of THORP in Relation to UK and US Limits and to ICRP Recommendation

Those Exposed	Estimate of Exposure from THORP operation (millirem per year)	DOSE EQUIVALENT LIMITS TO WHOLE BODY (millirem per year)		
		UK (Public exposure limits)	USA/EPA Nuclear fuel cycle	ICRP Publication 26
Individual members of Critical Groups	25*	500**	25	500
Whole population	0.03* (average)	5***	-	50 (average)

* based on derivation made earlier

** NRPB advice that a limit of 500 mrem in a year will in most cases result in an average dose rate equivalent to a critical group of less than 100 mrem per year of life-long whole body exposure from all sources of radiation, including waste disposal operations.

*** NRPB considers that the annual average dose equivalent resulting from all UK waste management practices is not likely to exceed 5 mrem.

Estimated exposure of individual members of the critical group associated with THORP operations at Windscale is shown in the previous table to be 5% of the permitted limit of exposure in a year recommended for the UK and by ICRP. It is also equal to the USA standard specified by EPA as applicable over the nuclear fuel cycle. Those individuals of the general public who are exposed at these levels as a consequence of living in the vicinity of Windscale will not receive significant additional exposure from other fuel cycle operations in the UK, or indeed elsewhere. If account is taken of the contribution from magnox fuel reprocessing at Windscale the estimated exposure remains under 10% of the UK and ICRP yearly limits and also not substantially different from the USA standard. Since the estimate of the environmental impact from THORP is believed to be a conservative one then the exposure of individuals living in the vicinity of Windscale is likely, in practice to be close to 5% of the national and ICRP limits and almost identical with the USA standard. The exposure of the UK population from THORP is only 0.6% of that suggested by NRPB as likely to be associated with all UK waste management practices and less than 0.1% of that suggested by ICRP in Publication 26. Magnox reprocessing will have a similar impact but the contribution from all other fuel cycle operations combined will be relatively small. Table 10 considers the atmospheric discharges.

Table 10. THORP Discharges related to Energy Produced by the Fuel

Radionuclide	Estimated Annual Discharge to Atmosphere from THORP (including margins)*	Curies per gigawatt year of electrical energy produced**	
	curies	THORP	EPA Limit
Krypton 85***	1.4×10^7	3.5×10^5	5×10^4
Iodine 129***	0.06	0.0015	0.005
Plutonium 239**** and other long-lived trans-uranic alpha emitters	0.1	0.0025	0.0005

The atmospheric discharges of krypton 85 from THORP are, on the current maximum estimate, 7 times the EPA standard. However since provision is to be made in THORP for a krypton 85 removal system to be installed, if a cost-effective study justifies this, then it would likely produce a decontamination factor of about 10. In this case the discharge level of krypton 85 from THORP

* Based on 1200 tonnes U/year, irradiation of 37,000 MWD(th)/tonne and a pre-reprocessing period of 1 year.

** It is assumed that THORP handles spent fuel which has generated 40 gigawatt years of electrical energy for each year of operation.

*** The atmospheric discharges of Krypton 85 Iodine 129 are based on the conservative maximum discharge values. Consideration is being given to providing additional clean up facilities for these radioisotopes.

**** The discharge of plutonium 239 and other long lives transuranic elements is again an upper figure with a substantial margin. In practice the discharge is likely to be considerably lower.

would be within the US standard specified by EPA. The Iodine 129 discharges are well within the EPA figures and when account is taken of the margin in the alpha-emitter figure and the intention as stated to the Windscale Inquiry, there is unlikely to be any major difference.

Non-nuclear Impact:

Apart from the nuclear implications, the impact of a reprocessing plant on its local environment is little different from that of any other modern plant in a clean, high-technology industry. Where there are differences of degree, they tend to favour the reprocessing plant. A plant of this kind is sited away from concentrations of industrial or business premises. This may mean that more attention is paid to its visual impact than would be the case if it were located on an industrial estate. To the extent that it necessarily includes several bulky and fairly high buildings, up to about 30 metres, and one or more chimney stacks of about 60 metres, it may look out of place against a rural background, and need careful siting and landscaping to preserve amenity. But, in other respects, it fits readily into such an environment. Furthermore, bearing in mind that it provides the total reprocessing capacity required for spent fuel from 40 gigawatts of installed nuclear power, it occupies a small area, not more than comparable to the area of one of the nuclear power stations. It produces no smoke or fumes and little noise and relative to its size and the value of its throughput little traffic. Once the plant has been built the deliveries of irradiated fuel and of process materials are not likely to average more than 20 to 30 consignments a week for the 1200 tonne U/year plant consequently, since rail access will be provided, the impact on local amenity will be small.

In terms of demands on local resources, the most important requirements are water, steam and electricity. Water usage will, of course, depend on plant design and the extent of water recycling, but is likely to be substantial. The THORP plant of nominal capacity 1200 tes/year is estimated to need some 15 - 20 megalitres/day of water for all purposes. Maximum electrical demand is expected to be some 12 MW, and steam consumption up to 18 tes/day.

The employment pattern and social impact of a reprocessing plant are similar in principle to those of a large unit in the chemical industry. It has been estimated that the THORP plant will require some 1000 extra workers at Windscale, but a similar plant on a new site might well require a total work-force of more than 2000. A greater than average proportion of those employed will be professionally qualified staff and skilled tradesmen. But there will also be a substantial number of jobs for workers who need no prior qualifications, but can be trained on the plant. The employment provided is balanced therefore, and if a large proportion of the jobs can be filled by local recruitment, no special infrastructure problems are likely to be created in the area.

To the extent that any such problems do arise through the necessity to bring in a substantial number of people from outside the area, there will be the normal strains associated with a rapid increase in population. Experience indicates that such movements, apart from the direct economic spin-off, can bring social and cultural benefits to an isolated area because those imported tend to be of comparatively high quality and attainment.

ENVIRONMENTAL IMPACT OF ACCIDENTS

The risk associated with any untoward event is the probability of that event occurring multiplied by the consequences of the event.

RISK = PROBABILITY x CONSEQUENCE

In the UK this approach was developed by Farmer in considering the criteria to be adopted in the siting of nuclear power stations. The well known "Farmer Curve" shows frequency of release versus quantity of radio-iodine released. Events lying above the curve are judged not to be acceptable. A report published by the UKAEA (1974), shows how this curve can be adapted to releases of other substances. This approach is to be applied in the safety assessment of THORP.

Rasmussen's reactor safety study (1975) on assessment of accident risks in US commercial nuclear power plants is a major contribution on the assessment of risks from nuclear power reactors. The work uses the above probabilistic approach most effectively. There has not been any similar study published as yet with reference to reprocessing plant but the same approach can be applied to individual cases as will be seen below. Before proceeding to consider possible accidents in reprocessing plant it may be useful to compare the nature of such plant with that of nuclear power reactors with regard to potential for release of significant quantities of radioactive material to the general environment. While the total radionuclide inventory of a reprocessing plant complex (including storage of spent fuel and highly active fission product) waste may well exceed that of a reactor, the following characteristics of a reprocessing plant act in its favour when compared with a reactor -

i. temperatures are in general much lower

ii. pressures are in general much lower

iii. the non-nuclear potential energy of the plant much lower.

The inventory of radioactive material in a nuclear fuel reprocessing plant is considerable. It is necessary, therefore, to design and operate the plant in such a way as to prevent accidents which could result in a release of more than a small fraction of the inventory. Discussion, here, will be confined to possible accidents in reprocessing plant which can be expected to result in a release of radioactive material beyond the site boundary.

Types of Reprocessing Plant Accident with Significant Potential for Release to the Environment:

In evidence given at the Windscale Inquiry three cases were presented and examined in some detail. These were:-

i. Release of activity from a highly active liquid fission product storage tank (HAST) due to loss of cooling.

ii. Release of activity from a spent fuel storage pond due to loss of cooling.

iii. Arbitrary release of plutonium to atmosphere.

Each of these will be considered in turn.

Loss of Cooling Water to Highly Active Storage Tanks: The storage tanks were chosen because they have a bigger inventory of fission products than other elements of the reprocessing plant complex, and also because the largest release of fission products could in theory occur. Heat generated by the process of radioactive decay of the fission products makes it necessary to provide means of cooling of the highly active liquor and to ensure that the cooling capability is always available. Cooling is carried out by means of cooling coils and jackets through which water is passed. The number of cooling components available is greater than that actually required so that failure of one or more cooling components would not constitute a failure of the system. An additional precaution is that water may be supplied in four independent ways. The chance of all these cooling methods failing is negligibly small (it has been assessed as about one chance in 1,000,000 years). Nevertheless if it is assumed that the cooling capability fails, and cannot for some reason be restored, and that it proves impossible to transfer the contents of the affected tank to spare tanks held in readiness, then after some hours the liquid would boil. This would result in the transfer of some activity from the boiling liquid to the vapour above the liquid with consequent discharge of activity from a stack via the vessel ventilation system.

The fraction of the total activity in the liquid which could enter the vapour phase had been determined (by laboratory expriments on inactive simulates and by measurements made on actual plants where radioactive liquids are evaporated) as being not more than one part in ten thousand (10^{-4}). Thus of the 10^8 curies in the liquid not more than 10^4 curies would be available for release in this way. It was pessimistically assumed that despite the presence of an efficient off-gas clean-up system, the whole of the 10^4 curies was released via the stack.

The consequences of such a release were assessed by the Safety and Reliability Directorate (SRD) of the UKAEA making use of a computer programme (1976). The imputs to the programme were the quantity and type of material released, the duration of the release, the stack height, weather conditions, deposition velocity and population distribution. These calculations indicated a possible need for evacuation, for up to a few days, of the small number of people who live within a mile of the Windscale plant, and a temporary ban of up to a few weeks on the consumption of foodstuffs (milk, free range eggs, etc) produced within a 10 mile radius. The programme was also used to obtain an estimate of the number of people who might, in the most unfavourable circumstances contract cancer as a consequence of the postulated release. In the worst case this might be about 10.

Further evidence was given on the calculated times to reach defined conditions assuming complete loss of cooling with no remedial action. These are indicated in Table 11.

Table 11

		PWR FUEL 1 YEAR COOLED	PWR FUEL 5 YEAR COOLED
Fuel and Waste liquid Parameters:-			
Irradiation	MWD/T	33,000	33,000
Rating	MW/T	37	37
Cooling time	years	1	5
Concentration	l/t*	888	200
Heat content	MW	1.5	1.1
Fission product content	expressed in tU from which FPs extracted	146	650
Initial bulk liquor temp.	^{o}C	60	60
Time in days at which:-			
Boiling point is reached		0.4	0.6
Boil-away is complete		2.6	3.5
Conversion to oxides is complete (Temperature taken as 1000°C)		3.6	8.0
Melting point of stainless steel is reached (Temperature taken as 1450°C)		3.9	8.5

Further evidence was given regarding the possible sequence of events which might follow those set out above. Three possibilities were postulated:-

i. The concrete cell in which the storage tank is housed continues to provide a barrier against release of activity.
ii. The cell walls and roof collapse but the base remains intact.
iii. The cell walls and roof remain intact but the base is penetrated by molten mass.

Calculation of temperature profiles indicated that, of these hypothetical events the third was the least improbable. The extent to which radioactivity might be released in this was discussed further. It was suggested fission product oxides would become intimately mixed with the molten steel and concrete, the degree of binding being quite high. It was not possible to quantify either fraction or rate of subsequent leaching by ground water resulting in conveyance to the nearby river or to sea but it was suggested that such release would certainly be less than the quantity of 50 mega-curies of radio-caesium considered by the RCEP

* The concentration of the stored liquor is expressed in litres of liquor per tonne of uranium in the spent fuel from which the fission products were extracted.

which they estimated might lead eventually to 50 deaths. As the hot mass penetrated the ground it would drive off ground water as steam which, it might be argued, could carry away radioactivity which might reach the surface and be released to atmosphere. It was difficult to see how a greater amount of activity could be released via this route than that postulated above for the case of boiling of the liquor (10^4 curies). The problem was likely to remain one of ground contamination.
With regard to aerial discharges via the stack it was concluded that the activity release was unlikely to be very different from the case of straightforward boiling of the liquor.

The low probability (10^{-6} per year) of the initiating event, (taking into account the system diversity and redundancy) must be emphasised. The probability that no remedial action would be taken is also low resulting in an extremely low overall probability. It is this element of the risk equation, when considered against the consequences indicated above, that makes the risk acceptably low.

Consequences of Aircraft Crash on Highly Active Storage Tank Cells at Windscale: The probability of an aircraft crash at any specific location in the UK on a random basis is less than 10^{-7} per year if it is assumed that the area affected is about $10^4 m^2$. This is based on data from both military and civil aircraft crashes. The probability will be less at Windscale which is away from airfields and recognised flight paths and which is also a restricted area. Furthermore the Major Hazards Advisory Committee in the UK (1976) has concluded that "there is generally little to be gained by taking into account and guarding against events which have a very remote probability - although they could happen tomorrow. For example in the UK the probability of an aircraft crash affecting a site is less than once per million years per site". Despite the low probability of such an event the possible consequences of an aircraft crashing on to the cells housing the highly active storage tanks were considered in evidence given at the Windscale Inquiry. Consideration was given both to the 'punch' penetration and also to the transfer of kinetic energy to the structure. In the first case material would break away from the point of impact, and also in some cases a 'scab' would be formed on the rear face which would fall away from the wall. In the second case the absorbed energy would result in a vibration response, the damage caused being dependent on the ability of the structure to absorb such energy. An NRDC (1976) formula was used to estimate both the penetration produced and the dynamic response of the structure. If it is assumed that an aircraft engine impacts on either the wall or roof of the cell, calculations show that it could just penetrate a wall of 1.1 m thickness. The walls of the cells are 1.7 m thick which provides an adequate margin of safety. Calculations also show that the box structure of the cell would ensure that the wall or roof would not collapse, and only local damage would occur. If it is assumed that the engine impacts on the roof in such a manner as to cause the breaking of all the external features of the cooling systems on the top of the cell, the possible consequences and possible remedial actions would be as considered earlier.

Loss of Cooling to a Spent Fuel Storage Pond: The possible consequences of such an event, again of low probability, were also presented in evidence at the Windscale Inquiry. Assuming complete loss of cooling and no remedial action the pond water temperature would increase to an equilibrium level and evaporation of the water would eventually expose the fuel element storage bottles (of stainless steel). Later the water in the bottles would boil and in turn expose fuel elements. This would lead to puncturing of fuel element pins (due to increasing pressure of the helium within the pins) resulting in the release of gaseous fission products. Such a puncture would not however cause sufficient damage to the pins to permit fuel pellets to fall into the bottles.

Times were calculated as follows:-

	Case 1 Heat Load 11 MW	Case 2 Heat Load 7.4 MW
Cumulative time (in days) to -		
- reach equilibrium temperature	4.3	7.0
- expose top of bottles	15.9	24.0
- puncture first fuel (approximately 700°C)	16.7	25.0

Case 1 assumed a storage pond containing 1000t of PWR fuel with an average cooling period of 1 year. This case was considered to be unrealistic because it bears no resemblance to programmed deliveries. Case 2 assumed that a 1000t capacity pond would be filled at the rate of 620 t/y (i.e. filled in approximately 1 2/3 years) which corresponds to the maximum expected rate of delivery. It was further assumed that the pond would contain one-third CAGR fuel (average cooling 13 months) from UK CAGR stations and two-thirds PWR fuel (average cooling 16 months).

Calculations were also made to establish whether the fuel element reaches an equilibrium temperature before the fuel cladding melts, if the loss of cooling persists beyond the stages considered above. The equilibrium temperature of a Zircaloy clad PWR fuel element was calculated as 1072°C which is to be compared with the melting point of Zircaloy of 1800°C. In the case of stainless steel clad PWR fuel element the equilibrium temperature would be less than 1072°C which is to be compared with the melting point for stainless steel of 1400°C.

With regard to the impact of an aircraft engine similar considerations to those out-lined above for highly active storage tank cells apply to fuel storage pond. The ponds walls are adequate to withstand the punch effect and it was calculated that the transfer of kinetic energy would be such that a collapse of a wall would not occur.

<u>Consequences of an Accidental Release of Plutonium from the Windscale Site</u>: The results of calculations were given in evidence at the Windscale Inquiry. Assumptions were as follows:

i. An <u>arbitrarily</u> assumed release of 15 kg of Pu 239 as oxide at ground level from the centre of the Windscale site. This assumption does <u>not</u> imply that such a release is possible.

ii. 1% of the plutonium is in respirable form.

The calculations indicated the following consequences assuming the duration of the release is between 0.5 and 2 hours:-

Probability of one or more cancers for the general public		0.04 - 0.3
Area/distance over which the initial airborne concentration due to resuspended activity exceeds the maximum permissible level for the general public.	Area* km^2	0.3 - 5.8
	Downwind distance km	1.1 -18.6
Downwind distance for which the Emergency References level for various organs is exceeded	km	
	lung	< 0.5 to 6.2
	Whole body	< 0.5 to 1.2
	Thyroid	< 0.5
	Bone marrow	< 0.5 to 5.2
	Bone cells	0.5 to 13

CONCLUSION

The environmental impact of reprocessing irradiated fuel is primarily the impact on man through the pathways in the biota leading to the radiation exposure of man. It has been demonstrated in this paper, that operation of the THORP reprocessing plant gives rise only to minimal impact. On the basis of the estimated maximum discharge rates for radioactive liquid and airborne wastes from normal function of the plant, the radiological detriment to any individual, who is a member of a critical group, is 24 millirem per year, equivalent to a maximum individual risk of 2.4×10^{-6} per year. This is well within the range stated in ICRP publication 26 as likely to be acceptable to any individual member of the public. Furthermore the NRPB has expressed the view that the long term average dose rate equivalent to any critical group will in practice be less than 100 millirem per year. The estimated maximum environmental impact of THORP is well within this predicted level.

The annual average dose equivalent to the UK population as a whole due to normal plant operation is estimated at 0.03 millirem which is only 0.6% of that suggested by NRPB as likely to be associated with all waste management practices and less than 0.1% of ICRP's estimate (on the assumption that the practices exposing the public are few and cause little exposure outside the critical groups).

---

* excluding area within the site (site radius 0.5)

The range of figures reflect a range of weather conditions and different sources of data used in the calculations.

Despite the differences in approach in controlling radioactive waste discharges in the environment, due to variance in the numerical limits described as standards in different countries, there is not necessarily a significant difference in the actual doses received from a given practice. This is confirmed in the paper.

Evaluation of the potential environmental impact of an accidental release of radioactivity from the plant indicates that for a loss of coolant on a highly-active liquid waste concentrate tank about ten thousand curies might be released to atmosphere. The probability of such an event is low, assessed to be a frequency of 10^{-6} $year^{-1}$ (one chance in 1,000,000 years). The short term consequence of such a release would be a need to evacuate a small number of people who live within 2 km of the Windscale Site for up to a few days and a temporary restriction of up to a few weeks on locally produced foodstuffs. On the assumption of the most unfavourable weather conditions of poor mixing and minimal dispersion the long term consequence might lead to 10 deaths due to cancer induction. Additionally, if the loss of coolant continues with no remedial action then there is the possibility that the base of the storage tank cell would be penetrated by a molten mass of steel containing fission products bound in the steel and concrete. Calculations suggest that such a release into the ground would certainly be less than the 50 MCi of radiocaesium considered by the Royal Commission on Environmental Pollution, which they estimated might lead eventually to 50 deaths. This type of accident was the worst envisaged. It must be stressed that the probability of the initiating event is low ($10^{-6}y^{-1}$) and further that the probability of being unable to take any remedial action is also low, resulting in an extremely low overall probability. It is this element of the risk equation, when considered against the consequences, that makes the risk acceptably low.

It can therefore, be concluded that operation of the THORP reprocessing plant would lead to acceptable levels of environmental impact both under normal and accident conditions.

References

Cmnd (1959): The Control of Radioactive Wastes. Cmnd 884.

EPA (1973): Environmental Analysis of the Uranium Fuel Cycle Part III - Nuclear Fuel Reprocessing. US Environmental Protection Agency, Washington, D.C.

EPA (1977): Environmental Radiation Protection Standards for Nuclear Power Operations, Environmental Protection Agency, Federal Register, January 13 1977, Part VII, Washington, D.C.

European Communities (1976): Official Journal of the European Communities, 19, No. L187. 12 July 1976.

Health and Safety Commission (1976): Advisory Committee on Major Hazards.

ICRP Publication 26 (1977): Recommendations of the International Commission on Radiological Protection, Pergamon Press, Oxford.

NAS (1972): The Effects on Populations of Exposure to Low Levels of Ionising Radiations, National Academy of Sciences, Washington, D.C.

National Radiological Protection Board (1974): Radiation Exposure of the Public - The Current Levels in the United Kingdom, NRPB Report 24.

National Radiological Protection Board (1977): Advice dated 9 August 1977 in relation to radiological protection standards ASP 1.

NRPB (1977): UK National Radiation Protection Board, Radiological Protection Bulletin No. 21.

Nuclear Engineering and Design 37 (1976), 183 - 203.

Rasmussen, N.C. (1975): Reactor Safety Study, An Assessment of Accident Risks in US. Commercial Nuclear Power Plants, WASH - 1400 (NUREG 75/014), United States Nuclear Regulatory Commission, Washington, D.C.

Royal Commission on Environmental Pollution (1976): Sixth Report Cmnd 6618 United Kingdom.

UKAEA (1974): Chemical Plants and Laboratories. Safety Assessment, Safety and Reliability Directorate, United Kingdom Atomic Energy Authority, SRD Report 8.

UKAEA (1976): A Guide to the Use of TIRION - a Computer Programme for the Calculation of the Consequences of Releasing Radioactivity to the Atmosphere, Safety and Reliability Directorate, United Kingdom Atomic Energy Authority, SRD Report 63.

UNSCEAR (1972): Ionising Radiation : Levels and Effects. A report of the United Nations Scientific Committee on the Effects of Atomic Radiation to the General Assembly. United Nations 1972.

Windscale Inquiry (1978): Reported by the Hon Mr Justice Parker.

7

NUCLEAR WASTE DISPOSAL: TECHNOLOGY AND ENVIRONMENTAL HAZARDS

F. Kenneth Hare and A.M. Aikin
Institute for Environmental Studies,
University of Toronto, Toronto M5S 1A4, Canada

INTRODUCTION

The wastes from the nuclear industries require comprehensive treatment if man and nature are to be protected against radioactivity. Though nuclear wastes are not alone in being dangerous to life, they have acquired special visibility in the past few years. Their proper management has long been a matter of concern, and one in which considerable experience and skill has now been gained, after some initial errors. The final stage of such management, however - the ultimate disposal of long-lived radioactive wastes - has not yet been successfully achieved.

Radioactive substances are those in which the atomic nuclei are unstable. The nuclei decay at precise rates, and emit radiation of several kinds, all of which may be harmful to man, plants and animals. Such substances are widespread in nature, and also occur within the human body. Some human cancers and birth defects arise from this natural radiation. They are the price we pay for living in a world whose atoms are not all stable.

Much more strongly radioactive substances are generated, however, when the nuclei of certain metals are split, either naturally or by man. One abundant natural element is fissile uranium-235 (U-235), which is a small but universal fraction of natural uranium. The earth's crust contains about four grams of uranium per tonne, and the sea about 3 grams of uranium per thousand tonnes. A constant 0.715 per cent of this uranium is fissile U-235. Other fissile nuclei occur, but are extremely rare in nature. They include plutonium-239 (Pu-239) and uranium-233 (U-233).

Nuclear technology, whether for weapons production or power-generation, depends on the harnessing of fission. The reverse process of fusion, whereby light atomic nuclei combine (as in the sun), is at present used only in weapons. Power generation depends on the controlled fission of U-235 in reactors, and on Pu-239 formed in those reactors from U-238, the most abundant type of uranium.

It may later be possible to use thorium-232 (Th-232) as the source of energy, after it has been changed to U-233. All these forms of fission create two product streams:-

(i) an enormous amount of heat, which is the source of energy for the electric power to be generated; and

(ii) fission products, which are lighter nuclei formed by the splitting of the original nuclei of U-235, Pu-239 or U-233.

The first of these streams is the desired product from the reactor. The second is the price we pay. It could be a high price, since the fission products are mostly intensely radioactive, and hence dangerous. They constitute most of the nuclear waste that has to be dealt with, in terms of total radioactivity.

The operation of reactors also produces other wastes. Part of the uranium is converted into other heavy nuclides*, called actinides, of which plutonium-239 is the prime example. Others include americium, curium, neptunium and other isotopes* of plutonium. These are also radioactive. Very few are useful aside from Pu-239, which is fissile. Still other radioactive wastes are formed from the materials of which the reactor is built, such as iron and cobalt, or from the air and water contained within the reactor.

Power-reactor operation thus creates many different radioactive substances, each with its characteristic (and unalterable) rate of decay. Some nuclei decay in seconds. Others linger for millions of years. Most are useless, and are hence wastes of a dangerous and awkward kind. Reactors used to manufacture plutonium for weapons create similar wastes, which have been accumulating ever since World War II. Nuclear waste management and disposal must protect man and nature against the hazards presented by all these unwanted by-products.

Wastes also occur at other stages of the fuel cycle, to use the usual if paradoxical title. Large amounts of radium, radon and thorium are released, for example, in the mining, milling and refining processes whereby uranium oxide fuel is manufactured for nuclear reactors. These heavy, radioactive substances occur in the tailings and waste heaps that accumulate round mines, mills and refineries, from which they may be leached into water supplies, streams and lakes. In addition radon gas, a radioactive decay daughter** of radium or thorium, occurs in the mine, and diffuses out of tailings and waste heaps. The jargon calls all these products front-end waste, to distinguish them from the wastes from reactors, the back-end of the fuel cycle.

---

* all types of atomic nuclei are collectively called nuclides. A radionuclide is any species of nuclide that emits α, β or γ radiations, or neutrons. All the different nuclides of a single element, which differ only in the number of neutrons present, are called isotopes of the element.

** radionuclides decay into other nuclides, called daughter products. Four decay chains are known, only three of which occur in nature.

Back-end wastes accumulate in all countries that operate reactors, for whatever purpose. Currently some twenty-six nations are known to have the problem. Front-end wastes, however, are accumulating only in the smaller list of countries that mine and refine uranium. Nevertheless these wastes are a problem for all mankind. Many are deadly, and pose a threat to future generations. Some are capable of world-wide dispersal, so that they imperil all people.

The management of these wastes has had a long history, going back to the first realization that they were threats to human health. In more recent times, the wastes from reactors have been reasonably well managed in most countries. What has been neglected, however, is the question of the <u>disposal</u> of the long-lived wastes that will not simply decay in storage. This problem now looms large in people's minds, and has become a subject for heated argument. A solution will have to be found - and the solution will have to satisfy a sceptical and alarmed world. Disposal means the planned, permanent placement of radioactive wastes in some sort of repository, with no intention of subsequent recovery. All highly radioactive waste with a long life expectancy should, if possible, be disposed of in this fashion. The present report reviews the options available for this process, and examines the possible environmental problems associated with these options. It also looks at measures that may be taken to offer long-term protection of man and biota against the hazards involved. The report begins with an analysis of the nature and origin of the wastes, and a review of disposal technology.

THE NATURE AND ORIGIN OF WASTES

The Fuel Cycles.

The nuclear fuel-cycle begins with the extraction of uranium ores from rock, and ends with the disposal of wastes. It is not, in this sense, a cycle but a sequence, with well-defined stages. Each stage produces wastes, many long-lived, and thus candidates for disposal. In Table 1 we have identified the main stages, and the wastes associated with each. Before we discuss these we must first define the classes of fuel cycle to which it refers. For practical purposes we have to deal with

- <u>the once-through cycle</u>, in which uranium or uranium oxide fuel is used in the reactors once, and then removed. The fuel rods are fabricated from mined uranium ore. Once-through cycles are of two main families, those using (a) heavy water moderated and cooled and graphite moderated and gas-cooled, both using natural uranium, with 0.715 per cent U-235; and (b) light water, boiling light water, pressurized light water and advanced gas-cooled reactors, all of which require enriched uranium (usually about 3 per cent U-235). Most power reactors now in use are of the light water kind (LWRs), and need enriched uranium.

- <u>advanced fuel-cycles</u>, now under intensive international study, which are based on recycling fissile materials derived from the reprocessing of irradiated fuel. These materials include plutonium-239 and uranium-235, both of which are present in irradiated fuel extracted from reactors. These cycles may be based on the recycling of plutonium and uranium, or on thorium-232, which can be converted to fissile uranium-233. A special case that has received much attention is the

Table 1. Process Stages and Wastes in the Nuclear Fuel Cycle

Process	Radioactive Waste
Uranium Mining →	Waste rock (low grade ore)
	Mine waters, (contain Ra-226 but are used in mill)
	Ventilation air, (contains Rn-222)
Milling →	Mill tailings Th-230, Ra-226, Rn-222 and sometimes Th-232
	Waste waters, mainly Ra-226
	Ventilation air, mainly Rn-222
Refining →	Impurities from U, mainly Ra-226
UF$_6$ conversion	Wastes are recycled
Enrichment	Wastes are recycled
Conversion to oxide	Wastes are recycled
Fuel Fabrication	Wastes are recycled
Reactor Operation →	Reactor Wastes
Irradiated Fuel → (If no processing)	Irradiated Fuel for storage and disposal
Fuel Reprocessing →	High level waste) Actinide wastes) for disposal

THE NUCLEAR CYCLE: Front End (Uranium Mining to Enrichment); Back End (Conversion to oxide to Fuel Reprocessing)

U and Pu (recycled from Fuel Reprocessing to Conversion to oxide)

- breeder reactor, in which natural uranium or depleted uranium from other reactors is mixed with plutonium-239 from fuel processing in such a way that more fissile material is produced (mainly plutonium) than is consumed by fission. In effect such reactors breed their own fissile fuel from the irradiated fuel derived from non-breeder reactors, or from natural uranium.

The wastes that result from these cycles are all much the same in character, but there are significant differences in the details of waste management. We have laid major stress on the highly radioactive wastes (HLWs) derived from fuel reprocessing, and on irradiated fuel that comes from once-through or advanced fuel cycle reactors.

Character of the Wastes.

Radioactive wastes are produced at all stages of the nuclear fuel cycle, though the major quantities (in bulk and total radioactivity) occur at the front and back ends of the cycle. The front end includes the mining, milling and refining stages, and affects those countries and regions actively exploiting uranium ores. The back end refers to the materials that arise from reactor operation. The middle parts of the cycle produce less waste. To guide the reader we have prepared Table 1, which lists the stages of the cycle, and briefly indicates the nature of the wastes associated with each. They are of five main kinds:-

(a) naturally occurring radionuclides rejected during the mining, milling and refining stages. These include isotopes of uranium, thorium and their decay daughters, such as radium and radon gas;

(b) heavy radionuclides, called actinides, formed from uranium and other heavy metals by neutron absorption in the fuel rods of the reactors;

(c) fission products, i.e. lighter nuclides formed by the fission of uranium-235, plutonium-239 or uranium-233 in the fuel rods;

(d) solid radionuclides formed in the reactor when neutrons are absorbed by the reactor structure, such as iron-59 and cobalt-60; and

(e) gaseous or liquid radionuclides formed by neutron absorption (e.g., argon-41; tritium, H-3; carbon-14) in and around the reactor.

Items (a) to (d) are mostly solid elements that can be contained and treated as wastes. Radon gas under (a), and the gaseous and liquid products under (e), are hard to contain, and most of them are deliberately dispersed to the hydrosphere or atmosphere. In this report we are concerned only with disposal, i.e., the permanent isolation from the biosphere, of dangerous, long-lived radionuclides that are retainable. These occur in (a) to (d) inclusive, though the major hazard for man is in (a), (b) and (c).

The Mining and Milling Operations: Waste Generation and Management.

Uranium is mined in many countries, the largest current exporter being

Canada. The ores contain uranium in oxides from UO_2 to U_3O_8, or in more complex mineral molecules such as carnotite or austenite, which contain other metals. Uranium content varies from very high (~ 40 per cent) in pitchblende to as little as 0.1 per cent by mass. Much commercial mining involves ore containing less than 1 per cent uranium, so that extraction produces large volumes of bulky wastes. Before the uranium can become nuclear fuel it must be extracted and purified, which includes separating it from its own radioactive decay daughters. These natural radionuclides generally stay with the bulk of the ore, and are discharged with the very large quantities of mine and mill waste, the principal nuclides of concern being thorium-230, radium-226 and radon-222. Uranium itself is of low radiotoxicity, but it has a chemical toxicity comparable to other heavy metals such as lead.

Uranium-bearing ores normally occur deep enough that they are not a hazard to life, but there are exceptions. In some areas of the world, radioactive minerals are close enough to the surface to raise the background radiation well above normal, both by direct radiation and by the release of radon gas. The radium content of some drinking waters are also above recommended levels. Statistically it is difficult to determine whether there has been a decrease in life expectancy in these areas, but it is safe to assume that it would not be good to multiply these occurrences. Potentially, that is what uranium mining is doing. It brings into the biosphere large quantities of these naturally occurring but immobile radioactive materials, and changes them into forms in which escape and dispersion are possible. This problem is well recognized, and measures have been taken to keep it under control. The International Atomic Energy Agency (IAEA) has issued a code of practice for the management of these mine and mill wastes. Many national agencies have also established regulations. The problem of long-term isolation of the radium-226, however, is still not adequately solved. Radium is rather easily dispersed by water, and ingested by plants and animals.

In the production of nuclear fuels the uranium is first mined, then milled, usually at the mine site, to an impure yellow cake. It is then brought to nuclear purity in a separate refinery. Thereafter, since the radioactive daughter products have all been removed, and grow back very slowly into the uranium, the subsequent enrichment and fuel fabrication stages are relatively non-hazardous. So it is the wastes from mining, milling and refining that must be managed with care.

Wastes from the mining operation consist:

- of waste rock, or very low grade ore, which can be disposed of in the mine if it presents an exposure hazard;

- of liquid waste, a combination of mine drainage and water used in the mining operation, and which is generally all used in the mill;

- and of gaseous wastes, the exhaust ventilation, which contains radon gas as well as ore dust.

The mine wastes present little hazard, as they are generally at low radioactivity levels, and management techniques are available and in general use.

The milling operation consists of grinding the ore to a very fine powder, which is leached with carbonate or sulphuric acid solution to dissolve the uranium. The uranium is then purified by a combination of ion-exchange, solvent extraction and precipitation processes, the details depending on the nature of the original ore. The milling wastes, which are generally collected into one stream, consist of a slurry of the pulverized ore containing almost all the uranium daughter products, together with a small fraction of unrecovered uranium. Most of the chemicals used in the process are also present. Radium-226 is the most hazardous nuclide in the slurry, a typical concentration being 800 pCi*g^{-1} of dry tailings, or less than one gram in 1000 tonnes of tailings. (In secular equilibrium, production of 1000 tonnes of uranium would produce approximately 300 g of radium-226 as waste, compared with about 1 tonne of plutonium after irradiation in a nuclear power plant). This slurry is discharged into a basin where the solids are retained behind a natural or man-made dam, and the liquid overflows to a waste treatment lagoon. Here the contained radium is co-precipitated with barium sulphate. Time is allowed for this precipitate to settle before the supernatant liquid is discharged into the natural water systems (rivers, lakes, sea or groundwater). This liquid often still contains some radium-226, but should be below the levels prescribed by IAEA and national regulations. As long as the mill is in operation the processing of the liquid waste will continue, and there must be control over the discharge of effluents. It is known, however, that continued leaching of the finely-divided solid mill wastes by rain water will remove radium that will then appear in the seepage from the piles. Radium concentration in such seepage is likely to exceed the recommended levels. If such leaching occurs, the seepage will have to be continuously collected and processed to remove the radium, and this operation might be needed for centuries to come. The alternative is tolerance of higher radium levels in the surrounding waters. Neither of these alternatives is acceptable. Thus methods of stabilizing these mill-tailings must be found. Many methods are under active study and testing, but because of the very large size of the tailings deposits they are costly and difficult. All methods aim at physical stabilization of the wastes, and at a decrease of wind and water erosion to near zero. Some thought has been given to putting the solid wastes back into the mine from which they came, but only a little more than half could be handled in this way because of the increase in volume brought about by the pulverization. Revegetation of the tailings piles appears to be the optimum method of stabilization, but little is known yet about how effective this will be in decreasing radium leaching. Some consideration is also being given to removal of the radium and thorium from the ore during mill processes, and to their separate disposal. If removal were complete enough, the large volume of mill wastes would no longer present a radioactivity hazard. For the removal to be effective, however, over 99 per cent of both these elements would need to be recovered. Given the nature of many of the ores, this would be very difficult and costly to achieve. Evaluation of this possibility is continuing, but it is unlikely to be the chosen waste management process.

The sheer bulk of the mining and milling wastes stands in the way of ready treatment. United States estimates, for example, give a waste-to-product ratio of 1,300 to 1. Tens of millions of tons of these low-grade but by no means harmless materials surround the mining and milling areas of producing countries. At present these wastes are under control; but a long term solution that will not require continual surveillance is not yet fully developed. The main hazard is the slow and continued leaching of radium into nearby water systems.

---

* 10^{-12} Curie, the Curie being 37×10^9 disintegrations per seconds.

The Middle Stages of the Cycle - Refining, Hexafluoride Conversion, Enrichment, Oxide Conversion, Fuel Fabrication.

These middle stages comprise:

- refining, to remove residual impurities (including radium-226 from the natural uranium);
- hexafluoride conversion, in which the yellow-cake (U_3O_8) from milling is converted into uranium hexafluoride (UF_6), a readily volatilized compound;
- enrichment of uranium-235 in the UF_6 from the original 0.715 per cent of natural uranium to between 2.5 and 3.5 per cent, for use in light-water and similar reactors;
- conversion of UF_6 to UO_2, or oxide conversion; and
- fuel fabrication, using the UO_2 to form ceramic pellets, rods to contain the pellets, and bundles to contain the rods.

Heavy water moderated reactors, such as CANDU in Canada (and under development elsewhere) and the Magnox type reactors, enable one to by-pass the hexafluoride conversion and enrichment stages, because such reactors use natural uranium. The middle stages produce wastes, but these are of low radioactivity, and are generally recycled or handled on-site. Hexafluoride conversion produces calcium fluoride (CaF_2) ash, which is drummed and buried at a suitable facility. Enrichment plants have large volumes of liquid waste that are locally ponded, the radioactive sludges being buried on site. Fuel fabrication from UF_6 produces a further quantity of CaF_2, a waste that is avoided in heavy-water technology. The CaF_2 is usually uranium-contaminated, and is typically stored on-site. Under modern practice these stages are quite tractable from the waste management standpoint. Local problems do occur, chiefly because of current public anxiety about all aspects of waste management. But by comparison with the front and back ends of the cycle the middle stages present no difficult problems, and do not raise questions of ultimate disposal on the same scale.

Irradiated Fuel from Reactor Operation.

In a nuclear reactor, heat is produced from the fission process that occurs when a neutron is absorbed by the fissile nuclides uranium-235 and plutonium-239. The U-235 isotope occurs naturally in uranium, and the Pu-239 is formed in the reactor by neutron absorption in U-238 to form U-239, which decays radioactively to Pu-239. This fission produces a spectrum of fission product nuclides that accumulate in the fuel. Many of these are unstable, and decay radioactively. Fission also produces neutrons, about 2.3 per fission on average. One of these neutrons is absorbed by a fissile nuclide to keep the fission process going, and the others are absorbed by non-fissile nuclides in the fuel, the fission products and the construction materials of the reactor core. As the concentration of fission product nuclides increases, they absorb more and more neutrons, so that ultimately a fuel bundle becomes a burden to the reactor and must be removed. Also in the fuel elements there is a build-up of isotopes of heavy elements. These are formed through a series of neutron capture reactions and radioactive decays. They generally go under two names: the transuranic elements, or TRUs, i.e., elements with atomic numbers above that of uranium; or actinides, with atomic numbers higher than actinium. The important members of this group, as well as the longer-lived fission products of interest in waste management, are listed in Table 2.

The actinides are generally long-lived alpha particle emitters, whereas the fission products are shorter-lived beta and gamma emitters. The radiological properties of the irradiated fuel are effectively determined for the first few hundred years by the fission products, and thereafter by the actinides. This is illustrated for typical fuel in figure 1. Though there are differences in production rates of individual nuclides in different nuclear fuel systems, it matters little as regards waste management whether the principal fissile material is U-235, Pu-239, the thorium U-233 cycle, or even if fast fission is used. All systems produce radioactive fission products and actinides that require careful storage, handling and ultimate disposal.

Table 2. Significant Nuclides in Radioactive Waste Management

Element	Isotope	Half-life	Mode of Decay	Generation Mechanism
Hydrogen	H-3	12.3y	β	Fission and neutron capture
Carbon	C-14	5.7×10^3 y	β	Neutron capture
Argon	Ar-41	1.8 h	β, γ	Neutron capture
Cobalt	Co-58	72 d	β, γ	Neutron capture
Cobalt	Co-60	5.3 y	β, γ	Neutron capture
Krypton	Kr-85	10.8 y	β	Fission
Strontium	Sr-89	54 d	β	Fission
Strontium	Sr-90	28 y	β	Fission
Yttrium	Yt-91	61 d	β	Fission and neutron capture
Zirconium	Zr-93	1.5×10^6 y	β	Fission
Zirconium	Zr-95	66 d	β, γ	Fission and neutron capture
Niobium	Nb-95	35 d	β, γ	Fission and daughter of Zirconium-95
Technetium	Tc-99	2.1×10^5 y	β	Fission
Ruthenium	Ru-106	1 y	β	Fission
Iodine	I-129	1.7×10^7 y	β	Fission
Iodine	I-131	8 d	β, γ	Fission
Xenon	Xe-133	5.2 d	β	Fission
Cesium	Cs-134	2.1 y	β, γ	Fission
Cesium	Cs-135	3×10^6 y	β	Fission
Cesium	Cs-137	30 y	β, γ	Fission
Cerium	Ce-141	33 d	β, γ	Fission
Cerium	Ce-144	285 d	β, γ	Fission
Promethium	Pm-147	2.6 y	β	Fission
Samarium	Sm-151	93 y	β	Fission
Europium	Eu-154	16 y	β, γ	Fission
Lead	Pb-210	21 y	β	Daughter of Polonium-214
Radon	Rn-222	3.8 d	α	Daughter of Radium-226
Radium	Ra-226	1.6×10^3 y	α, γ	Daughter of Thorium-230
Thorium	Th-229	7.3×10^3 y	α, γ	Daughter of Uranium-233
Thorium	Th-230	8×10^4 y	α	Daughter of Uranium-234

Table 2 (continued)

Element	Isotope	Half-life	Mode of Decay	Generation Mechanism
Uranium	U-234	2.4×10^5y	α	Daughter of Protoactinium-234
Uranium	U-235	7.1×10^8 y	α, γ	Natural source, daughter of Plutonium-239
Uranium	U-238	4.5×10^9 y	α	Natural source
Neptunium	Np-237	2.1×10^6 y	α	Neutron capture and daughter of Americium-241
Plutonium	Pu-238	87 y	α	Neutron capture and daughter of Curium-242
Plutonium	Pu-239	2.4×10^4 y	α	Neutron capture
Plutonium	Pu-240	6.6×10^3 y	α	Neutron capture
Plutonium	Pu-241	15 y	β	Neutron capture
Plutonium	Pu-242	3.87×10^5 y	α	Neutron capture
Americium	Am-241	433 y	α	Neutron capture and daughter of Plutonium-241
Americium	Am-243	7.37×10^3 y	α	Neutron capture
Curium	Cm-242	163 d	α	Neutron capture
Curium	Cm-244	18 y	α	Neutron capture

Source mainly OECD (1977)

Thus the discharged fuel still contains useful quantities of fissile uranium-235 and plutonium-239 but to be used again they must be chemically separated from the neutron-absorbing fission products. This chemical reprocessing to recycle the fuel is expensive, and must compete economically with the use of fresh fuel; but it would allow much more energy to be produced per unit mass of mined uranium.

Fuel Reprocessing : Waste Generation and Management.

By 1985 there will be about 10,000 tonnes of irradiated fuel in storage at nuclear power plants around the world. Most of these storage systems were designed with the intent of storing fuel only for a few years before shipping it to a chemical reprocessing plant. There have been delays in building such commercial plants, so additional fuel storage capacity is needed, and is being built. It is still assumed in most countries using nuclear power that fuel reprocessing will be profitable and acceptable; but it will be at least another decade, if not longer, before enough reprocessing plant capacity (and recycle fuel-fabrication plants) will be available to deal with the ever increasing supply of irradiated fuel. If nuclear power is to be a provider of large quantities of electrical energy during the next century, there is little doubt that fuel recycling will be needed to extend the usefulness of the available uranium. This is the basis for the assumption that, in the long term at least, all irradiated fuel will be reprocessed to recover the unused fissile and fertile material. Thus the programmes to develop long term safe methods for waste storage and isolation have generally concentrated on the waste streams from fuel reprocessing plants. It is only lately that any thought has been given (as we do later in the report) to methods for permanently disposing of irradiated fuel.

The common method of fuel reprocessing consists of cutting the fuel elements into short pieces (1 to 2 cm), dissolving the uranium oxide fuel in nitric acid, extracting the uranium and plutonium from this solution by solvent extraction, and recovering these two products separately by selective backwash conditions. Two major waste streams result, the undissolved metal fuel cladding, usually zirconium alloy (Zircaloy) or stainless steel, and the acidic solutions containing the fission products, the actinides and some unextracted uranium and plutonium. This acid solution is often referred to as high level waste, (HLW), since it is indeed highly radioactive. As a result of military programmes large quantities of this solution have been produced. So far it has been concentrated to reduce the volume, sometimes neutralized, and stored in large tanks. Containment has not been perfect, and the few leaks have been widely publicized, though they have done little environmental damage.

More permanent methods of storage or disposal of this high level waste are now being developed, with most work concentrated on converting the waste to an insoluble solid. Glass has become the preferred form, and several countries are testing processes at the pilot-plant stage. A typical glass block will be a metal clad cylinder, 1500 mm long and 400 mm in diameter, containing the waste from 1 to 2 kg of irradiated fuel. These glass canisters will be stored, with cooling and shielding, pending their placement in a permanent repository. Similarly the fuel cladding wastes will be compacted and incorporated into an isoluble medium.

These processes have yet to be used on a commercial scale, and so there are still doubts concerning their ability to deal in a fully satisfactory manner with all the toxic wastes that arise in fuel reprocessing. These doubts have resulted in delays in authorization of the construction of fuel reprocessing plants. They will only be answered by further research and development, and ultimately by demonstration on a commercial scale. Sufficient international attention is being given to this waste management problem that an acceptable solution should be found.

Other Wastes From Reactor Operation (Reactor Wastes).

During the operation of a nuclear power plant many of the neutrons from the fission process are absorbed by the materials in the reactor core. Thus radioactive nuclides are produced in structural components, such as pressure vessels, pressure tubes or control mechanisms. In addition the coolant carries small quantities of corrosion products around the circuits, and as these pass through the reactor core they become radioactive. They are eventually deposited throughout the piping system, in the boilers and pumps, thus causing radiation fields around this equipment. The principal radionuclides formed in this way are cobalt-60, iron-59 and manganese-54. Of these, cobalt-60 has the longest half-life - 5.26 years. In addition to the radioactivity due to these corrosion products, some fission products can, on occasion, be embodied in the coolant circuit through small holes in the fuel cladding. Such cladding defects are rare, and the fuel is removed from the reactor at the first opportunity; but the fission product contamination of the coolant circuit remains.

During operation, the radiation fields around the process equipment are not a hazard, since they are located behind shielding. Like all such systems, however, occasional maintenance is required and station personnel must enter these areas to work on such items as pumps and valves. During these procedures gloves, rags, clothes and other such items become contaminated radioactively. The clothes are generally laundered at the plant, giving rise to potentially radioactive solutions. The gloves, rags and other disposable objects are bagged, and become solid radioactive waste. Any replaced items of equipment, such as valve packings and pump seals, are also bagged or wrapped in plastic for disposal. To keep the radiation fields around the process equipment down to tolerable levels, the radioactive corrosion material and fission products are continuously removed from the coolant circuits by filtration and ion-exchange purification of a side-stream. These filters and ion exchange resins become another form of solid wastes, since they are generally used only once, and then removed when loaded.

Some radioactive gases are also formed in the reactor, and many are released directly to the atmosphere. The significant gaseous radionuclides are tritium (half life 12.3 years, H-3) carbon-14 (half life 5730 years) and krypton-85 (half life 10.8 years). They are formed by various nuclear reactions or as fission products, and because of their long half lives are of environmental concern. Other radioactive gases that are released, such as argon-41 or nitrogen-13, are shortlived and thus have little environmental impact. These gaseous wastes are usually released from stacks after various forms of treatment, such as filtration and delay systems to allow the shortlived radioactivities to decay. More elaborate decay or collection systems, such as cooled charcoal traps or cryogenic distillation, have been developed, but they have not been widely used, since their small benefits have not so far been thought to warrant their cost.

The liquid wastes are treated in two ways. When the amount of contained radioactivity is low they are often dispersed by blending with the condenser water discharge. Alternatively they are purified by ion exchange, or on occasion concentrated by evaporation. The concentrate is then converted to a solid, either by incorporation in cement or, in some experimental cases, bitumen. The function of these processes is to convert the radioactive waste into solid form.

Various methods for the immobilization of the solid reactor wastes are in use, and others are being investigated. At some nuclear plants the combustible wastes are put through a special type of incinerator to achieve a large volume reduction. At other plants they are mechanically compacted. This reduces the magnitude of the waste problem, but does not provide immobilization for long-term disposal. In most countries these solid wastes are stored in concrete bunkers. West Germany has taken the lead in setting up a national disposal system. The evaporation concentrates from their power plants are incorporated in cement, and the ion exchange materials in thermosetting resins. These immobilized solids are then placed underground in the Asse salt mine.

Most of the radioactivity in the reactor wastes will decay to non-toxic levels in a few decades. For this reason there is still some controversy as to whether it will be necessary to bury and/or dispose of these wastes with the same requirements as for high level waste. Engineered structures could contain them for at least 100 years, which is probably long enough. As the nuclear industry grows, however, the quantities of these wastes will become large, and a central disposal site serving several nuclear plants would make environmental control much easier. The situation today is not dangerous, but it is also not tidy. Continued and more extensive planning and regulation are needed for the future.

Some consideration has also been given recently to the decommissioning of nuclear facilities. Some research reactors have already been shut down and decontaminated, and the equipment sent to waste burial sites. However, the dismantling of a nuclear power plant will present problems on a much larger scale. As few studies of this potential environmental problem had been made, IAEA convened a Technical Committee to discuss this topic in October 1975, and a report was subsequently issued. Other national studies have also been recently published, and estimates have been made of the volumes of radioactive waste that will be produced. No problems of feasibility have been found, but most plans will allow for the nuclear plant to sit idle for a number of years before dismantling is started. Much of the radioactivity in the structures is relatively shortlived, and a few years' decay will make the job much easier. No detailed studies of the decommissioning of fuel processing plants have been published. This disposal of nuclear facilities at the end of their useful life is not an insurmountable problem, or a big hazard to the environment if done carefully; but regulations and procedures have yet to be developed. It is beyond the scope of the present report.

DISPOSAL TECHNIQUES

The Objectives of Disposal Systems.

Growing uneasiness about the accumulation of wastes in the nuclear industry has led in the past few years to strong pressure for demonstrations of the feasibility of safe disposal. In several countries there has been a slowing-down of the industry's growth because of this uneasiness. One of the most drastic reactions is that of the Government of Sweden, whose 1976 Conditions Act says that permission to commission a nuclear reactor... "may be granted only providing that the reactor owner

1. has produced an agreement which adequately satisfies the requirement for the reprocessing of spent nuclear fuel and *has demonstrated how and where an absolutely safe final storage of the high-level waste obtained from the reprocessing can be effected*, or

2. has demonstrated *how and where an absolutely safe final storage of spent, un-reprocessed nuclear fuel can be effected*." (the italics are ours).

Similar pressure is being exerted on the industry in many other countries. Accordingly there is under way a major study of the technology of disposal systems. Several technical studies have been published. Of special value to the present report are studies by the Swedish Nuclear Fuel Safety Project (Kärn-Bränsle-Säkerhet, KBS), 1978; by the American Physical Society (1977); by the U.K. Royal Commission on Environmental Pollution (1976); and by OECD (1977). Reference has also been made to a report from Canada's Department of Energy, Mines and Resources (Aikin, Harrison and Hare, 1977).

The distinction between ultimate disposal and long-term storage is important. Storage implies retrievability; one seeks temporary containment and isolation of the materials, until one finds a further use for them (for example, the decision to reprocess irradiated fuel). Storage also implies monitoring and surveillance, the maintenance of security and safeguards, and vigilance for health and environmental protection. Disposal implies the reverse : the materials are put away with no intention of recovery, and with the purpose of isolating them from the biosphere for an exceedingly long period - in effect, for all time. It implies the intention ultimately to relax monitoring and surveillance, since these are obligations we cannot put on future generations. The objective of permanent isolation from the biosphere arises from the very long lives of many of the radionuclides. It will be necessary to isolate the fission products for several hundred years, and the actinides for some tens of thousands of years, even to reduce levels of radioactivity to those typical of high-grade uranium ore bodies. In practice, current planning aims at much longer isolation, on the scale of hundreds of thousands of years. Even the shortest of these periods is longer than that elapsing since the end of mediaeval times in Europe. The Neolithic revolution and the beginnings of urbanization are less far back than the 10,000 years needed to reduce even reprocessing wastes to something comparable to the radioactivity of rich uranium ore. It is this long isolation time, and the difficulty of long-term guarantees of performance, that makes many critics sceptical of disposal technologies.

In this report we shall not challenge the need for such disposal. It may well be that present perception of the hazard is exaggerated, and that public health and the natural environment could be protected with less drastic measures. Many calculations suggest that this is the case. But the level of public concern is high enough to justify defence in depth against the hazard. At some subsequent date, experience and public perception may call for less draconian measures. For the present, however, very elaborate, fail-safe technologies for disposal are necessary. All present plans for disposal are based on the multiple barrier principle, as in the case of reactor safety itself. In brief, the plans call for

- maximum possible immobilization of the wastes or irradiated fuel, especially against solution or suspension in water;

- encapsulation of the immobilized wastes in long-lasting, corrosion-proof canisters or capsules, often multiple;

- incorporation of the encapsulated wastes in buffering material designed to retard water penetration, while freely permitting heat release;

- choice of site for the engineered repository in deep rock, or in the ocean floor, so as to minimize water migration past the wastes, and to retard the migration of radionuclides in such migrating water;

- choice of site for the repository remote from earthquake or volcanic hazards, possible economic resources, and with good access to the required transportation links.

The major objective of the present report is to consider the potential environmental impact of such repositories. Clearly, if the above specific requirements are met, there will be no impact, apart from that associated with the preparation of the wastes for disposal, their transportation to the repository, and the operation of that repository until it is sealed off (as it will be when full). But it is vital to consider the possibility of miscalculations or accidents leading to failure. The ruling principle is to design the facility and the handling chain so as to minimize the likelihood of such mishaps. It is also vital to estimate their consequences, should they occur. In what follows we shall concentrate on the disposal of high-level reprocessing wastes, or of irradiated fuel. It may also be necessary to dispose of waste from earlier stages of the fuel cycle, such as reactor waste containing transuranic elements, or materials derived from the decommissioning of reactors. But the big and urgent problem is to learn how to cope with the intensely radioactive fuel or reprocessing wastes that are being or will be produced in growing volumes every year. A solution to this problem can readily be adapted to look after other wastes.

Disposal of Irradiated Fuel.

We have seen that fuel rods are removed from all kinds of reactors after one to five years of irradiation because of the accumulation within them of fission products that absorb too many neutrons. One tonne of uranium typically yields 25,000 to 35,000 MW(t) days of heat in light water reactors (much less in natural uranium reactors). This irradiated fuel contains wastes (which, if reprocessed, would yield about 150 litres of glass, or 4 m^3 per 1,000 MWe y^{-1}). This fuel is intensely radioactive when removed from the reactor, but the radioactivity is mostly due to short-lived fission products that decay very rapidly. The fuel is then stored in water-filled cooling bays at the reactor site to shield workers and the public from the radiation, and to remove the great amount of waste heat generated. When removed from a light-water reactor it contains

- about 3 per cent by weight fission products (mostly β and γ emitters)
- about 95 per cent unconsumed uranium, of which 1 per cent is fissionable U-235
- about 1 per cent plutonium (mostly fissionable Pu-239)
- and less than 1 per cent other actinides (mostly α emitters).

Fuel from heavy water reactors is poorer in fissionable U-235 and Pu-239.

After two years or more in the cooling bays at the reactor site the fuel needs to be moved, because the bays become full. Most nuclear engineers advocate storage at a central interim storage facility, also under water - though Atomic Energy of Canada Limited (AECL) is experimenting with dry concrete canister storage, which renders the constant filtration of cooling water unnecessary. In interim storage the fuel continues to cool, and to lose radioactivity as the radionuclides decay to stable nuclides. Throughout this period of interim storage the fuel must be rigorously isolated from human beings, and protected against accident, wilful sabotage or theft. There is division of opinion as to how long the fuel should remain in interim storage. A recent Swedish study suggests a limit of 10 years, but longer periods have already been achieved.

A decision must then be taken as to reprocessing. If the plan is to reprocess the fuel to recycle its fissile contents (U-235 and Pu-239), it will be transported to a reprocessing plant - which may well be nearby. We shall treat this option later. At present we shall consider the course of action to be taken if it is decided to dispose directly of the fuel.

Fig.1 shows the hazard presented by irradiated light-water reactor fuel (3.1 per cent U-235) per tonne, relative to that presented by the uranium ore that is equivalent to a tonne of such fuel, together with the contributions

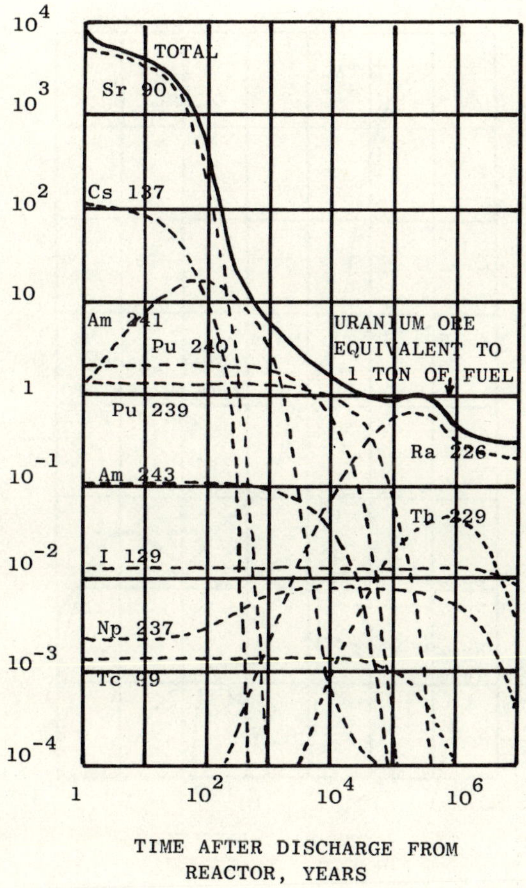

Fig. 1. Hazard indices of various radionuclides, and total hazard, presented by one tonne of irradiated light-water reactor fuel, as a ratio of the hazard presented by uranium ore equivalent to one tonne of fuel. Burn-up 33,000 MWd(t) per tonne of uranium. Enrichment 3.1 per cent U-235. (After KBS, 1978).

made by the chief radionuclides (compare with Table II for identification of nuclides). Fig.2 shows the same data for high-level-wastes from the reprocessing of one tonne of fuel. In each case the fuel is assumed to have undergone a burn-up in the reactor of 33,000 MW(t) days per tonne of uranium (data from KBS, 1978).

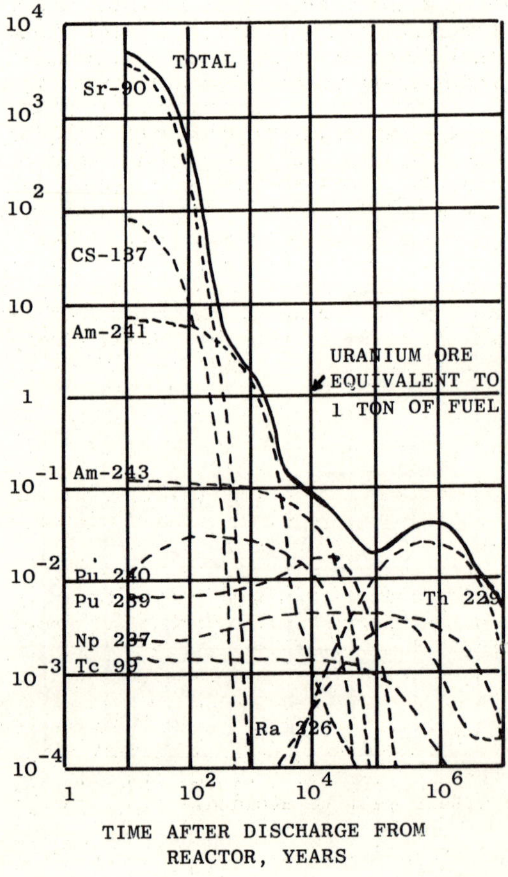

Fig. 2. Similar to figure 1, but is for the equivalent immobilized high-level waste. (After KBS, 1978).

The hazard is assumed to be the radioactivity associated with each nuclide multiplied by that nuclide's relative toxicity to man. The details of such curves depend on the original fraction of uranium-235, the burn-up achieved, and the type of reactor. Canadian CANDU reactors, for example, have much lower burn-ups (about 7,500 MW(t) d per tonne) and the evolution of the waste is rather different. The following broad properties manifest themselves:-

- In the first 600 years, it is the fission product nuclides that present the greatest hazards. The main activity is due to the very toxic nuclides strontium-90 and cesium-137, both having half-lives near 30 years, and both readily ingested by man. Other fission products are rather less hazardous. Most have short half-lives, but iodine-129 (half-life 17×10^6 years) is an exception that escapes to the atmosphere (being volatile). Other long-lived nuclides include technetium-99 and cesium-135. The fission products are mainly beta and gamma emitters.

- After 600 years, the hazard is primarily due to the actinides and radium. For 50,000 years americium-241 and plutonium-239 and -240 dominate. These nuclides are alpha-emitters. After that the main hazard is due to radium-226, which also emits hard γ-radiation.

- Certain heavy radionuclides not present in the irradiated fuel at the time of its withdrawal grow into the fuel as decay daughters of other heavy nuclides. Examples include radium-226 and thorium-229.

- Irradiated fuel's total hazard drops to that typical of uranium ore (i.e., to naturally occurring radiation hazards) in about 30,000 years, and then remains near that level for about another 150,000 years. The time taken to reach these levels depends, of course, on the burn-up of fuel achieved in the reactor.

It is generally assumed from analyses of this kind that isolation of spent fuel from the biosphere is required for about 600 years in the case of the dominant fission products, and at least long enough for the longer-lived activity (chiefly due to the actinides) to decline to ore-body levels. Another figure often quoted is ten half-lives of plutonium-239 (244,000 y). Precise figures are not very logical. In practical terms the presence in the fuel of long-lived actinides like plutonium-239 means that isolation must be planned to be permanent. Any escape from confinement while activity levels are still above those typical of nature itself must be avoided.

Though the direct threat to the biota (including man) is posed by the radioactivity of the wastes, it is the release of heat that poses the major problem from the point of view of their storage and subsequent disposal. This heat must be dispersed, to avoid damage to the containment. Fig.3 shows the heat released per tonne of uranium from irradiated fuel from pressurized water reactors, as a function of time out of the reactor (with 3.15 per cent uranium-235, and 33,000 MW(t) days per tonne burn-up). Curves are shown for both irradiated fuel and reprocessing wastes. Fig.4 shows the equivalent cooling of Canadian CANDU fuel after 7,500 MW(t) days per tonne, with natural uranium. It follows from these and other data that much heat has to be dispersed from the irradiated fuel, especially in the first few years, when the fuel is stored under water

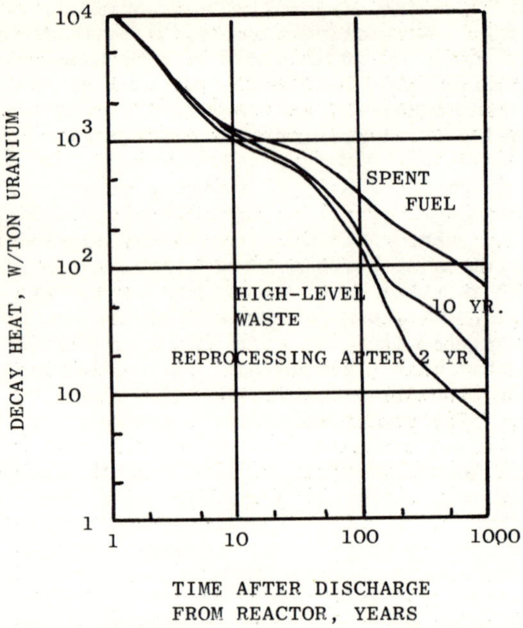

Fig. 3. Decay heat generation, in Watts per tonne of uranium, from irradiated fuel and immobilized high-level wastes (99.5 per cent uranium removal) from a pressurized water reactor using 3.25 per cent U-235, after a burn-up of 33,000 MWd(t) per tonne or uranium. (After KBS, 1978).

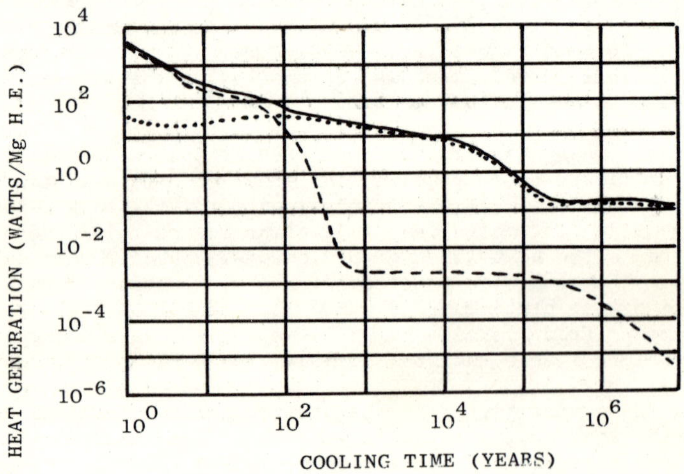

Dashed : Fission Products
Dotted : Actinides
Continuous : Total

Fig. 4. Decay heat generation, in Watts per tonne (megagram), from irradiated fuel from a Canadian CANDU reactor, natural uranium, 7,500 MWd(t) per tonne of uranium burn-up.

(whose high heat capacity helps in the dispersal process). After the fuel has been installed in a repository, the heat must disperse itself through the surrounding medium by conduction. The rate of heat generation decreases rapidly at first. In the first century the fission products dominate the release, but the actinides take over thereafter. If we assume that irradiated fuel is disposed of in a repository after fifty years in surface storage, its heat output is likely to be about 600 W T^{-1} for LWRs and PWRs, and 100 W T^{-1} for CANDUs. We return to this subject later.

The immobilization of irradiated fuel should not pose severe problems. The fission products, actinides and unused uranium remain within the ceramic pellets of the fuel rods, whose Zircaloy cladding will help to confine any that escape. The cladding, however, after irradiation is somewhat brittle. Hence it must be encapsulated in some form, to minimize the migration of radionuclides. Little study was given to this problem until recently, since the direct disposal of irradiated fuel was considered undesirable for economic reasons. One recent published study (KBS, 1978) recommends encapsulation of the rods in pure copper sheaths to ensure prolonged containment. Within the next few years it is to be expected that the major nuclear powers will publish designs and safety requirements for repositories for irradiated fuel.

Disposal of Reprocessing Wastes.

As shown above, reprocessing wastes will be disposed of in the form of immobilized solid blocks, probably consisting of glass containing the fission products, small residual amounts of uranium and plutonium not removed in the reprocessing, and the remaining actinides. The glass is likely to be clad in a resistant metal canister. The wastes differ from irradiated fuel in lacking the radioactivity and heat release associated with the uranium and plutonium that have been removed. As Fig.2 shows, the effect of this removal is to permit a more rapid decay in radiological hazard than in the case of spent fuel. For the first three hundred years strontium-90 and, to a lesser extent, cesium-137, dominate the hazard index. Thereafter, for a period of about 50,000 years, the actinides americium-241 and -243 and the residual plutonium-239 and -240 dominate. Thorium-229 (a decay daughter) builds in to dominate the hazard index from 50,000 years out to well over 10 million years. Overall hazard drops to that of uranium ore equivalent to 1 tonne of fuel after about 25,000 years. As in the case of irradiated fuel, these rough figures vary a little according to the type of fuel, the burn-up achieved, and the type of reactor. Fig.3 shows the accompanying heat release. The vitrified high-level wastes cool more rapidly than fuel, especially if reprocessing is carried out promptly after removal from short-term storage. Reprocessing wastes hence differ in waste-form from irradiated fuel, and pose rather less long-term radiological hazards. The differences are not great, however, and are not in themselves a good argument for reprocessing. The demands made on the repository's capacity to contain the wastes and to disperse heat are not very different. In our judgement repository design should permit the disposal of either irradiated fuel or vitrified high-level wastes. It is impossible to judge at this time how extensive future reprocessing will be.

SITING OF REPOSITORIES

The Options.

The objective of disposal is the permanent isolation of the wastes from all contact with living organisms, especially from human beings. To do this, one must design an engineered site called, by convention, a repository. Options that have been considered for sites for repositories include

- surface or shallow sub-surface permanent storage, involving permanent surveillance and monitoring;
- burial in glacial ice, in Antarctica, or in Greenland;
- dispatch to space by rocket vehicle;
- disposal in canisters on the sea-floor, probably in deep oceans, or burial of the canisters in holes in the sea floor, in each case in stable abyssal plains. A variate is disposal near subduction zones between tectonic plates;
- geological containment on land
    - in sediments such as salt, clay, shale or volcanic tuffs
    - in massive igneous rocks such as granite

The first of these options is not, in our judgement, valid for disposal as normally defined. It has nevertheless been espoused by some leading authorities in Canada and elsewhere. Ocean disposal clearly has much merit, but technical and jurisdictional uncertainties are handicaps. At present opinion usually favours on-land burial in an excavated rock repository within the home territories of the country. This is the method receiving most attention today.

Current Choices for Geological Containment.

The sites being actively considered differ from country to country, depending on the geology, density of population and the nuclear technology being employed. The kinds of rock being examined include clays, shales, salt beds and dunes, tuffs and massive igneous rocks.

Salt beds or domes (diapirs) are being actively explored and tested in several countries, notably the United States, the Federal Republic of Germany (where storage in salt at Asse is a well-developed technique), and the Netherlands. Salt has many advantages. Its mere existence is proof that groundwater circulation is absent, since it is highly soluble. It may, however, contain entrapped water that will tend to migrate towards a heat source such as a repository, but this would not lead to the escape of radionuclides. Salt tends to flow under high pressure, and hence self-heals any damage to excavated structures. It is an excellent heat conductor.

Igneous rocks, which are often massive and crystalline in form, are being examined in other countries, especially Canada, the United Kingdom, France and Sweden. Such rocks are often fractured and fissured, and water movement follows these lines of weakness. In Canada the preference is for small quartz-poor, felspar-rich intermediate or basic intrusions, which are abundant in the Laurentian Shield, the ancient core of the country. Similar rock-bodies are being explored in Sweden, where granites and granite-gneisses are the choice. These rocks contain quartz, but are also rich in felspars.

POTENTIAL ENVIRONMENTAL IMPACTS

General Considerations.

Disposal aims to eliminate environmental impact due to the radiotoxicity or chemotoxicity of the wastes. In this section we examine the likelihood and magnitude of such impacts associated with the disposal technologies now available or under development. We also consider how the impacts can be minimized. Desirable levels of environmental and health protection have long been established by the International Commission for Radiological Protection (ICRP), whose standards form the basis for most national codes. The emphasis has naturally been upon human health, rather than on that of other organisms. Apart from our desire to protect ourselves, we appear to be more sensitive to ionizing radiation than any other animal or plant species so far tested. The United Nations Scientific Committee on Atomic Radiation (UNSCEAR) has recently (1977) published detailed estimates of the radiological exposure of human populations from all parts of the fuel cycle. Hence this subject will not be treated here. Environmentalists are likely to insist, however, that the impact of disposal on other organisms be considered, especially as regards the tendency of specific nuclides to be concentrated by certain species. Effects on overall ecosystem functions will also need to be specified, if these are detectable.

The potential environmental impacts of waste disposal can be considered in relation to specific stages of the process:-

- preparation of the wastes for disposal;
- transportation to the repository site;
- emplacement in the repository; and
- after emplacement.

In this report we shall examine only the fourth of these stages, since this is the vital concern. The impacts of reprocessing and immobilization technology, and of transportation, have been covered elsewhere - for example in Flowers, 1976; UNSCEAR, 1977; OECD, 1977; KBS, 1978. The risks associated with emplacement have not been examined in as much detail, but these are likely to be confined to small occupational exposures, similar to those experienced in earlier parts of the fuel cycle.

Impacts After Emplacement in a Rock Repository.

The irradiated fuel or vitrified high-level wastes will be installed in excavated structures in salt, igneous rock, shale, clay or tuff. The repository must be at sufficient depth for the rock to absorb all the radiation emitted, and to disperse the heat so that there is no significant rise of soil and surface temperatures. If there is to be impact on living organisms there will hence have to be either

- a breaching of the containment by earthquake, vulcanism (volcanic activity), erosion*, meteorite impact or faulting (catastrophic effects); or

- dispersion of radionuclides from the repository by migrating groundwater, essentially a long-term gradual process.

Proper location of the repository, and its design, will minimize each of these possibilities. We shall consider then in turn.

Catastrophic Effects.

Catastrophic processes are considered very unlikely to breach the containment of wastes. The following can be visualized:-

(a) bedrock-movement due to warping. Loading of the land-surface by glacial ice can downwarp the crust, in some areas by as much as 1,000 m. Subsequent recovery can be quite rapid (of order 1 cm y^{-1}). Vertical warping of this kind will not breach a repository, since it is uniform over large areas. The districts being explored by both Swedish and Canadian authorities have been repeatedly subjected to such movements in the past million years, and may well be downwarped again. Renewed glaciation within 10,000 years is possible.

(b) bedrock-movement due to faulting. In the choice of repository site it must be a first requirement to avoid major fault-planes or fracture-zones that have a history of recent movement. In rigid Shield areas like the Laurentian Shield of Canada, the Angara Shield in the Soviet Union, and the Fenno-scandian Shield in Norway, Sweden, Finland and parts of the Soviet Union, there are extensive areas in which little faulting has occurred in the past several hundred million years. The KBS study for Sweden, for example, estimates that the probability of a fault movement affecting a 1 km^2 repository is less than 1 in a billion per annum. Many other countries, however, are tectonically** more active. Japan, Italy and New Zealand are good examples. In such countries special care will be needed in locating areas with minimum likelihood of faulting. New faults are rare. Movement is nearly always along pre-existing fault-planes that can be readily detected, mapped and monitored.

---

* erosion is normally slow, and hence not catastrophic. But it is important only if accelerated by some drastic change in environmental conditions.

** tectonic refers to the tendency for the rocks of the crust to be folded or faulted due to internal compressive stresses, or to tension.

Even in tectonically active areas, however, movement along a fault-plane rarely exceeds a few centimetres per movement, and this would be most unlikely to cause rupture of the repository, or damage to the encapsulated wastes. Open faulting, in which new, open fissures appear, is essentially a surface phenomenon.

(c) earthquake-shocks. The seismic effects of faulting of the above kind include substantial shock-wave movement into the rock around the fault, extending out to several hundred kilometres. Damaging earthquakes occur in association with fault-movement along well-known axes (for example, the St. Lawrence Valley in Canada; the Pacific Coast of the United States). Minor earthquakes or tremors occur in many areas, occasionally with damage to surface structures. These shocks have maximum effect at the surface, and rarely damage mines, subways and other engineered sub-surface structures. There is a long history of mine management in seismically active areas from which these effects can be confirmed and analysed.

(d) meteorite impact. Analysis of meteorite impacts over the past million years (KBS, 1978) suggests that the odds against a direct meteorite impact causing a 100 m crater in a given square kilometre are about 10 trillion (10^{13}) to one per annum. Canadian estimates are closer to 10^{11} to 1. Both are quite negligible hazards.

(e) In hard rock areas stream <u>erosion</u> does little work in periods of a hundred thousand years, except in mountainous terrain. Glacial erosion may overdeepen existing valleys. In the Laurentian Shield of Canada, successive glaciations in the past million years have locally excavated pre-existing valleys by 1,000 m. In flat plateau areas, glacial erosion is minor, and consists chiefly of removal and redistribution of soil and subsoil. A total regional erosion of 10 to 100 metres per million years probably overestimates the effect.

(f) vulcanism. Active volcanic areas will be avoided in the choice of site. New vulcanism is usually in areas with a previous history of such activity. Impact of such new vulcanism on a well-chosen repository should be nil.

These and other possible catastrophic effects are best avoided, or at least minimized, by choice of a repository that is

- remote from active faults, and seismically or volcanically active areas

- deep within the rock

- remote from deep valleys, especially those aligned on faults or fracture zones.

These conditions are readily met in most areas, but are harder to satisfy in countries like Italy, Japan, New Zealand and Turkey that have no seismically inactive areas. In such cases self-sealing rock types such as clay and salt are probably better than rigid rocks - but more research is needed to verify this. It is unnecessary to avoid areas potentially open to continental-scale glaciation (such as Canada and Scandinavia). The growth of a new ice-sheet would exclude man, plants and animals. Moreover it would provide a thick, extra covering for the repository, and would still further compress any existing fracture or fault planes. It would probably endure some tens of thousands of years.

Dispersion Processes.

Dispersion of radionuclides by migrating groundwater is the chief hazard in waste disposal. Avoidance of such dispersion provides most of the site-selection and design criteria for the repository.

Groundwater originates from rain or melted snow at the land surface. In a few areas, such as the Sahara and the interior basins of Australia, the groundwater may be fossil, i.e., derived from previous epochs with a wetter climate. Elsewhere the year's surplus rain tends to sink into permeable rocks in upland areas, and to emerge from springs in valleys, having followed a curving path through the rock. Certain rocks, called aquifers, permit ready groundwater movement, and store large quantities in pores and fissures. Others are highly impermeable. Groundwater may occur under high pressure below an impermeable cap layer - for example in the sedimentary limestones of southern England, Northern France, the Low Countries and in the Münster basin of Germany. Groundwater movement is easy in porous and fissured rock, and is hence most rapid near the surface. At greater depths the pores and fissures are narrower, or may be virtually absent. Motion is so slow that it may be unmeasurable. Calculations of flow at 500 and 1000 m depth in crystalline, fissured igneous rocks give velocities of the order of 0.1 to 0.2 litres per square metre per annum, for areas of low surface relief, corresponding to permeabilities much below 10^{-9}m s^{-1}, possibly as low as 10^{-12} m s^{-1}.
Confirmation that groundwater movement is very slow at even moderate depths in level terrain can be derived from age determinations of water sampled at those depths. This is difficult to do, since present-day water has to be injected into the hole during the drilling. Hence determinations should underestimate the true age. Using the carbon-14 technique Swedish experiments (KBS, 1978) at one site have given ages of 4,400 years at 291 m; of 4,275 at 510 m; of 11,055 at 407 m; and of 8,205 at 493 m, the pure experimental error being in each case $\pm$100 years. Finnish measurements at one site have given an age of 4,010 years at 136 m. The rapid local variations are probably real, reflecting the fact that movement is local in fractured igneous rock with a hillocky surface. The rock has rapid variations of fissure-density, and hence of permeability. Age determinations of groundwater around a chosen repository will be valuable but not conclusive evidence of the slowness of movement (and hence desirability of the site).

The repository itself will be sited and designed so as to minimize the risk that such groundwater movement will leach radionuclides out of the wastes. The latter are in ceramic or vitreous form, encased within low-corrosion metal canisters. These are likely to be embedded in highly absorbent buffer material that entirely fills the excavated repository and backfilled shafts. Even rapid groundwater movement through such a medium will be greatly retarded. Great care will have to be taken, however, to minimize the risk that the actual construction of the repository does not open up new paths for more rapid circulation.

If, nevertheless, groundwater movement does penetrate the repository, it will eventually cause some leaching of the radionuclides in the irradiated fuel or vitrified wastes, insoluble though they are. It is necessary to attempt some estimate of how rapid this leaching and outward transport may be. If there is any danger that radionuclides may be transported to the surface before their radioactivity has decayed, it will be necessary to compute the degree of dilution that will occur, and to estimate their subsequent progress through the drainage basin and its ecosystems. A low rate of leaching will depend upon the following succession of barriers:-

- a low rate of groundwater migration;
- a high integrity of the encapsulation;
- a low rate of solution or suspension of radionuclides from the glass or fuel;
- a high capacity of the buffer material to retard the passage of the radionuclides;
- a high capacity of the bedrock to sorb the radionuclides.

As we have just seen, the rate of groundwater migration decreases with depth. Hence a deep repository is less likely to suffer leaching. It is also more remote from hazards due to seismic activity, erosion, meteorite impact, or human interference. Swedish plans call for a repository at 500 m in igneous rock, where the present groundwater is at least of order 4,000 to 11,000 years of age. It would be reasonable to assume an equivalent period for the groundwater around the repository to return to the surface. In fact, their plans assume the improbably fast return time of 400 years (KBS, 1978, Vols. II and IV). A Canadian study calls for a repository at 800 to 1,000 m (Aikin, Harrison and Hare, 1977), where the age of groundwater is probably considerably more, and the rate of flow even slower. The greater the depth, the greater the security against leaching and the other hazards mentioned above - this is the ruling assumption. But the assumption needs to be qualified with the warning that the structural integrity of the repository may be less easily maintained at these greater depths. The optimum depth for a given site should be experimentally determined from local field measurement, coupled with modelling of groundwater flow.

The metal canisters in which the wastes will be encapsulated can be made corrosion-resistant by a suitable choice of metal - for example titanium or copper - with known long-term resistance properties. If, as in some plans, a cladding of lead is included within the canisters, resistance will be even higher. It is assumed in several studies that such canisters are likely to be breached in about 1,000 years, and effectively removed within the next few thousand years, gradually exposing the wastes to circulating waters. This is probably conservative. In calculating the effects of canister failure, allowance must also be made for accidental damage, either during installation, or as a result of rockfalls or earth movements.

The rate of solution or suspension of the radionuclides within the glass or fuel has been studied both theoretically and empirically. It is not certain that this leaching is simply equivalent to the solution rate for the glass or ceramic, since some radionuclides may migrate towards the interface with the groundwater. Initial leaching rates at 20°C, using borosilicate glass, are reported to be about 10^{-4} g cm^{-2} y^{-1}. The nepheline syenite glass developed by the Chalk River Nuclear Laboratories of Atomic Energy of Canada Ltd. started at 10^{-5} and decreased after ten years of continuous leaching to 2×10^{-8} g cm^{-2} y^{-1}. Thus under prolonged solution, leaching rates decrease with time, because of the accumulation of a coating of corroded material, which decreases diffusion rates. All authorities agree that the rate of leaching increases rapidly with rising temperature - perhaps thirty- or forty-fold by 100°C. Hence it is important to avoid excessive rises of temperature due to the heat release. Equivalent leaching rates for UO_2 fuel have not been as well determined, but they, too, are likely to be very low.

The retardation of movement of dangerous materials away from the wastes towards the surface depends on the capacity of the buffering material and the rock along the migration path to sorb the radionuclides that have entered solution. Even though groundwater may move quite freely through fissures or porous rock, it is common for the movement of dissolved or suspended substances to be retarded by ion exchange or adsorption, mechanical filtration, precipitation and mineralization; this is how, in the remote past, many ore-bodies were formed. The combined process is called sorption (with sorb as the verb).

Research in several countries, most notably France (de Marsily, Ledoux, Barbreau and Margot, 1977), suggests that high sorption capacity is the key to optimum performance by deep rock repositories. In rock with good sorption capacity, the radionuclides mostly move much more slowly than the groundwater itself. The same should be true of well-chosen buffer materials (especially <u>along</u> the boreholes, shafts, and drifts, which might otherwise act as migration pathways). The radionuclides differ as regards retardation:-

- technetium-99 and iodine-129, both with very long half-lives, are poorly sorbed, or not at all. It can be assumed that they will move with the same velocity as the groundwater, and hence ultimately escape. Strontium is also poorly sorbed, moving at 1-3 per cent of groundwater velocity; its shorter half-life, however, means that it will decay long before it has travelled far.

- the actinides, especially americium, are moderately or strongly sorbed, and will move much more slowly than the groundwater. Neptunium is less uniformly sorbed than the other actinides.

Clearly, then, a series of barriers exists between the wastes and the surface ecosystems. Their probable overall effectiveness has been calculated by means of numerical simulations, using conservative values of the large number of parameters that govern the transfer processes (e.g. ERDA, 1976; KBS, 1978). These models predict the length of time required for the main radionuclides to reach the earth's surface, and to have impacts on critical population groups. They may also be used to compute the radiation doses received by these groups as functions of time. The use of models is essential, since no empirical test can be devised on so long a time-scale. The models are simplified and crude, but one can have some confidence in their results, since fail-safe values of the parameters are used.

KBS in Sweden has applied such a model for an installed capacity of 10,000 MWe (KBS, 1978) to a critical group consisting of a population drawing its main water supply from a well near a repository that undergoes leaching of its wastes after the encapsulation is breached (1,000 years from emplacement). Fig. 5 shows the calculated impact, in radiation doses to individuals within the critical group, in rems per thirty-years (i.e. per generation). The diagram compares the dose likely to be experienced with other levels of exposure typical of a Swedish population - natural background, and the dose due to radium-226 in drinking water. The maximum impact will be deferred until between 100,000 and 1,000,000 years hence. The peak dose is likely to be of the order of 0.1 rem (100 millirem) in 30 years, compared with typical background doses of 100 mrem per annum, or 3 rem in 30 years.

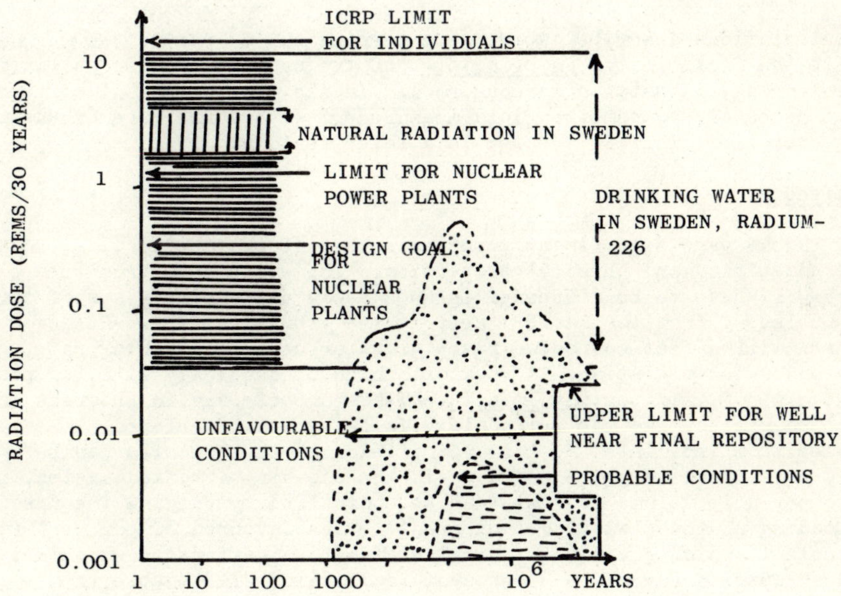

Figure 5. Calculated upper limiting and expected radiation doses to persons living near a well close to a deep geological repository in Sweden, from KBS, 1978. Dose loads from natural sources, and established Swedish dose limits, are also plotted. (After KBS, 1978).

The same model was used to compute the collective dose to the world's population due to the operation of the proposed Swedish repository. In the most unfavourable 500 years, the dose is expected to average .007 manrem* per MWe per annum, which is well below the recommended limit for the entire nuclear fuel cycle (1 manrem per MWe y^{-1}). The assumptions and numerical values used in these calculations were highly conservative, and the actual impact is expected to be substantially less than those given above. It therefore appears that the effect of the multiple barriers introduced into the disposal system should

- defer the impact of the leaching of the wastes until a very remote period of the future, when much of the radioactivity will have decayed;

- reduce the doses to individuals in critical groups to extremely low values; and

- similarly reduce lifetime impacts to the entire world population to very low values.

It is conceivable that catastrophic processes of the sort described before could accelerate and magnify these impacts, but the probability of such catastrophes is exceedingly low.

All the calculations described above were carried out as part of exercises to demonstrate the feasibility, in principle, of the concept of disposal in deep rock repositories, in salt or igneous rock. In the case of an actual repository it will be necessary to combine field measurement with a far more exhaustive modelling exercise. This is treated in a later section.

Thermal Effects.

Rocks vary significantly in their thermal characteristics, both as regards conductivity and physical properties. The waste heat from the repository will have to be dispersed in such a way that undue rises of temperature occur at no level, from the waste itself to the ground surface. Rising temperatures will affect corrosion rate, glass or ceramic leaching rate, gas formation and water migration. A rock with high conductivity will, if possible, be chosen, after thermal testing in situ. Tests are already in progress in mines in several countries, and the methodology is being well developed. The rate of heat release from irradiated fuel or vitrified high-level wastes can be accurately calculated (Figs. 3 and 4). In the case of vitrified wastes, the rate of release per unit volume of waste can be controlled by varying the fission product loading of the glass. French practice has favoured 20 per cent fission product content, which gives substantial heating in the first few centuries. Some Swedish opinion favours a 9 per cent load, which will enable them to keep core temperatures in the glass below 90°C, and surface temperatures of each canister below about 60°C. These figures depend critically on the duration of pre-disposal storage, on the reprocessing date in relation to withdrawal from the reactor, and on the geometry of the actual emplacement of waste in the repository; one can space out the waste to optimize heat disposal, subject to considerations of cost and physical integrity. The same applies to fuel. The heated area around the repository will gradually enlarge. Maximum temperatures in the waste or fuel will be reached within a century, and the repository itself should then cool. The heated area will continue to expand for many centuries, and will ultimately affect surface temperatures slightly - possibly by as much as +1°C. These effects will need to be accurately calculated for the specific site chosen.

---

* i.e., one rem impacting one man.

Future Security.

We shall not treat the problem of present-day security in the nuclear cycle, since this does not seem to us to be an environmental issue. We must, however, at least raise the problem of future security. It is highly unlikely that future generations will deliberately seek to re-enter a sealed-off repository. But it is possible that they may do so inadvertently. The possibility exists of a future Dark Ages, in which there is a loss of technological skills, and perhaps a rupture of information memory.

The only precautions that we can advocate to guard against such an event are

- to leave as many and as varied records in other forms as we know how;

- to avoid using sites which may in future be of value in mineral or geophysical exploration.

It is the second of these measures that seems to us to argue against the use of salt formations. Salt has a high economic value, and is often associated with other valuable materials, such as oil and gas.

Environmental Survey, Monitoring and Modelling.

Following the choice of an actual repository, it will be necessary to conduct exhaustive field surveys and testing, combined with hydrological, geochemical, ecological and ecotoxicological surveys and modelling exercises. The environmental impact analysis will have to predict, as far as is humanly possible, the consequences of potential failure of containment by the repository, at various future dates. A first requirement will be the establishment, through the entire drainage basin in which the repository is to be located, of background levels of the chief radionuclides, and of radiation dose. This includes water-bodies, soils, selected species of the biota, rocks and atmosphere. The background will include contributions from nuclear weapons testing fallout (which is still in progress), as well as from naturally occurring radionuclides. The establishment of these baseline data is necessary if future increases due to migration of radionuclides from the repository are to be detected. Such migration is not likely within the first millennium, so that a major purpose of the exercise is to bequeath to posterity an adequate record of pre-existing levels of radioactivity. Another purpose, however, is to provide for the unlikely contingencies of failure in handling at the repository itself, of sabotage, of catastrophic disturbance - and of the calculations made by the designers.

A second, and equally vital, requirement must be exhaustive hydrological, geochemical and ecological surveys of the drainage basins. The most accurate possible picture of water movement through the whole hydrologic cycle should be sought, especially as regards groundwater movement, soil moisture relationships and the rôle of the chief reservoirs (for example, the great northern bogs). This should be coupled with an adequate review of fluid geochemistry, and of fluviatile, lake and bog ecosystem function. The latter should include, if possible, the identification of potential bio-accumulators or food-web multipliers within the biota. There is already much experience of heavy metal behaviour in aquatic ecosystems, including the vital questions of metal speciation, complexing and accumulation within species.

The design of adequate experiments of this kind will be a vital exercise for the various national licensing agencies. It is a new kind of environmental impact analysis, in three main senses. First, it aims at the identification of potential impacts that are very improbable, and may be literally inconceivable with present experience. Second, it aims at very distant epochs. And thirdly it aims at the design of monitoring systems not so far visualized. Ecotoxicological methods of today, as exemplified in the brilliant modelling exercises of UNSCEAR, aim much more specifically at impacts on contemporary man.

Finally, one must stress that all such effort should aim at satisfying widely-felt public anxieties that have been aroused by alarming statements about the impacts of nuclear wastes on future generations. Whether these statements are justified or not, it is indeed a heavy responsibility to pass on to posterity large amounts of very toxic material. This alone seems to us to justify the large effort and considerable expenditures represented by the above suggestions, at least to the extent needed to allay these concerns - and to protect future generations.

CONCLUSION

We have arrived finally at what appears to be a comforting conclusion - that the ultimate disposal of nuclear wastes should be technically feasible, and very safe. We find that the environmental and health impacts will be negligible in the short-term, being due to the steps that precede the emplacement of the wastes in the repository. Disposal itself, once achieved, offers no short-term threat - unless an unforeseen catastrophe of very low probability occurs. The risks appear negligible by comparison with those associated with earlier stages of the fuel cycle.

Ultimately - millennia hence - a slow leaching of radionuclides to the surface might begin. But it would be so slow that great dilution of each nuclide will occur. This phase is likely to be reached somewhere in the period 100,000 to 1,000,000 years hence. We cannot visualize what the world will be like at that time. Past experience suggests that it will pass through a sequence of ice ages and warmer interglacials. We do not believe that these will of themselves disrupt a repository. But they will certainly disrupt human society. Man will by that time probably have evolved dramatically. We cannot guess what he will be like, or what his capacities and needs will be. In the end - millions of years hence, when most of the radionuclides will have decayed into harmless daughters - the remains of the nuclear age will accumulate in two main reservoirs. These will be the ocean, and the continental rock bodies.

We are dissatisfied with the situation at the front end of the nuclear cycle. No satisfactory technique has yet been reached for the disposal of the heavy radionuclides that accumulate in the tailings from the milling stage. Isotopes of radium and thorium are present in these tailings, and are leached outwards by percolating rainwater or snowmelt. These heavy nuclides pose the same sort of threat to mankind as the actinides contained in the high-level wastes from reactors. It does not make sense to contain the one without the other.

We are convinced that the environmental and health effects of waste disposal are small beside those posed by the accumulation of fissile materials in storage (mostly plutonium-239 for military purposes at present, but soon to include Pu-239 and other fissile nuclides from fuel reprocessing). They are also small by comparison with the maintenance of adequate security and safeguards for such materials.

REFERENCES

Aikin, A.M., J.M. Harrison and F.K. Hare, The Management of Canada's Nuclear Wastes, Ottawa, Energy, Mines and Resources Canada, Report EP77-6.(1977).

American Physical Society, Nuclear Fuel Cycles and Waste Management, Reviews of Modern Physics.(1977).

Energy Research and Development Agency (ERDA), Alternatives for Managing Wastes from Reactors and Post-Fission Operations in the LWR Fuel Cycle. Springfield, Virginia, National Technical Information Service, 5 Vols. (1977).

Flowers, B., Nuclear Power and the Environment, Sixth Report of the Royal Commission on Environmental Pollution, London, H.M.S.O. (1976).

Federal Republic of Germany, Radioactive Abfälle in der Bundesrepublik Deutschland, Bonn, Bundesministeriums für Forschung und Technologie, 5 Vols. (1976).

International Atomic Energy Agency (IAEA), Management of Radioactive Wastes from the Nuclear Fuel Cycle, Vienna, 2 Vols. (1976).

Effects of Ionizing Radiation on Aquatic Organisms and Ecosystems, Vienna, Tech. Report 172.(1976).

Keeny, S.M. Nuclear Power, Issues and Choices, Report of the Nuclear Energy Policy Group, Ford Foundation and Mitre Corporation, Cambridge, Mass., Ballinger.(1977).

Kärn-Bränsle-Säkerhet (KBS), Handling of Spent Nuclear Fuel and Final Storage of Vitrified High Level Reprocessing Waste, Stockholm, 5 Vols.(1978).

Marsily, G. de, E. Ledoux, A. Barbeau and J. Margot, Can the geologist guarantee isolation? Science, 197, pp. 519-528.(1977).

Organisation for Economic Cooperation and Development (OECD), Report on Objectives, Concepts and Strategies for the Management of Radioactive Waste, Paris, 273 pp. (1977).

Pigford, T.H., Environmental aspects of nuclear energy production, Annual Review of Nuclear Science, 24, pp. 515-559.(1974).

Pochin, E., Estimated Population Exposures from Nuclear Power Production and other Radiation Sources, Paris, Nuclear Energy Agency, OECD.(1976).

United Nations Scientific Committee on the Effects of Atomic Radiation (UNSCEAR), Nuclear Power Production, UN General Assembly Document A/AC.82/R.243.(1977).

8

IMPACT SUR L'ENVIRONNEMENT DE LA GESTION DES DECHETS DU CYCLE DE COMBUSTIBLES NUCLEAIRES

Y. Sousselier
Institut de Protection et Sûreté Nucléaire
Commissariat à l'Energie Atomique
92260 Fontenay-Aux-Roses, France

INTRODUCTION

Toute action industrielle a - ou peut avoir - à un degré divers un impact sur l'environnement. Cet impact peut se situer :

- en "amont" de l'action : les matières premières nécessaires pouvant entraîner un épuisement ou une diminution des ressources pour l'homme, la construction des installations nécessaires pour l'action envisagée entraînant une modification du cadre de vie local pouvant être, en certains cas, important (barrage, endigage, dérochage de falaises, etc.).

- pendant l'action : certaines nuisances sont entraînées directement par cette action (bruit, fumée, odeur). D'autres ne se produisent qu'en cas d'accident (explosion, incendie).

- en "aval" de l'action : pollution du milieu. En réalité, ce type d'impact se produit "pendant" l'action mais les conséquences n'apparaîtront souvent "qu'après" l'action par suite de phénomènes de relâchement différé (pollution à partir d'un stockage de déchets se dégradant avec le temps) ou de phénomènes lents de reconcentration.

Toute étude d'impact doit analyser ces différents points et essayer de les quantifier. Mais une telle étude n'a d'intérêt et de justification que si elle vise à comparer les impacts de deux actions industrielles alternatives ou à entrer comme un des éléments de l'analyse coût bénéfice dans l'analyse de justification d'une action industrielle. Or, toute comparaison d'impact est difficile et peut être même sujette à caution. Si on peut, relativement, facilement comparer les impacts de deux installations fabriquant le même produit selon le même procédé mais avec des moyens différents de traitement des effluents, avec des appareillages différents, il sera déjà plus difficile de faire la comparaison quand il s'agira de produits différents (mais dont l'usage serait le même) : comment comparer l'impact d'un bruit ou d'une odeur, celui d'un rejet continu d'effluents peu toxique et d'un rejet accidentel de plus grande quantité d'effluents plus toxiques? Comment surtout comparer une nuisance qui va se produire tout au long de l'action et une nuisance qui risque de se produire seulement dans quelques siècles voire dans quelques millénaires?

Toute étude d'impact de l'énergie nucléaire et de ses différents composants va se trouver confrontée à ces difficultés. Par contre, les études seront facilitées par le fait que tous les impacts possibles ont commencé à être étudiés, dès les premiers développements de l'énergie nucléaire, bien avant les premières réalisations industrielles à grande échelle. Les études du type coût-bénéfice ont été menées dans nombre de domaines de l'énergie nucléaire, en particulier, dans celui de l'irradiation du public, suivant les recommandations de la Commission internationale de protection radiologique.

Une telle étude a été implicitement faite dans le cas des déchets radioactifs dès la prise de position vis-à-vis de l'option fondamentale en matière de gestion des déchets :

"Rejeter-diluer" ou "concentrer-confiner" option qui se traduira souvent par produire des effluents ou produire des déchets et qui se traduira aussi, dans la prise en compte du premier terme de l'alternative, par la question des faibles doses et par celle du risque à long terme, voire même à très long terme, dans la prise en compte du second terme de l'alternative.

Cette étude se propose donc d'analyser les différents impacts possibles de la gestion des déchets en essayant de quantifier chacun de ses impacts mais sans chercher à les comparer : une irradiation potentielle d'une population a-t-elle un impact plus important que la stérilisation d'un terrain? On verra, à de nombreuses reprises, que les impacts ne sont que potentiels. La question des déchets radioactifs a été traitée avec une telle prudence que, dans de nombreux cas, la probabilité est grande que l'impact réel soit nul.

Le développement de l'énergie nucléaire est un fait, la continuation de ce développement une quasi certitude. Par contre, son rythme et ses modalités sont beaucoup moins définis. Or, les conséquences sur la production, voire sur la nature des déchets seront très importantes. L'utilisation de la filière des réacteurs à neutrons rapides, le non retraitement des combustibles irradiés se traduiront de manière très différente en ce qui concerne les déchets. Ce rapport essaiera d'examiner non seulement la gestion des déchets telle qu'elle se présentera vraisemblablement dans le développement le plus probable de l'énergie nucléaire mais aussi l'incidence d'autres développements de cette énergie.

GESTION ACTUELLE ET FUTURE DES DECHETS RADIOACTIFS

Origine et caractéristiques des déchets radioactifs

Notion de déchet radioactif

Le terme de déchet radioactif couvre, suivant les auteurs, des choses très diverses. La notion de déchet en elle-même n'est pas appréhendée par tous de la même façon. Etymologiquement, un déchet est "ce qui tombe d'une matière que l'on travaille" (Dictionnaire français Larousse du XXème siècle, édition 1929) mais est de plus en plus considéré comme synonyme de "résidus inutilisables, en général sales et encombrants" (Dictionnaire français le Petit Robert, édition 1973). La définition qui nous paraît la meilleure est celle qui a été retenue dans le récent rapport publié par l'Agence pour l'énergie nucléaire de l'OCDE intitulé "Objectif, concept et stratégie en matière de gestion des déchets radioactifs" (OECD/NEA, 1977) et qui est la suivante : "toute matière contenant des radionucléides ou contaminée par des radionucléides en concentration supérieure aux valeurs que les autorités compétentes considéreraient comme admissibles dans des matériaux propres à une utilisation sans contrôle ou au rejet et pour laquelle aucun usage n'est prévu".

Les éléments qui feront que les matières seront des déchets radioactifs seront que ces matières contiennent des radionucléides, que cette quantité de radionucléides excède certaines valeurs et qu'aucun usage n'est prévu. Il est bien évident qu'il y a un certain arbitraire dans la valeur admissible de la radioactivité et qu'en fonction des procédés de récupération et de l'évolution des conditions économiques, ce qui est considéré comme déchets aujourd'hui, pourra, dans un avenir plus ou moins lointain, ne plus être considéré comme déchets.

Des exemples classiques sont constitués par l'uranium appauvri et surtout par les combustibles irradiés. L'uranium appauvri résultant des opérations d'enrichissement du combustible a longtemps été considéré par certains comme déchet du fait qu'on n'en voyait pas d'utilisation. Puis, on a envisagé l'utilisation dans les couvertures de réacteurs surrégénérateurs. Maintenant, on estime que cet uranium pourrait encore servir de matière première, pour récupérer l'uranium 235 encore présent, avec des procédés de séparation isotopique avancés comme le laser. Nous ne considérerons pas cet uranium appauvri comme déchet dans ce rapport.

Les combustibles irradiés n'avaient jamais été considérés comme déchets dans le passé, sauf par certains tenants de la filière des réacteurs à eau lourde. Depuis deux ans le Gouvernement des Etats-Unis d'Amérique a pris une position, liée aux problèmes de prolifération nucléaire, qui l'entraîne à considérer ces combustibles irradiés comme déchets. Bien que, dans l'ensemble de ce rapport, nous ne considérions pas ces combustibles irradiés comme déchets, nous serons amenés, vu l'importance de la question, à envisager les conséquences d'une telle prise de position.

Nous engloberons dans ce rapport les effluents radioactifs. En fait, il s'agit de déchets rejetés dans l'environnement même si la définition que nous avons proposée ne s'applique pas. De plus, les conséquences éventuelles que peuvent entraîner le stockage de déchets radioactifs contenant des émetteurs de longue période sont une migration d'ions radioactifs dans l'environnement, ce qui correspond à un rejet différé d'effluents radioactifs.

Déchets et sous-produits

Quelle que soit la stratégie de développement de l'énergie nucléaire, la production de déchets radioactifs, pour un cycle donné, ne sera pas un absolu et dépendra des techniques utilisées, des procédés mis en oeuvre. Ce qui sera un absolu est la quantité de produits de fission formés, liée par des lois physiques à la quantité produite d'énergie obtenue par fission nucléaire. Ces produits de fission vont être séparés au premier cycle d'extraction des usines de retraitement, on va y retrouver environ 99 % de l'activité formée au cours du fonctionnement du réacteur. Dans le courant de ce rapport, ces produits seront appelés "les produits de fission" ou les "déchets de très haute activité".

Un point important est d'examiner si les produits de fission dont la production est inéluctablement liée à celle de l'énergie nucléaire constitueront obligatoirement ou non des déchets radioactifs. Des études - et même des réalisations - de valorisation de ces produits de fission ont été faites; en particulier, on s'est orienté vers l'utilisation du strontium 90 dans les générateurs isotopiques et de celle du césium 137 comme source d'irradiation au lieu du cobalt. En fait, il faut bien voir qu'en ce qui concerne les déchets, le problème ne serait pas changé. Ce n'est pas parce qu'on utiliserait du strontium pendant des durées égales à une période et même deux périodes radioactives que l'on changerait vraiment la question du stockage à long terme des déchets.

Par contre, les produits de fission donnent, comme tous les éléments radioactifs, à plus ou moins long terme des produits stables dont certains sont intéressants. C'est en particulier le cas pour les métaux nobles, le palladium, le ruthénium et le rhodium qui au bout d'un certain nombre d'années resteront à l'état d'isotopes inactifs et qui sont en quantités non négligeables dans les combustibles irradiés. Des études ont été faites sur l'intérêt de la récupération de ces produits (Sousselier, 1978), néanmoins, dans l'état actuel des choses, on ne peut que considérer que la totalité des produits de fission devra être considérée comme déchets.

La quantité de transuraniens formés est également un absolu pour un fonctionnement donné d'un réacteur. Par contre, la réutilisation de ces transuraniens dépendra beaucoup de la stratégie suivie (recyclage du plutonium dans les réacteurs de la filière à eau ordinaire, développement de la filière surrégénérateurs).

Mais les grandes variations dans la production de déchets viendront en particulier :

- des conditions de fonctionnement des réacteurs et du comportement des combustibles. Les périodes d'intervention en zone active sont génératrices de déchets, de même que le fonctionnement avec des gaines ruptées.

- des schémas de procédés utilisés dans le retraitement et aussi de la fiabilité de ces usines. Si le recyclage, quasi systématique, des effluents liquides est réalisé, les productions de déchets seront très limitées. Enfin et de même que dans le cas des réacteurs, les périodes d'intervention en zone active sont très génératrices de déchets et sont, en particulier, responsables de la production de la presque totalité des déchets dits "technologiques".

En résumé, si pour un réacteur de type donné, les productions de produits de fission et de transuraniens sont des données physiques, leur répartition dans les déchets, leur volume et leurs caractéristiques dépendent des procédés utilisés et du fonctionnement des installations. On peut donc raisonnablement escompter une diminution des quantités produites, ne serait-ce que parce que l'intérêt évident des producteurs est d'améliorer le facteur de charge de leurs installations.

<u>Caractéristiques fondamentales des déchets</u>

Toute étude concernant ces déchets et, en particulier, l'étude des impacts sur l'environnement ne pourra pas être faite globalement mais devra l'être pour les différentes catégories de déchets. De nombreuses classifications de déchets ont été proposées, en général adaptées à un but déterminé. En ce qui concerne les études d'impact, on devra examiner les critères caractéristiques fondamentaux et, en fonction de cela, procéder à une classification valable pour le but que nous poursuivons.

Le premier critère caractéristique des déchets est bien évidemment la radioactivité - ou plus communément l'activité - contenue qui s'exprime en curie. La gamme est très large. Si on prend la valeur inférieure, on peut la fixer à peu à 10^{-2} Ci/t et même pour certaines règlementations à $2x10^{-3}$ Ci/t, ce chiffre est, dans les règlements des transports des matières radioactives, la valeur en dessous de laquelle les matières ne seront plus considérées comme radioactives. En terme de comparaison on peut rappeler qu'un granite, sans qu'il soit question de granite uranifère, a une activité qui varie entre $2x10^{-5}$ à $2x10^{-4}Ci/m^3$. A l'autre extrémité de la gamme, les produits de fission concentrés, séparés dans les usines de retraitement et qui seront vitrifiés ont des activités de l'ordre de 10^3 ou 10^4 Ci/l.

Le second critère caractéristique sera la période radioactive. Certains produits de fission ont des périodes tellement courtes qu'à leur sortie du réacteur, ils n'existent déjà plus que comme isotopes stables, d'autres ont des périodes excessivement longues. Il est bien certain qu'en ce qui concerne les problèmes de long terme, les problèmes de stockage définitifs ou d'évacuation, la période est un critère essentiel.

Un autre critère important est la radiotoxicité. En première analyse, on pourra la mesurer par les concentrations maximales admissibles (C.M.A.) des radionucléides contenus dans les déchets. Ce critère interviendra, en particulier, dans les évaluations de capacité du milieu, c'est-à-dire dans les études destinées à calculer les radioactivités susceptibles d'être rejetées dans l'environnement par une installation nucléaire.

D'autres critères jouent également un rôle, par exemple, la forme physique et chimique. Mais le dernier critère important est le volume, surtout quand il s'agit de volume à stocker et spécialement quand le mode de stockage ou son lieu sont sophistiqués.

En fonction de ces critères, on va pouvoir, avec bien entendu un certain arbitraire, déterminer les catégories essentielles de déchets pour le but qui nous intéresse.

1) Les déchets de haute activité contenant des éléments à période longue. Ce seront les produits de fission tels qu'on les sépare actuellement dans les usines de retraitement et dans lesquels sont concentrés environ 99 % de la radioactivité contenue initialement dans les barreaux. Ce sont ces déchets qui font l'objet des conditionnements les plus élaborés comme la vitrification.

2) Les déchets de faible activité et ne contenant que des éléments à courte période. Cela est à peu près le cas pour tous les déchets provenant de l'exploitation normale des réacteurs de puissance.

3) Les déchets de faible activité contenant des éléments à période longue, c'est-à-dire des émetteurs alpha. Et pour cette raison, les déchets de cette catégorie sont souvent appelés "déchets alpha".

4) Les déchets de haute activité ne contenant que des éléments de courte période. Ce sont essentiellement les déchets du démantèlement.

5) Les stériles venant des opérations d'extraction de minerais et de traitement de minerais. Il s'agit de déchets de faible activité et contenant des émetteurs radioactifs à longue période (^{226}Ra en particulier) mais, compte tenu de leur mode particulier de gestion et de la toxicité élevée du radium, il est préférable d'en faire une catégorie spéciale.

Origine des déchets radioactifs

A parler rigoureusement, l'origine de la radioactivité contenue dans les déchets provient de deux sources :

- la radioactivité de l'uranium naturel et de ses descendants,
- les réactions nucléaires qui se produisent dans les réacteurs : réactions de fission conduisant aux produits de fission, réactions de capture conduisant aux transuraniens, activation sous flux neutronique.

Une première distinction importante est à faire : celle entre les déchets contaminés par de l'uranium naturel et ceux contenant des nucléides provenant des réactions nucléaires. Il est certain que dans l'un et l'autre cas, il s'agit toujours de la même radioactivité. (Du point de vue de la physique, il n'y a pas de radioactivité "naturelle" et de radioactivité "artificielle") Mais la différence se situe quand l'uranium a été séparé de ses descendants, du radium en particulier. C'est le cas pour les déchets provenant de la conversion, de l'enrichissement et de la fabrication de combustibles à l'uranium. On pourrait objecter qu'à terme, le radium se reformera mais d'abord cela ne se produira que dans un temps assez long - le radium se retrouvera en équilibre dans plus de 10 000 ans - et ensuite les déchets correspondants ne seront jamais que des déchets de faible teneur en alpha. Des fûts de 200 litres de déchets provenant des installations du cycle avant le réacteur ne contiennent généralement que quelques dizaines de grammes d'uranium, même s'ils en contenaient 200 g, cela ne correspondrait à terme qu'à un déchet alpha de 0,7mCi/m3 alors que les déchets alpha venant des installations de retraitement ou de fabrication de combustibles au plutonium en contiennent couramment 100 à 1 000 fois plus. L'impact de ces déchets ne peut donc être que faible.

Une deuxième distinction importante est à faire entre les déchets qui contiennent intentionnellement des éléments radioactifs et ceux dans lesquels on en retrouve d'une manière fortuite, c'est-à-dire entre ce que nous avons appelé les produits de fission (séparés au premier cycle d'extraction du retraitement) dans lesquels on essaie de rassembler le maximum de radioactivité et les autres déchets dans lesquels on essaie d'en avoir aussi peu que possible.

Cela étant dit, on peut commodément rassembler les déchets suivant cinq origines :

1) Les produits de fission que les procédés de retraitement essaient de séparer en totalité.

2) Les déchets de procédés. Les différentes opérations de traitement chimique ou de séparation mécanique (telles que les filtrations par exemple) n'ont jamais un rendement qui soit rigoureusement de 100 %. Cette perte (et l'on retrouve la première définition citée plus haut) est souvent appelée "déchets de procédés". Les éléments radioactifs se trouvent mélangés avec un effluent chimique (solvant, solution acide, etc.) ou fixés sur un produit solide (résines échangeuses d'ions par exemple).

3) Les déchets technologiques. Toutes les interventions d'entretien, de réparation faite dans les parties actives d'une usine de traitement chimique ou d'un réacteur vont entraîner une certaine contamination des gants, blouses, plastiques de recouvrement. Les matériaux enlevés de ces installations (joints, pièces mécaniques, etc.) seront le plus souvent contaminés par de la radioactivité déposée lors du fonctionnement de l'installation. S'il n'est

pas possible - ou pas intéressant économiquement de décontaminer - les pièces correspondantes constitueront alors des déchets.

4) Les déchets de fonctionnement du réacteur pourraient être assimilés aux déchets de procédés et aux déchets technologiques. S'il s'agit des concentrats d'évaporation des effluents radioactifs et des résines échangeuses d'ions en fin de leur vie, il s'agit bien de déchets de nature intermédiaire entre les déchets de procédés et les déchets secondaires de traitement. S'il s'agit des filtres actifs, ce sont des déchets technologiques mais constitués par du matériel dont la vie prévue est limitée et le remplacement programmé à l'avance.

5) Les matériaux de démantèlement des installations nucléaires sont de même nature que certains déchets technologiques mais n'apparaissent qu'après le déclassement des installations et par contre, constituent des volumes beaucoup plus importants.

Il est évident que dans deux de ces classes, déchets de procédé et déchets technologiques, on peut avoir comme contaminant de l'uranium naturel et/ou des nucléides provenant de réactions nucléaires. Il est clair aussi que cette seconde classification ne se superpose pas à celle donnée dans le paragraphe précédent, mais les deux sont nécessaires dans l'étude d'impact et l'exposé des résultats.

Schéma de gestion actuel et schéma possible dans le futur

Déchets primaires et déchets secondaires

Le fonctionnement des installations nucléaires entraîne la production de déchets qui peuvent se présenter sous forme gazeuse, liquide (effluents gazeux ou liquides) ou solide.

L'option fondamentale, mentionnée dans l'introduction de ce rapport "rejeter-diluer" ou "concentrer-confiner" se présente alors. L'étude faite pour servir de base à la décision est une étude du type coût-bénéfice basée, en particulier, sur une étude d'impact ou plus précisément sur une étude des conséquences radiologiques des rejets, en suivant le chemin critique des divers radionucléides et en déterminant le groupe critique de population le plus exposé (AEN, 1971). De telles études, pour être entièrement significatives, demanderont de plus en plus à être faites non seulement au niveau local mais aussi au niveau régional et même mondial.

Dans le cas où ces études - ou d'autres considérations - conduisent à la décision de confiner, les déchets feront l'objet de traitements appropriés dans le but de diminuer leur volume (concentrer) et de les mettre sous une forme telle que le confinement soit garanti compte tenu du mode de stockage (confiner). Il y a donc, en général, deux stades de traitement que l'on appelle souvent traitement (évaporation des liquides, filtration des gaz, décontamination des solides) et conditionnement (mise en matrice résistante : vitrification, bitumage, etc.).

Ces opérations peuvent conduire à la production d'autres effluents radioactifs qui devront, à leur tour, être traités et, là aussi, une étude devra être faite de façon à juger quel est le mode optimal de gestion.

Schéma actuel de gestion des déchets

Le développement actuel de l'énergie nucléaire est essentiellement basé sur l'utilisation des réacteurs de la filière à eau ordinaire. Les réacteurs de la filière gaz-graphite à uranium naturel ont constitué une part non négligeable (avec leur prolongement AGR), enfin ceux de la filière eau lourde ont un certain développement. En fait, l'étude de la gestion actuelle des déchets peut être faite en se basant principalement sur l'étude du cycle de la filière à eau ordinaire. Les principales différences sont que pour la filière gaz-graphite, la production de déchets au niveau du réacteur est sensiblement inférieure et que pour toutes les filières à uranium naturel, il n'y a pas d'enrichissement ni de déchets correspondants.

Le point fondamental est le devenir du combustible irradié. Dans ce rapport, nous considérerons comme gestion actuelle, le retraitement et prendrons comme hypothèse que le plutonium récupéré sera utilisé comme combustible (soit dans la filière à neutrons rapides, soit en recyclage dans la filière à eau ordinaire).

Les schémas de gestion de déchets, utilisés ou prévus font appel à des procédés divers mais qui sont suffisamment proches pour que l'on puisse ne pas les considérer tous en détail :

- traitement des effluents liquides de faible activité par évaporation, coprécipitation ou échange d'ions, le résidu radioactif obtenu étant conditionné par bétonnage, bitumage, etc.

- traitement des produits de fission par calcination-vitrification,

- traitement des gaz par filtration, absorption, adsorption, etc.

- traitement des solides par tri-compactage-incinération, les résidus étant bétonnés ou bloqués par des techniques similaires. Une description schématique complète et précise en est donnée dans un article publié dans la Revue de l'Energie (Gauvenet, 1977).

Ces diverses opérations ne se font pas toujours dès que les déchets sont produits. Dans certains cas, on a intérêt à laisser décroître la radioactivité pour faciliter les opérations. Par ailleurs, certains procédés de conditionnement n'ont pas encore atteint le stade industriel (fusion des déchets de gaines de combustibles irradiés) ou viennent seulement de l'atteindre (vitrification), enfin des stockages définitifs n'ont été, à ce jour, réalisés que pour des déchets de faible et moyenne activité.

La gestion des déchets va donc comporter un certain nombre de stockages répondant à des buts différents. Nous distinguerons :

- les stockages opérationnels, destinés à conserver un déchet en attente d'une opération qui le transformera en produits différents (par exemple, stockage de produits de fission liquides avant vitrification).

- les stockages intérimaires où sont conservés les déchets dans leur conditionnement définitif mais pour lesquels on ne considère pas qu'il s'agisse d'un stockage définitif.

- les stockages définitifs (final storage) correspondant à l'évacuation (disposal) d'un déchet. Les déchets y sont mis avec l'intention de les y laisser définitivement. Mais ces stockages pourront comporter une phase pendant laquelle la reprise (retrievability) sera possible.

Schémas possibles de gestion dans le futur

Il est bien connu que l'impact le plus important que serait susceptible d'avoir la gestion des déchets radioactifs sur l'environnement est un impact à long terme, voire à très long terme. Comme on l'a déjà souligné dans l'introduction, la comparaison est très difficile à faire entre des risques potentiels actuels et des risques - potentiels aussi - futurs.

Par ailleurs, le problème de l'action des faibles doses reste sujet de controverse. Bien qu'il semble peu vraisemblable que de très faibles doses puissent avoir une action quelconque sur l'homme et les espèces biologiques, même si ces très faibles doses s'appliquent à un nombre très important d'individus, il ne semble pas que cela ait pu être, jusqu'à présent, scientifiquement démontré.

Il est donc important d'étudier les conséquences d'autres schémas de gestion des déchets, en particulier de ceux diminuant les rejets d'effluents dans l'environnement et de ceux susceptibles de diminuer les impacts à très long terme.

Les premiers correspondent à la philosophie des rejets nuls (zero release). Il est actuellement techniquement possible dans un certain nombre de cas de le réaliser. On connaît par exemple des procédés de piégeage de krypton 85 dans les usines de retraitement, même si le développement industriel reste à compléter. On est un peu moins avancé pour le tritium mais des procédés de récupération existent (Pautrot et Arnault, 1975).

Les conséquences qu'aurait la mise en oeuvre de telles techniques seraient de deux ordres : économiques et risques comparés d'irradiation. Sur le plan économique, il s'agirait de comparer le coût supplémentaire des traitements et stockages correspondants par rapport à l'avantage de la diminution de risques dus aux très faibles irradiations. Mais il faudra aussi comparer ce risque avec celui des irradiations qui résulteraient d'un accident lors des opérations de traitement et de stockage entraînant un relâchement important de nucléides. Il s'agira d'évaluer la probabilité d'un tel accident et d'en estimer les conséquences. Il semblerait peu réaliste, compte tenu de ces comparaisons, d'imposer à l'ensemble des installations nucléaires des rejets nuls. Par contre, ce pourrait être le cas, pour les usines de retraitement, avec les rejets de krypton 85 et, pour les usines de retraitement non situés en bord de mer, pour les rejets de tritium.

Les seconds schémas possibles de gestion des déchets sont ceux qui visent à une solution "finale" du problème des déchets alpha, les deux voies semblant actuellement possibles étant : la transformation des radionucléides à longue période en radionucléides à période courte en moyenne (méthode appelée couramment "transmutation des déchets alpha") et l'évaluation spatiale de ces déchets.

On va se trouver aussi devant une comparaison très délicate : augmentation d'un risque immédiat (irradiation des travailleurs et risque d'accident lors des opérations de séparation, transmutation, évacuation spatiale) et des coûts de gestion en face de la diminution d'un risque à très long terme. Il semble que la décision relèvera plutôt de critère philosophique (est-il raisonnable de prendre en compte le très long terme?).

ORDER NUMBER

PO NYAATP08

BOOK NUMBER

9393052

BAY NUMBER

4035-7

AUTHOR	El-Hinnawi, E.E. (ed)
TITLE	Environmental Sciences and Applications: Nuclear Energy and the Environment (Volume

DESCRIPTION

Volume 11. This is an ex-library book and may have the usual library/used-book markings inside.This book has hardback covers. In poor condition, suitable as a reading copy. No dust jacket.

Production de déchets suivant les modes actuels de gestion

Compte tenu de ce qui a été dit plus haut sur le développement de l'énergie nucléaire, on peut, en première approximation, pour estimer les productions de déchets radioactifs se baser sur le cycle d'un réacteur de la filière à eau ordinaire. Il est commode de prendre un réacteur type de 1 GWe et d'envisager son fonctionnement pendant un an.

Le tableau 1 synthétise la production de déchets associée au fonctionnement de ce réacteur ainsi que de son cycle associé.

TABLEAU 1

PRODUCTION ANNUELLE DE DECHETS ASSOCIEE AU CYCLE DE COMBUSTIBLE D'UN REACTEUR DE 1 GWe DE LA FILIERE EAU ORDINAIRE (1)

STADE DU CYCLE	NATURE DES DECHETS	VALEURS m3/an	ACTIVITE Ci/an (3)
Concentration	Stériles de minerais	80 000 t(2)	60
Conversion, Enrichissement	Essentiellement déchets technologiques	40	0,3
Fabrication combustibles uranium	- idem -	50	négligeable
Réacteurs	Filtres, résines usées	10	1 000
	Concentrats d'évaporation	240	100
	Déchets technologiques	250	10
Retraitement	Produits de fission vitrifiés	3	150×10^6
	Déchets de gaine	15	2×10^6
	Déchets de procédé	30	2 500
	Déchets technologiques	100	1 000
Fabrication combustible plutonium	Déchets de procédé	2	
	Déchets technologiques	25	

(1) Production hors effluents gazeux - Les déchets sont considérés conditionnés suivant les techniques actuellement en vigueur (béton, bitume, etc.) à l'exception des déchets de gaines pour lesquels les chiffres correspondent aux déchets non compactés, non conditionnés.

(2) Tonnes pour stériles, le reste en m3.

(3) Curies β γ pour les lignes réacteurs et retraitement,
Curies radium pour les stériles de minerais,
Curie uranium pour les déchets conversion et fabrication de combustibles.

On peut apporter quelques précisions à ce tableau :

- le chiffre concernant les stériles de minerais est basé sur une teneur moyenne de minerais de 2 ‰. L'activité indiquée est celle du Ra-226.

- les chiffres concernant l'enrichissement et la conversion sont basés sur une teneur de soutirage pauvre de 0,275 %, ce qui correspond à une consommation annuelle de 160 t d'uranium naturel pour 30 t/a d'uranium enrichi contenu dans les combustibles. (Ces deux valeurs sont des valeurs intermédiaires à celles du BWR et PWR : nous considérons comme réacteur LWR type, celui qui résulterait d'une pondération moyenne des caractéristiques de ces deux types en considérant que le développement se fera alors en proportion de 2 PWR pour 1 BWR).

- les déchets venant de la conversion, de l'enrichissement et de la fabrication des éléments combustibles au plutonium ne contiennent comme nucléide radioactif que de l'uranium séparé de ses descendants.

- les activités indiquées pour les déchets provenant du réacteur et du retraitement sont des activités β γ. En fait, l'activité α des déchets du réacteur est nulle ou très faible. L'activité alpha pour les déchets de retraitement n'a pas, en elle-même une grande signification : car les déchets contiennent du plutonium dont la composition isotopique varie avec l'irradiation du combustible et le plutonium 241 n'est qu'un émetteur β mais il donne naissance à de l'américium 241 qui, lui est émetteur α. Cependant, on peut prendre, en première approximation et pour fixer les idées, <u>pour les activités alpha</u> :

  - verres .... 600 Ci/m3 (contient l'ensemble des Am et Cm)
  - déchets de gaine 4 Ci/m3
  - déchets de procédé 1-10 Ci/m3
  - déchets technologiques $<$ 100 m Ci/m3
  - les chiffres indiqués pour la fabrication de combustibles au plutonium sont ceux résultant de l'utilisation du plutonium récupéré dans le traitement des combustibles irradiés du réacteur considéré (30 t/an à 10 kg/t soit 300 kg de Pu total) quelle que soit l'utilisation des combustibles ainsi obtenus (recyclage dans les réacteurs à eau ordinaire ou envoi dans les réacteurs à neutrons rapides).

Nous n'avons pas fait figurer les références d'où sont tirés ces chiffres car nous avons préféré prendre les dernières données disponibles dans chaque domaine. C'est ainsi que pour les réacteurs nous nous sommes basés sur le guide de l'AIEA, en cours de publication, sur la gestion des déchets dans les centrales de puissance (AIEA, in press).

Pour les déchets de retraitement et de fabrication de combustibles au plutonium, nous avons utilisé les chiffres que prend le CEA et qui sont tirés des projets de ses nouvelles installations de retraitement ou de fabrication de combustibles. (pas de publication à ce sujet). Les autres chiffres sont tirés du rapport ERDA 76-43 (1976).

On peut souligner que d'autres données sont assez sensiblement différentes. C'est le cas pour l'étude publiée par CULLER (1977). Cette étude estime à 2 400 m3/a les déchets de faible et moyenne activité non alpha (c'est-à-dire les déchets des réacteurs et des installations situées avant le réacteur) alors que notre étude évalue cet ensemble à 600 m3/a. On peut supposer deux raisons à cette divergence : la première est que l'étude CULLER peut considérer les déchets "bruts de production" sans utilisation de techniques de réduction de volume (compactage, incinération, etc.) qui sont de plus en plus utilisées pour ces genres de déchets et qui conduisent, dès maintenant, à des facteurs de réduction de volume de l'ordre de 2. La seconde est que cette étude est probablement basée sur les résultats obtenus dans les premières années d'exploitation des réacteurs; or, on sait que plusieurs maladies de jeunesse ont affecté les réacteurs de la filière à eau ordinaire mais depuis deux ou trois ans, les facteurs de charge des réacteurs nouvellement mis en service ont sensiblement supérieurs. Des chiffres de 1 000 m3/a de déchets par réacteurs sont certainement été atteints à de nombreuses reprises : ils correspondent aussi à un facteur 2 sur les chiffres retenus par le groupe d'experts de l'AIEA qu'il semble plus raisonnable de suivre.

En ce qui concerne les déchets issus du retraitement, les chiffres donnés dans l'étude CULLER sont en très bonne concordance pour les verres et les déchets de gaine (pour les verres 3,1 m3/GWe x an et pour les gaines considérées compactées 2,7 m3/GWe x an) ce qui est normal puisqu'il s'agit, comme on l'a souligné précédemment, de sous-produits. (déchets fatals).

Par ailleurs, il faut bien remarquer que, dans une étude générale d'impact sur l'environnement, un facteur 4 n'est pas réellement significatif puisque les impacts sont, en fait, très faibles.

Enfin, on peut souligner que le groupe de travail No 7 de l'INFCE, consacré aux déchets radioactifs, a retenu comme hypothèses de travail des valeurs très proches de celles proposées dans cette étude (80 m3/GWe xa pour les déchets des usines avant réacteur, 502 m3/GWe xa pour les déchets de réacteurs). Les différences les plus notables concernent les déchets de fabrication des combustibles au plutonium : 56 m3 au lieu de 27 m3; ce qui correspond peut-être à des résultats obtenus mais ne paraissent pas correspondre à ce qu'on est, dès maintenant, capable de faire.

Incidences des différents paramètres

Les chiffres que l'on vient de donner correspondent donc à des acquits actuels même si dans certains cas, on peut estimer que ce qui a été effectivement obtenu jusqu'à ce jour correspond à des valeurs un peu plus élevées. Mais le développement de l'énergie nucléaire en est encore à une phase de pleine croissance. Même si au cours des dernières années la plupart des programmes ont été révisés en baisse, les prévisions de mise en service de nouvelles centrales au cours des 5 à 10 années qui viennent - et qui correspondent à des programmes engagés - correspondent à une forte augmentation des programmes installés et on peut raisonnablement prévoir que toutes ces centrales fonctionneront au moins pendant une trentaine d'années. On doit donc autant considérer quelle sera l'évolution probable de la production de déchets au cours des décennies à venir.

Il a été souligné plus haut que la production de tous les déchets n'est pas inéluctable. En particulier, celle des déchets liés au fonctionnement des réacteurs dépend dans une large mesure de la tenue des gaines des éléments combustibles et de la corrosion des divers équipements du circuit primaire. La contamination par les produits de fission du circuit primaire se produit pratiquement uniquement par les ruptures de gaine (la seule exception notable étant la diffusion du tritium à travers les gaines dans le cas où celles-ci sont en acier inoxydable). En fait, le fonctionnement du réacteur n'est pas affecté par des ruptures de gaine et cela a probablement eu comme conséquence que les constructeurs de réacteurs, dans leurs calculs d'optimisation économique, ont visé à l'épaisseur minimale de gaine (et l'enrichissement minimal du combustible). La prise en compte de l'ensemble des conséquences des ruptures de gaine : coût du traitement des effluents, coût du transport et du stockage des déchets (celui du stockage étant susceptible d'augmentation avec l'accroissement des exigences pour ce stockage), contamination des châteaux de transport de combustibles irradiés et des piscines de réception des usines de retraitement, conduira peut-être à une optimisation différente, c'est-à-dire se traduira peut-être par la fabrication de gaines plus résistantes si du moins les gaines actuelles ne s'amélioraient pas. Mais cela semble déjà être le cas et le taux de rupture de gaines dans les réacteurs semble diminuer.

La seconde source de contamination du circuit primaire est constituée par les produits de corrosion qui s'activent sous rayonnement neutronique lors de leur passage dans le coeur. Mais ces produits peuvent se redéposer en divers endroits du circuit primaire, en pouvant entraîner aussi d'ailleurs les produits de fission ayant migré dans le circuit primaire par les ruptures de gaine. Cela sera une des causes importantes de la contamination de certains composants (échangeurs par exemple) et donc de l'irradiation du personnel lors des interventions. En vue de diminuer ces irradiations et aussi de réduire la durée des périodes d'entretien, on sera également incité à diminuer les corrosions en choisissant des matériaux appropriés.

En conclusion, il est vraisemblable que la production annuelle de déchets due au fonctionnement des réacteurs ira en diminuant, mais il est difficile de prévoir l'ampleur de cette diminution.

Des raisonnements analogues peuvent être faits pour la production de déchets des usines de retraitement : par exemple, celles-ci utiliseront certainement de plus en plus, dans l'avenir les recyclages systématiques des raffinats des deuxième et troisième cycles d'extraction.

Une autre cause qui devrait entraîner dans l'avenir, à court terme, la diminution de la production de déchets sera une utilisation plus systématique des techniques de réduction de volume et, en particulier, l'incinération des déchets combustibles. Cette technique a été relativement peu utilisée jusqu'à maintenant en fonction de son coût d'une part et des difficultés technologiques de mise au point. Mais le coût des stockages, les exigences plus contraignantes pour les conditions de stockage et les progrès technologiques déjà faits vont certainement entraîner une utilisation plus systématique de cette technique.

De nombreuses autres méthodes conduisant à des diminutions de volume des déchets sont actuellement étudiées et à des degrés de développement divers. On peut citer les méthodes de fusion, de décontamination des métaux, de tri, etc.

On peut donc, là aussi, conclure que les valeurs de production de déchets mentionnées dans le tableau 1 devraient aller en diminuant d'ici à quelques années, sans toutefois que l'on puisse chiffrer, dès à présent, cette diminution.

Evaluation de la production globale de déchets

Toutes les études de production globale de déchets se basent d'une part, sur des estimations de production unitaire rapportée en général au fonctionnement d'un réacteur de puissance donnée pendant un an et d'autre part, sur une estimation de développement de la puissance électronucléaire.

Peu d'études d'ensemble de la production globale de déchets radioactifs au niveau mondial ont été faites. De nombreuses études ont été faites au niveau américain, on peut citer, en particulier, le rapport 4451 d'Oak Ridge publié en 1970, la communication BLOMEKE (1976). Ces études sont évidemment intéressantes d'autant plus que l'on estime que la puissance électronucléaire installée aux Etats-Unis représentera environ 40 % de la puissance mondiale, tout au moins au cours des prochaines décennies. Au niveau mondial, une étude récente a été présentée à la conférence internationale de Salzbourg en 1977 par CULLER.

Compte tenu de ce qui a été expliqué, il semble plus adapté de prendre d'autres chiffres que ceux de CULLER et il est normal de prendre ceux retenus dans ce paragraphe. Comme nous recherchons une évaluation d'ensemble dans le but d'une étude d'impact, ce qui est important ce sont les catégories de déchets. Pour l'évaluation de la production globale, nous regrouperons les déchets en cinq catégories en considérant ensemble les déchets de faible et moyenne activité sans émetteur alpha (déchets des réacteurs et des installations avant le réacteur) et les déchets alpha de faible et moyenne activité β γ (déchets du retraitement, autres que produits de fission vitrifiés et déchets de gaine, et déchets de la fabrication des combustibles au plutonium).

En arrondissant les chiffres, ce qui est raisonnable compte tenu des incertitudes, nous nous baserons sur les chiffres suivants :

- stériles de minerais .... 80 000 t/ an x GWe
- déchets FA-MA non alpha .... 600 m3/an x GWe
- déchets alpha .... 150 m3/an x GWe
- déchets de gaine .... 15 m3/an x GWe
- P.F. vitrifiés .... 3 m3/an x GWe

Plus que les prévisions de production unitaire de déchets, les prévisions de puissance électronucléaire diffèrent entre elles. La figure 1 groupe trois évaluations récentes auxquelles nous avons ajouté celle figurant dans l'étude du N.R.P.B. britannique (UK National Radiation Protection Board, 1975), étude qui sera largement utilisée dans ce rapport. Les trois évaluations sont celles figurant dans la communication CULLER, celle faite par l'AIEA en 1977 et enfin celle figurant dans le rapport connu sous le nom de rapport WAES (1977). Les deux dernières prévoient une hypothèse haute et une hypothèse basse.

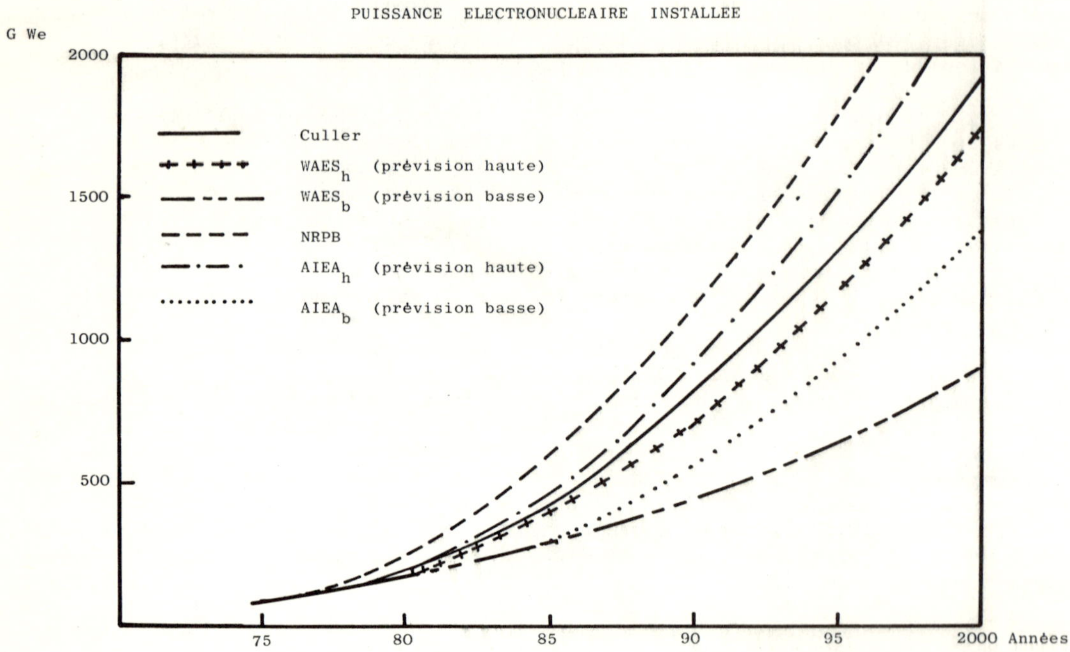

Fig. 1 Puissance Electronucleaire Installee

Si l'on se rappelle qu'au 31 décembre 1977, il y avait dans le monde une puissance électronucléaire totale correspondant à des réacteurs en fonctionnement de 109 GWe (dont 49,7 pour les Etats-Unis) et qu'à la même date, il y avait 212 GWe en construction dans le monde, chiffre auquel s'ajoutaient 153 GWe en commande, on peut juger les prévisions basses pour l'étude WAES et l'étude AIEA assez faibles. Par contre, l'étude NRPB correspond certainement à des chiffres trop élevés. Compte tenu de l'objet de cette étude, il semble préférable d'avoir des prévisions majorantes mais dans des proportions raisonnables et nous adopterons les chiffres de CULLER.

Pour l'évaluation des quantités cumulées, il est nécessaire de tenir compte des décalages qui existent entre les différentes opérations du cycle. Nous admettrons un décalage d'un an pour les opérations minières, c'est-à-dire que nous prendons en compte pour estimer la production de déchets à l'année n une puissance installée égale à celle de l'année n+1. Pour les autres déchets, le décalage est dans l'autre sens, un an pour les déchets de réacteurs (la première année correspond en fait à la montée en puissance, aux essais, etc.) quatre ans pour les déchets de retraitement et cinq ans pour la vitrification.

Les calculs ont été faits pour la période 1980-2000 et les résultats figurent dans le tableau 2.

TABLEAU 2

PRODUCTION CUMULEE DE DECHETS RADIOACTIFS PRODUITS DANS LE MONDE DE 1980 à 2000

Stériles de minerais	1 500 x 10^6 t
Déchets de faible et moyenne activité $\beta\gamma$ sans émetteur α	9 900 x 10^3 m3
Déchets α de faible et moyenne activité $\beta\gamma$	1 720 x 10^3 m3
Déchets de gaines	170 x 10^3 m3
Produits de fission vitrifiés	33 x 10^3 m3

Il est à souligner que n'y figure pas la production de déchets correspondant à la période avant 1980, ce qui est relativement limité compte tenu du développement de l'énergie nucléaire si l'on exclut les déchets provenant des programmes militaires. Il faut, bien entendu, considérer tous ces chiffres comme des ordres de grandeur.

Ces chiffres diffèrent bien entendu de ceux figurant dans les études citées mais les ordres de grandeurs demeurent les mêmes et nous pouvons les considérer comme des valeurs probables et des estimations suffisantes pour l'étude d'impact.

Conséquences d'autres schémas de gestion sur la production de déchets

Autres schémas du cycle

Le principal schéma à considérer est celui comportant le non retraitement. En effet, les différents schémas possibles, dans l'hypothèse générale considérée - et correspondant actuellement aux faits - d'un développement de l'énergie nucléaire basée essentiellement sur la filière des réacteurs à eau ordinaire sont :

- retraitement et recyclage seul de l'uranium,
- retraitement et recyclage de l'uranium et du plutonium,
- retraitement et utilisation du plutonium dans les surrégénérateurs,
- non retraitement.

En fait, les trois premiers schémas comportent tous le retraitement des combustibles irradiés et ne diffèrent que peu entre eux vis-à-vis de la production de déchets. Par contre, l'option du non retraitement change de manière très importante la production de déchets.

En effet, il n'y aurait plus, dans ce cas, de déchets de retraitement ni de fabrication de combustibles au plutonium. Par contre, les éléments combustibles eux-mêmes seraient des déchets et les quantités d'uranium nécessaire au fonctionnement des réacteurs étant supérieures, les quantités de stériles de minerais augmenteraient. Ce serait également le cas pour les déchets de l'enrichissement et ceux de la fabrication des combustibles à l'uranium, mais, comme on l'a vu plus haut, ces déchets sont de faible signification au point de vue puissance radiologique, on peut ne pas prendre en compte ces augmentations.

Le groupe de travail No 7 de l'INFCE a évalué l'incidence sur le tonnage de minerais. La quantité d'uranium naturel nécessaire au fonctionnement d'un réacteur de 1 GWe est de 153 t/a sans recyclage. On retrouve un chiffre équivalent à celui indiqué dans un paragraphe précédent où l'on considérait le plutonium utilisé dans les surrégénérateurs mais ce qui change entre les deux est le nombre de réacteurs de chaque type. Ce chiffre tombe à 85 t/a avec le recyclage et à 2 t/a pour les surrégénérateurs.

Les chiffres correspondant pour les quantités de stériles de minerais sont estimés, dans l'étude INFCE, à 80 000 t/a sans recyclage, à 45 000 t/a avec recyclage et à 1 000 t/a dans le cas de l'utilisation de surrégénérateurs.

Autres schémas de gestion des déchets

Les schémas de gestion étudiés et qu'il serait éventuellement possible d'appliquer dans le futur correspondraient, eux aussi, à des productions différentes de déchets. Ce serait le cas si l'on utilisait le piégeage total - ou au moins le piégeage aux usines de retraitement - de gaz radioactifs. Il n'y aurait bien entendu, dans ce cas, pas de "production" nouvelle de déchets mais les radionucléides en question, au lieu d'être rejetés dans l'environnement, seraient piégés, conditionnés et les produits obtenus devraient être stockés comme déchets.

Les quantités correspondantes dépendent évidemment du mode de conditionnement. Le mode le plus simple de stockage du krypton 85 est la mise en bouteilles de gaz comprimé. Bendixsen (1976) évalue de 25 à 75 bouteilles de 50 l. de modèle classique le nombre nécessaire pour stocker le krypton 85 provenant des combustibles retraités par une usine de 1 500 t/a, ce qui correspond à une activité de 10 MCi de Kr-85. La production cumulée dans le monde est évaluée à 12 000 MCi en l'an 2020, ce qui correspondrait à 60 000 bouteilles, en prenant le chiffre moyen de Bendixsen.

Les quantités d'iode à stocker dans le cas de piégeage total ne seraient pas non plus considérables. Thomas (1976) évalue à 330 kg/an la quantité d'iode 129 correspondant au retraitement de 1 500 t/a de combustibles irradiés. Il faut ajouter à ce chiffre 80 kg/an d'iode 127 stable, fin de chaîne de fission, et la quantité d'iode ajoutée avec les réactifs chimiques dans laquelle cet iode se retrouve comme impureté. Ce même auteur l'évalue à 175 kg/an. On aurait donc au total de l'ordre de 600 kg/an d'iode à stocker pour une usine de 1 500 kg/an qui correspond au retraitement des combustibles des réacteurs ayant une puissance installée de 50 GWe. En l'an 2000 et en tenant compte du décalage de trois ans entre puissance installée et retraitement, il y aurait l'équivalent de 1 600 GWe à retraiter soit 20 t d'iode à stocker. Même en admettant que cet iode soit à l'état d'iodure de plomb mis dans une matrice appropriée, les quantités correspondantes resteront minimes.

L'adoption du mode de gestion des déchets prenant en compte la séparation des déchets alpha et leur "transmutation" se traduirait aussi par des changements dans le chiffre total de déchets à stocker. Théoriquement, on devrait en avoir moins puisqu'il n'y aurait plus de déchets alpha. En fait, la situation sera un peu différente. La récupération des alpha devrait se faire à deux stades :

- récupération de l'américium et du curium dans les produits de fission,
- récupération du plutonium (et éventuellement de l'américium formé avant cette opération) dans les déchets alpha.

La première opération se fera par des opérations telles qu'extraction par solvant ou précipitation chimique, il est vraisemblable que ces opérations conduiront à la production de déchets très légèrement contaminés en alpha et contenant des émetteurs $\beta\gamma$, déchets qu'il ne sera pas forcément possible de recycler dans les produits de fission. Quant à la seconde, elle produira des déchets non alpha (ou éventuellement très peu contaminés en alpha) mais dont le volume sera probablement supérieur au volume initial des déchets. Enfin, les opérations liées à la "transmutation" (fabrication d'éléments combustibles correspondants) produiront aussi des déchets dont certains (déchets technologiques par exemple) ne pourront peut-être pas être recyclés. Il est donc vraisemblable, en définitive, que ce mode de gestion se traduira par une augmentation du volume total des déchets, augmentation difficile à chiffrer mais il serait peu vraisemblable qu'elle corresponde à un doublement du volume des déchets initiaux.

IMPACTS SUR L'ENVIRONNEMENT DES MODES ACTUELS DE GESTION DES DECHETS RADIOACTIFS

Différents impacts possibles de la gestion des déchets

Les différents stades de la gestion des déchets peuvent conduire à des impacts sur l'environnement, ce qui est le cas, comme il a déjà été souligné dans l'introduction de ce rapport, pour toute activité industrielle. Néanmoins, c'est essentiellement :

- par l'intermédiaire des rejets d'effluents radioactifs ou chimiques, lors de l'ensemble des opérations liées au cycle de combustibles,
- lors des stockages intérimaires ou définitifs,
- que les impacts pourront avoir lieu.

On peut remarquer que les impacts possibles peuvent être :

- immédiats,
- différés.

Les impacts différés pourront provenir de ce que l'action n'est pas suivie d'effets immédiats (par suite de phénomènes de reconcentration lente par exemple) ou parce que l'action ne se produira que dans un futur plus ou moins lointain (relâchement d'activité à partir d'un stockage).

On peut schématiquement regrouper en quatre classes les impacts :

Modification du milieu

Il s'agira essentiellement de la pollution radioactive (ou chimique) de l'air dans le cas de rejets atmosphériques d'effluents ou de l'eau dans le cas de rejets d'effluents dans les réseaux hydrographiques et aussi dans le cas de migration de radioactivité dans ces réseaux à partir des stockages. Les stockages et surtout les stockages définitifs ou les stockages intérimaires de longue durée pourront intervenir également en modifiant le réseau hydrographique, par exemple à la suite de travaux de drainage destinés à éviter que les eaux naturelles ne percolent à travers un stockage.

Utilisation des sols

Cet impact ne sera sensible que dans le cas des stockages définitifs ou intérimaires de longue durée et sera surtout important pour les stockages en surface ou à très faible profondeur (shallow burial disposal). Il devra être estimé en tenant compte de l'intérêt des surfaces stérilisées à court terme aussi bien qu'à long terme. On ne devra pas négliger ce facteur dans l'étude des stockages dans les mines, tunnels, ni même dans le cas de l'immersion en mer.

Incidence sur les ressources naturelles ou matières premières

On devra essayer de tenir compte, en étudiant ce point, de l'utilisation potentielle par les générations futures de matières premières dont l'utilisation n'est actuellement ni pratiquée, ni envisagée.

Contraintes socio-économiques

Elles peuvent être de nature diverse en englobant aussi bien :

- les nécessités de contrôle et d'entretien de stockage par les générations futures,
- l'effet psychosociologique de l'implantation d'installation de traitement ou de stockages de déchets qui peut conduire à des conséquences négatives (diminution de valeur de terrain, baisse du tourisme) ou positives (création d'emplois pour un site implanté dans une région peu peuplée et pas industrialisée).

Impact des rejets d'effluents

Mode d'action des rejets

Les rejets d'effluents radioactifs se traduisent bien évidemment par une pollution de l'environnement. Il peut s'agir d'une pollution à caractère global quand il y a diffusion du ou des radionucléides concernés dans le milieu (cas, par exemple, du rejet de gaz rares dans l'atmosphère ou de tritium dans le milieu marin) ou à caractère localisé quand il y a rétention des radionucléides dans une partie du récepteur (adsorption par les sédiments d'un fleuve, par exemple) voire reconcentration dans telle ou telle espèce biologique (concentration de l'iode dans le lait).

Tous ces phénomènes ont fait l'objet de nombreuses études de radio-écologie depuis un certain nombre d'années. Il est actuellement possible de suivre le devenir préférentiel du radionucléide dans le milieu (chemin critique) et d'estimer les conséquences sur le groupe de population le plus exposé (groupe critique). Les comptes rendus des symposiums consacrés à la question permettent de voir l'étendue des études faites. La publication 26 de l'ICRP explique de manière détaillée ces notions, de même que le rapport des experts de l'OCDE déjà cité.

On peut raisonnablement conclure :

- que les connaissances en radio-écologie sont extrêmement étendues. On connaît bien les facteurs de concentration des différents radionucléides dans les différents milieux. Il semble cependant (IAEA, 1975) que quelques lacunes subsistent encore en ce qui concerne certains transplutoniens.

- que certaines données géophysiques ou géohydrologiques sont moins connues. C'est semble-t-il le cas pour certains coefficients de diffusion dans les océans.
- qu'il n'y a pas à prendre en compte dans ces études l'action des rayonnements sur les autres espèces vivantes que l'homme. En effet, ce qui compte pour ces autres espèces est l'équilibre global et non les dommages possibles sur quelques individus. Le rapport Flowers (1976) concluait que les radiations peuvent jouer sur la fertilité, mais qu'il ne pourrait s'agir que d'une cause mineure parmi toutes les causes possibles. On devrait cependant examiner ce facteur si, dans le long de la chaîne, figure une espèce dont l'équilibre est très fragile.

Il y a, bien entendu, lieu dans ces études de prendre en compte l'ensemble des sources de rejets et d'examiner les impacts globaux, au plan local, au plan régional et au plan mondial.

Une importante question qui se pose est de savoir quelle est la durée à prendre en compte dans le calcul de l'intégration des effets des rayonnements. On calcule, en général, les effets en les exprimant en engagement de dose collective exprimé en hommes-rems. Si l'on fait le calcul mathématique des engagements de dose, on aboutira pour les radionucléides ayant des périodes radioactives longues comme,par exemple, le carbone 14 à des résultats élevés même pour des rejets très faibles. Il est cependant très probable que le même nombre d'hommes-rems délivré sur quelques décennies (ce qui est le cas pour les rejets de Kr-85) ou sur quelques millénaires (cas du C-14) n'ont pas les mêmes conséquences.

Il ne semble pas que la C.I.P.R. se soit clairement prononcée sur ce point. Plusieurs scientifiques ont recommandé que pour ces calculs, on ne prenne en compte qu'une période de 500 ans. De toute façon, cela n'aurait certainement pas de sens de prendre en compte des durées de l'ordre de la dizaine et surtout de la centaine de millénaires, ne serait-ce que parce que l'on ne peut faire aucune prévision sur la population du globe et sa répartition, pas plus que sur les climats, l'évolution de la faune et de la flore sur de telles périodes.

Enfin, une dernière inconnue dans ces études est la durée pendant laquelle l'humanité aura recours à l'énergie nucléaire de fission. Si l'on se rappelle qu'il y a cinquante ans, on ne connaissait pas le phénomène de fission et à peine le neutron, on peut à juste titre se demander si d'ici un siècle ou deux, des sources d'énergie nouvelles, peut-être inconnues actuellement dans leur principe, n'auront pas complètement supplanté l'énergie nucléaire. Néanmoins, il est probablement prudent, là aussi, de compter sur une durée d'utilisation de plusieurs siècles et le Comité des experts scientifiques de l'ONU a pris en compte une durée d'utilisation de 500 ans.

Impact des rejets pour les radionucléides à période longue ou moyenne

Des études globales prennent en compte la totalité des rejets, mais compte tenu de leur importance, des études particulières ont été consacrées pour quatre radionucléides figurant - ou susceptibles de figurer dans les effluents - et dont les périodes sont au minimum de l'ordre de la décennie. Ce sont le krypton 85, le tritium, le carbone 14 et l'iode 129.

Krypton 85 : ce radionucléide a constitué le principal composant des rejets gazeux des usines de retraitement de combustibles irradiés depuis le début du fonctionnement de ces usines. En effet, aucune d'entre elle n'a été jusqu'à présent dotée de systèmes de rétention pour cet isotope, la seule voie de rejet a toujours été la cheminée et les teneurs des combustibles irradiés n'ont jamais été négligeables pour ce radionucléide. De nombreuses études ont été consacrées à l'accroissement des teneurs en krypton 85 de l'atmosphère (ORNL, 1970). L'étude la plus récente parue concernant l'exposition aux rayonnements des populations est celle publiée en 1975 par le National Radiological Protection Board du Royaume-Uni pour le compte de la Commission des Communautés Européennes.

Le krypton est un gaz rare et, dans les conditions dans lesquelles il est susceptible de se trouver, aucun processus de fixation dans les espèces biologiques n'existe (on a cependant mentionné une certaine accumulation dans les graisses animales et aussi humaines (Kirk et al.,1976), mais ce processus paraît extrêmement limité et non susceptible en aucune façon de changer les conclusions indiquées). L'impact essentiel à étudier est l'irradiation potentielle des populations et l'on doit distinguer l'impact local ou régional dû au rayonnement entraîné par le passage du nuage rejeté et l'impact mondial dans lequel il faudra tenir compte du fait que le krypton aura diffusé d'une façon homogène dans l'atmosphère.

La figure 2 tirée du rapport cité du NRPB donne l'impact local et régional des rejets de krypton 85 d'une usine de retraitement de 1 500 t/an considérée comme usine type. On y voit clairement que l'impact local ne saurait être que très faible : la dose aux gonades à 1 km du point de rejet n'excédera pas 1 mrad/an en l'an 1995 et à 10 km du point de rejet, elle restera en l'an 2000 inférieure à 0,1 mrad/an. Les doses à la peau en l'an 1995 n'excéderont pas 60 mrad/an (organe pour lequel la limite annuelle recommandée par la CIPR est de 3 000 mrad/an pour les populations).

Le calcul de l'impact au niveau mondial est dépendant des hypothèses de développement de l'énergie nucléaire. Il doit par ailleurs tenir compte que, jusqu'à l'an 2000 du moins, toutes les usines de retraitement seront situées entre les 35° et 60° parallèles Nord et que les diffusions sont plus lentes entre parallèles que le long de celles-ci. Le rapport NRPB prenant en compte des hypothèses de développement de l'énergie nucléaire considérées comme les plus vraisemblables en 1975 (figure 1) mais que l'on considère généralement actuellement comme trop optimistes en déduit les valeurs d'irradiation pour l'ensemble de la population mondiale reportée sur la figure 3. On constate que, même avec des hypothèses probablement donc trop élevées de développement de l'énergie nucléaire, la dose aux gonades ne dépassera guère en l'an 2000 le centième de mrad/an et la dose à la peau sera de l'ordre du mrad/an : là aussi l'impact peut être considéré comme négligeable.

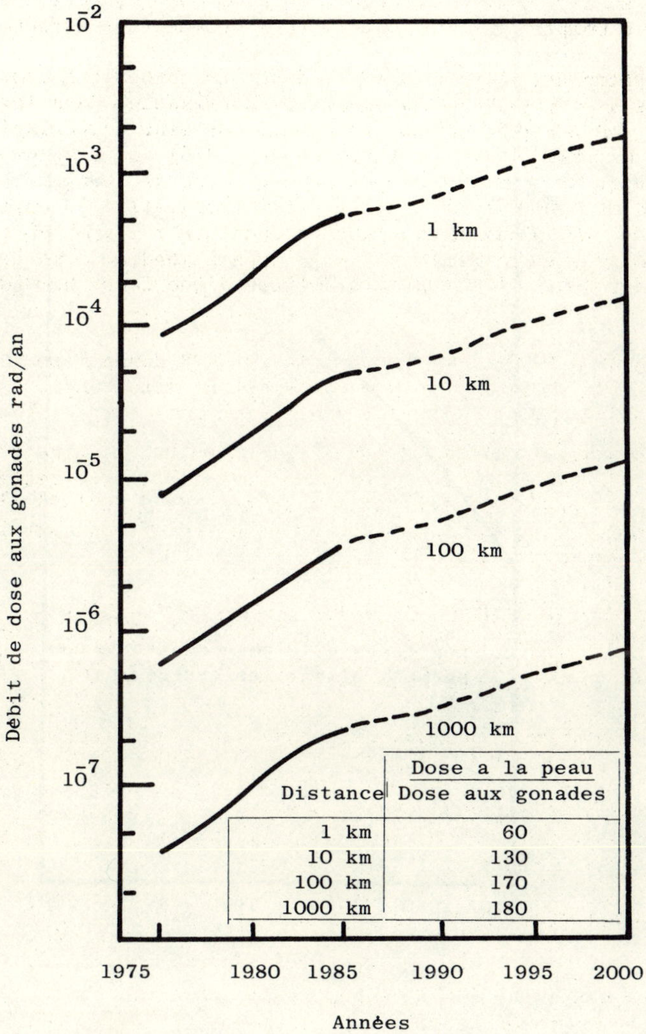

Fig. 2 Débit de dose, en fonction de la distance d'une installation de retraitement de référence, réultant du rejet de Kr-85 par une cheminée de 100 m de haut

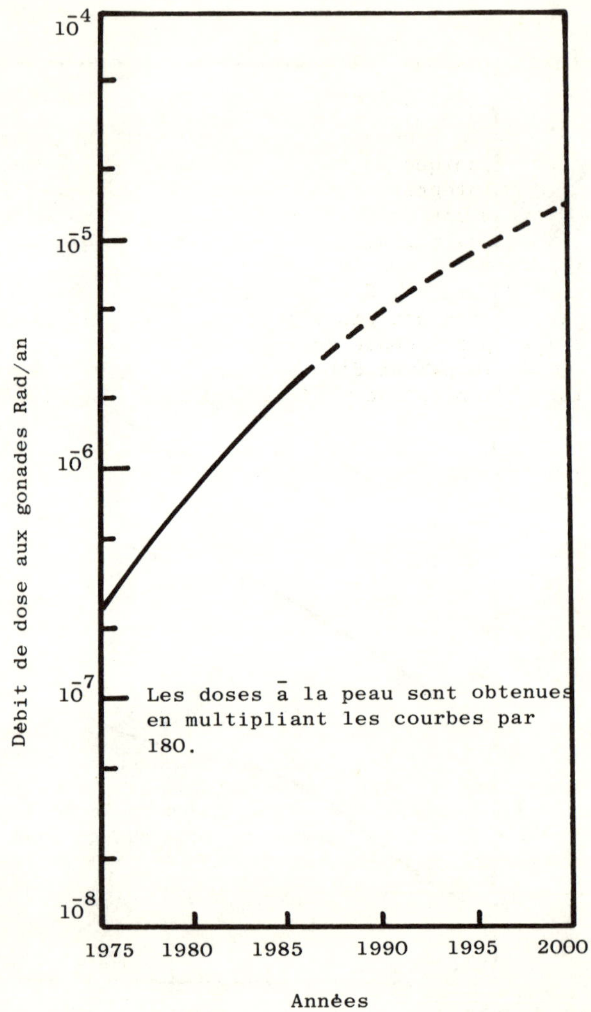

Fig. 3 Débit de dose aux gonades, entre les parallèles 35-60°N, résultant de la circulation globale de Kr-85 rejeté, compte tenu du programme de puissance nucléaire mondial.

On doit cependant considérer deux autres mécanismes possibles d'impact sur l'environnement pour ces rejets de krypton. Le premier est l'incidence sur le taux d'ionisation de l'atmosphère. L'ionisation se produisant naturellement dans l'air au niveau du sol ou à faible altitude a trois origines principales :

- l'ionisation tellurique
- l'ionisation due aux rayons cosmiques
- l'ionisation due au radon 222

Au-dessus des mers et compte tenu de la courte période de radon 222, les rayons cosmiques constitueront l'unique source de l'ionisation naturelle. L'énergie moyenne associée à la désintégration du krypton 85 produira 285 paires d'ions par m3 et par seconde pour une concentration de 1 pico curie de krypton 85 par m3. Boeck (1974) estime que vers 2020 il y aura 12 000 MCi de Kr 85 accumulés dans l'atmosphère et que la concentration à faible altitude sera de 3 nCi/m3 (ce qui représenterait 1 % de la CMA), ce qui correspondrait à la formation de 0,85 paire d'ion par cm3 et par seconde, pour une formation de deux paires d'ions par cm3 et par seconde due aux processus naturels au-dessus des mers et une formation de 10 paires d'ions au-dessus des continents. L'augmentation du taux d'ionisation ne sera donc pas négligeable, surtout au-dessus des océans où elle sera de l'ordre de 20 %. Il est cependant douteux que ces variations puissent se traduire par des conséquences mesurables, par exemple sur le plan météorologique. D'ailleurs même si cela était, il est possible que l'impact soit bénéfique.

Le second mécanisme possible d'impact résulte de l'augmentation de la teneur en krypton 85 du krypton naturel. On a estimé que cette teneur passerait à 1 μCi/g en l'an 2000. Il pourrait en résulter certaines conséquences au niveau des usines de liquéfaction de l'air produisant du krypton pour l'industrie des ampoules électriques. En effet, à la concentration mentionnée, le krypton obtenu serait considéré comme produit radioactif. En réalité, compte tenu d'une part des caractéristiques nucléaires (émission β d'une énergie maximale de 0,67 MeV avec un rayonnement γ associé de 0,515 MeV pour seulement 0,4 % des émissions β) et d'autre part de la conception des usines (risques de fuites très faibles) les conséquences radiologiques pour le personnel de ces usines ne pourraient être en pratique que négligeables. Il n'en reste pas moins qu'en toute rigueur, un certain nombre de précautions devraient alors être prises dans ces usines comme dans certaines usines manipulant de la radioactivité : contrôle atmosphérique, surveillance des agents, etc. L'impact sur les conditions de travail dans ces usines ne serait pas entièrement négligeable.

En fait, tout ceci suppose qu'aucun piégeage de krypton 85 ne sera effectué et montre que même dans ce cas, l'impact demeurerait limité. Mais pratiquement, les règles adoptées par un certain nombre de pays prévoient déjà la rétention de ce radionucléide dans les usines de retraitement dès la dernière décennie de ce siècle et il est hautement probable, justement du fait que les impacts de rejets ne seraient pas nuls (industrie de l'air liquide) ou mal connus (effet d'une augmentation d'une ionisation de l'air), qu'en l'an 2000 toutes les usines de retraitement seront équipées d'une installation de piègeage.

Tritium : A plusieurs égards, le cas du tritium se présente de façon assez différente de celui du krypton 85. Son mode de production dans les réacteurs nucléaires est plus complexe, de même que son rejet dans l'environnement et son mode d'action sur les individus. Le tritium se forme dans les réacteurs nucléaires d'une part, par fission ternaire et d'autre part, par activation neutronique par suite de l'interaction de neutrons avec certains éléments présents dans le réacteur et en particulier le bore, le lithium, le deutérium. Il s'ensuit que la production du tritium, contrairement à celle du krypton 85, varie sensiblement suivant les filières. Le tableau 3 synthétise les principales productions.

TABLEAU 3

TAUX DE PRODUCTION DE TRITIUM
(en Ci/GW_t x an)

FILIERE	MODE DE PRODUCTION	
	PAR FISSION	PAR ACTIVATION
gaz graphite	7×10^3	$1,9 \times 10^2$
eau sous pression	7×10^3	$3,3 \times 10^2$
eau bouillante	7×10^3	5
eau lourde	7×10^3	2×10^5
haute température	7×10^3	$1,4 \times 10^3$
neutrons rapides	1×10^4	$< 7 \times 10^2$

Compte tenu du fait que les réacteurs de la filière eau lourde et de la filière à neutrons rapides ne représentent - et ne représenteront très probablement jusqu'en l'an 2000 - qu'une faible proportion de parc des réacteurs en fonctionnement, on peut retenir comme valeur du taux de production de tritium, le chiffre de $7,2 \times 10^3/GW_t$ x an. Ce tritium produit va se retrouver, pour la plus grande partie à l'usine de retraitement. Le tableau 4 synthétise cette répartition des rejets.

TABLEAU 4

TAUX DE REJETS DE TRITIUM
(en Ci/GW_t x an)

FILIERE	POINT DE REJET	
	REACTEUR	RETRAITEMENT
gaz graphique	$2,6 \times 10^2$	$6,9 \times 10^3$
eau sous pression	4×10^2	$6,9 \times 10^3$
eau lourde	7×10^3	$6,9 \times 10^3$
haute température	$< 3,4 \times 10^4$	5×10^3
neutrons rapides	$< 10^4$	faible

Ces rejets se produisent à la fois dans les effluents liquides et dans les effluents gazeux. Pour déterminer l'impact sur l'environnement de ces rejets, il est nécessaire d'étudier le cheminement du tritium. Les rejets gazeux peuvent conduire à deux voies d'irradiation du public :

- l'inhalation et d'adsorption par la peau des vapeurs d'eau tritiée,
- l'ingestion d'eau ou d'aliments contaminés durant le passage du nuage radioactif. Cette contamination provient de l'entraînement du tritium par la pluie et des échanges en phase vapeur en l'absence de précipitation.

Les conséquences des rejets liquides sur l'irradiation du public dépendront de la nature du récepteur (rivière, fleuve, mer) de l'utilisation des eaux correspondantes. On doit remarquer à ce point de vue que le rejet direct dans la mer des effluents tritiés ne pourra avoir, au moins jusqu'en l'an 2000, qu'une incidence négligeable au plan radiologique. Ceci découle de l'énorme facteur de dilution apporté par la mer et de la faible contribution apportée par la mer à l'alimentation humaine dans de nombreuses régions du globe et spécialement en Europe occidentale, région dans laquelle se situera la principale concentration de réacteur nucléaires. Cependant une autre voie d'action est théoriquement possible, le passage du tritium dans l'atmosphère par évaporation suivi par sa reprécipitation avec les eaux de pluie. Ceci serait surtout à considérer dans le cas de rejet de tritium dans des mers relativement fermées et peu profondes. Le rejet dans les rivières peut avoir une importance plus grande, l'eau pouvant être utilisée comme boisson et aussi pour l'irrigation. Il existe enfin une dernière possibilité pour le devenir du tritium : les déchets solides. Ce peut être en particulier le cas pour la fraction du tritium qui diffuse dans les gaines des éléments combustibles si ces gaines sont conditionnées par des procédés ne comprenant pas de passage à température élevée. Cependant la fraction de tritium correspondante est faible et on peut négliger cette fraction.

On voit donc qu'un facteur fondamental dans l'étude de l'impact des rejets de tritium est la proportion de tritium rejetée dans l'atmosphère et celle rejetée dans le réseau hydrographique lors du retraitement. L'étude déjà mentionnée du National Radiological Protection Board du Royaume-Uni estime respectivement ces deux fractions à 25 % et à 75 %. Dans le projet de l'usine de Barnwell (Caroline du Sud) on prévoyait un rejet de 90 % dans l'atmosphère. A l'usine française de la Hague, la proportion rejetée dans l'atmosphère n'est que de 5 à 10 %, le reste étant rejeté dans la mer. Si l'on considère tous les rejets de tritium à l'usine de retraitement effectués par la cheminée (en fait sous forme d'eau tritiée) la figure 4 donne les débits de dose correspondant au rejet de 1 MCi/an; on remarque que les impacts sont faibles : à 10 km du point de rejet, le débit de dose global au corps entier n'est que de 2 mrad/an, comme on l'a vu plus haut, la plupart des usines rejetterait une grande partie du tritium dans les effluents liquides, ce qui entraînerait une diminution sensible de ce chiffre.

Les impacts globaux sont beaucoup plus difficiles à estimer, compte tenu de ces incertitudes sur les différentes voies de rejet, mais elles ne pourront être que très faibles. La figure 5 tirée elle aussi du rapport du National Radiological Protection Board donne une valeur, pour le débit de dose au corps entier, comprise entre 1 et 2 μrad/an.

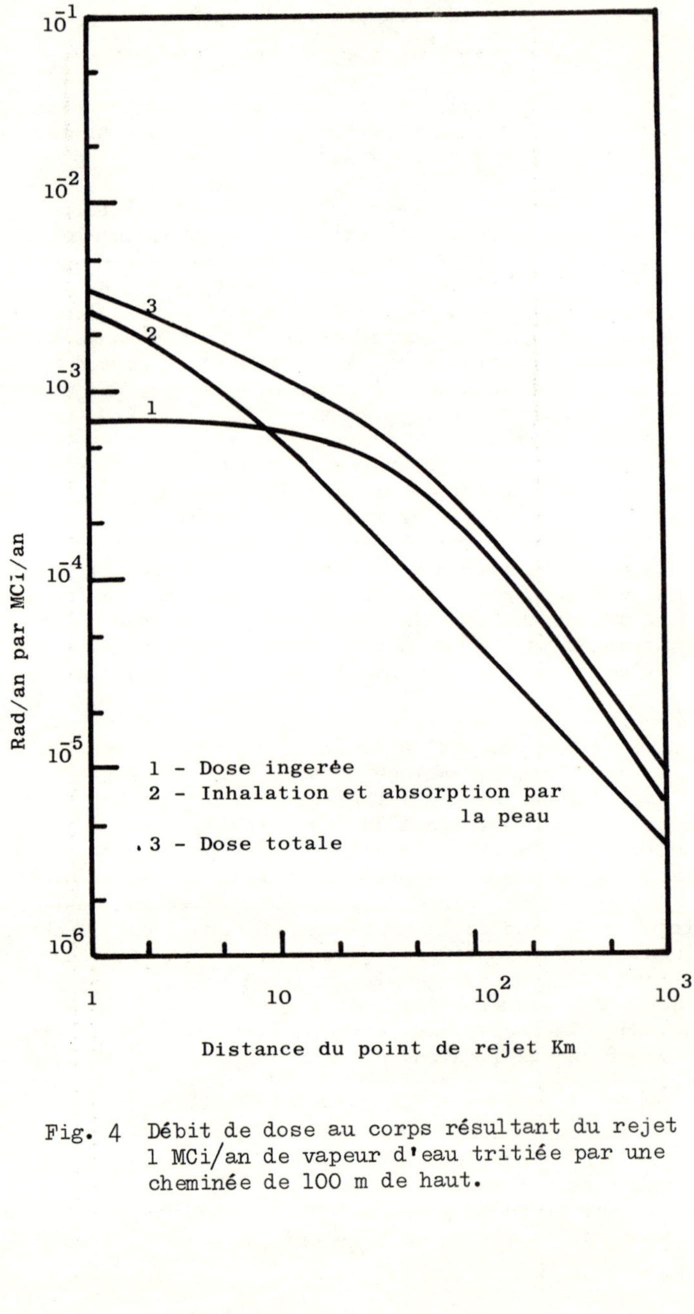

Fig. 4 Débit de dose au corps résultant du rejet 1 MCi/an de vapeur d'eau tritiée par une cheminée de 100 m de haut.

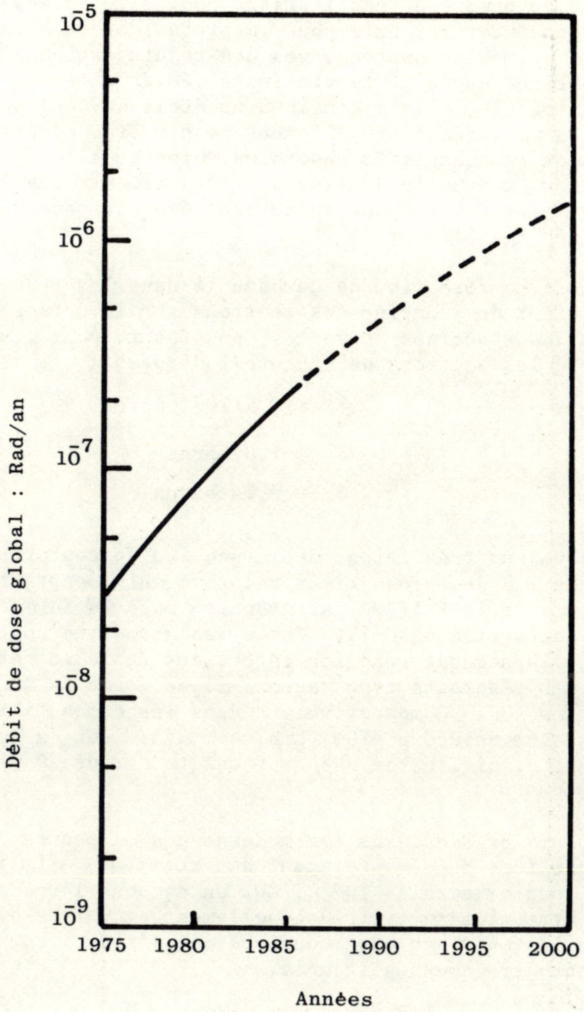

Fig. 5 Débit de dose au corps entier venant de la circulation globale de tritium rejeté compte tenu du programme nucléaire mondial.

En définitive, pour les rejets de tritium, l'étude la plus importante sera celle de situation particulière au niveau local ou régional. Ce pourrait être le cas, comme il a été mentionné plus haut pour des rejets importants dans des mers fermées ou dans des fleuves. Le cas du Rhin a fait l'objet d'études spéciales, compte tenu du nombre d'installations nucléaires prévu sur son parcours. La Commission internationale pour la protection du Rhin contre la pollution a récemment estimé les conséquences des rejets radioactifs jusqu'à l'an 2000. En se basant sur une capacité totale installée dans le bassin du Rhin de 100 GWe à la fin du siècle, elle a conclu à un débit de dose pour la population de 0,14 mrad/an. Cette valeur est probablement pessimiste, compte tenu des hypothèses sur les rejets et les quantités absorbées, bien qu'elle ne tienne pas compte de l'installation éventuelle d'usine de retraitement dans le bassin, mais le chiffre est suffisamment faible pour qu'en tout état de cause, l'impact puisse être considéré comme négligeable.

Carbone 14 : La formation de carbone 14 dans les réacteurs nucléaires est faible. Elle provient de l'action des neutrons sur le carbone 13, l'azote 14 et l'oxygène 17 selon les réactions suivantes, possédant, pour les neutrons thermiques de 0,025 eV, les sections de capture indiquées :

$^{13}C\ (n,\gamma)^{14}C \qquad \sigma = 9 \times 10^{-4}$ barns

$^{14}N\ (n,p)^{14}C \qquad \sigma = 1{,}8$ barns

$^{17}O\ (n,\alpha)^{14}C \qquad \sigma = 0{,}24$ barns

Malgré la valeur très faible de la section de capture, le premier mode de formation conduit à des productions relativement importantes dans le modérateur des réacteurs de la filière gaz graphite avec la formation intermédiaire du C-13 par activation de C-12. Comme les graphites contiennent en moyenne 10ppm d'azote, la seconde réaction intervient également et finalement dans les modérateurs des réacteurs type Magnox, il se forme 64 Ci/GW_t x an et 102 Ci/GW_t x an dans les AGR. Comparativement dans les combustibles eux-mêmes, les quantités formées sont beaucoup plus faibles : 4,5 Ci/GW_t x an dans les réacteurs à eau ordinaire, si l'on se base sur une teneur de 20 ppm d'azote dans les oxydes d'uranium.

Le carbone 14 présent dans les modérateurs ne pourra avoir un impact sur l'environnement que lors du démantèlement des réacteurs. Un réacteur de la filière gaz graphite aura emmagasiné 10^3 Ci/GW_t en carbone 14. Mais compte tenu du développement de cette filière limité actuellement et qui se développera certainement peu dans l'avenir, on peut conclure que l'impact correspondant sur l'environnement ne peut être que négligeable.

Il y a donc surtout lieu d'examiner le devenir du carbone 14 produit dans les combustibles irradiés et aussi celui produit dans le fluide de refroidissement. Dans ce dernier, il se produirait respectivement pour les réacteurs PWR et BWR 2 Ci/GW_t x an et 5,3 Ci/GW_t x an qui seraient rejetés dans l'environnement au site du réacteur. Il faut cependant remarquer que ces chiffres, comme d'ailleurs ceux relatifs à la formation de carbone 14 dans les combustibles eux-mêmes sont peut-être surestimés. Ils sont tous, comme on l'a souligné, basés sur des teneurs en impureté et des calculs de formation. Mais des analyses faites jusqu'à présent sur les combustibles irradiés ont conduit à des valeurs expérimentales inférieures. Compte tenu de la difficulté de mesures de quantités si faibles, il est plus prudent de prendre en compte les chiffres théoriques en étant conscient qu'il est très possible que l'on majore le problème.

Le devenir du carbone 14 formé dans les éléments combustibles est plus sujet à caution. Le rapport anglais déjà mentionné évalue à 10 % la quantité rejetée par les cheminées des usines de retraitement (UK Nat. Rad. Prot. Board, 1975), le solde restant confiné avec les diverses cagégories de déchets, mais l'on ne dispose pour l'instant d'aucun résultat mesuré, ce qui tendrait d'ailleurs à prouver que les chiffres réels de rejet sont peut-être inférieurs. L'impact du rejet du carbone 14 dans l'environnement découlera du fait qu'il va se retrouver incorporé dans tous les composés carbonés et en particulier dans les protéines. L'incorporation de carbone 14 à une activité spécifique de 1 pCi de carbone dans le corps humain se traduira par une dose moyenne à l'organisme entier de 0,17 mrad/an. Deux voies sont possibles pour cette incorporation : l'inhalation pendant le passage du nuage et l'ingestion d'aliments contaminés, cette contamination produirait des échanges isotopiques, les estimations qui ont été faites montrent que les doses moyennes au corps entier pour la population résultant de ces rejets demeureront très faibles. Comme le montre la figure 6 tirée toujours du même rapport, les débits de doses seront de l'ordre de 2/100 mrad/an en l'an 2000.

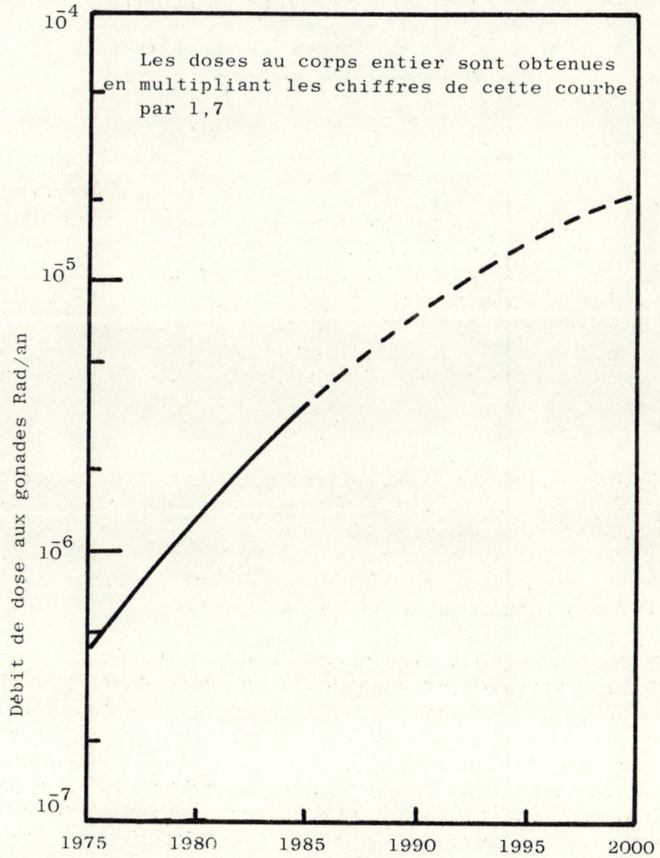

Fig. 6 Débit de dose aux gonades, entre les parallèles 35-60°N, venant de la circulation du C 14, compte tenu du programme nucléaire mondial

Iode 129 : L'iode 129 est un produit de fission de période extrêmement longue (16 millions d'années). Son taux de production, qui est sensiblement le même pour les différentes filières de réacteurs, est de 0,4 Ci/GW_t x an. L'iode 129 reste pratiquement totalement confinée dans le combustible pendant son séjour dans le réacteur et se trouvera libérée pendant le retraitement. Mais son devenir va dépendre des schémas utilisés pour le retraitement. A Windscale 75 % de l'iode 131 sont rejetés dans la mer et 0,1 % dans l'atmosphère, le restant se trouvant dans les déchets stockés sur le site. Le rapport du N.R.P.B. prend dans son étude d'impact 75 % comme rejet dans la mer et 1 % de rejet à l'atmosphère. L'expérience française conclut à des chiffres différents : 3 % de rejet à l'atmosphère, pratiquement pas de rejet dans les effluents liquides, la quasi totalité de l'iode reste donc confinée dans les déchets stockés sur le site.

En toute rigueur, l'on devrait dans une telle étude prendre en compte la fraction confinée dans les déchets, car à long terme on ne peut garantir l'intégrité totale du confinement, mais ce point ramène au problème des effets à très long terme qu'il ne semble pas raisonnable de prendre en compte, comme il a déjà été souligné dans ce rapport. De toutes façons, les ordres de grandeur des résultats demeureront inchangés. Les résultats des études sont synthétisés par la figure 7 et montrent que globalement les effets sont très limités : quelques centièmes de mrad/an pour le débit de dose à la thyroïde pour les populations.

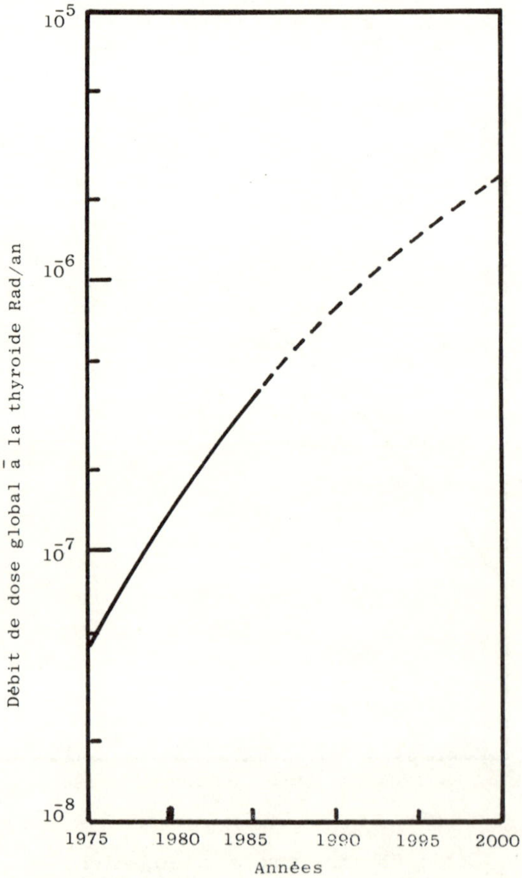

Fig. 7 Débit de dose à la thyroide venant de la production globale de l'I 129 rejeté, compte tenu du programme nucléaire mondial

L'impact sur l'environnement peut être sensiblement plus important au voisinage des rejets, soit que ceux-ci soient effectués dans des cours d'eau de débit relativement faible, soit qu'une proportion plus grande soit rejetée par la cheminée, compte tenu des possibilités de reconcentration de l'iode dans les chaînes biologiques. C'est ce dernier facteur qui conduit à équiper les usines de retraitement de dispositifs de rétention d'iode. Mais les quantités stockées ne seront pas négligeables. L'ERDA (1976) estimait qu'aux Etats-Unis en l'an 2000, alors que la puissance électronucléaire installée serait de 800 GWe, l'activité totale à stocker d'iode 129 serait de 4 800 Ci. Nous reviendrons sur ce point dans l'étude de l'incidence d'autres schémas de gestion.

Impact global des rejets d'effluents

A court terme, c'est-à-dire à l'échelle des deux décennies à venir, le principal impact sur l'environnement de l'énergie nucléaire au niveau mondial sera très vraisemblablement les conséquences potentielles de l'irradiation due aux rejets d'effluents. Ce qui montre l'importance des études faites sur l'impact global des rejets d'effluents.

Le Comité scientifique des Nations Unies sur les effets des radiations atomiques (UNSCEAR) a publié plusieurs études à ce sujet; en particulier son rapport de 1977 à l'Assemblée générale synthétise les résultats auxquels ce comité est arrivé. L'UNSCEAR a évalué les engagements de dose collective due aux rejets d'effluents radioactifs de tous les stades d'énergie nucléaire en l'exprimant en engagement de dose collective par unité d'énergie produite, en fait par MWe x an. La prise en compte d'une telle unité permet plus aisément l'évaluation de la fraction respective de chaque stade et surtout permet d'additionner ces fractions. Cette dose collective à la population mondiale comprend des contributions de quatre groupes : les travailleurs, la population locale, la population régionale et la population mondiale.

L'UNSCEAR dans ce rapport traite le cas des engagements de dose due aux nucléides de très longue période, en particulier celui de l'uranium 238 (période $4{,}5 \times 10^9$ an) et de l'iode 129 (période $1{,}6 \times 10^7$ an). Ce comité estime que l'accumulation de ces nucléides dans la biosphère n'entraînera pas un débit de dose de plus de 1 mrad/an, même si la production d'énergie nucléaire continue pendant 500 ans à un niveau de 2 000 GWe avec les technologies actuelles, mais il souligne que si l'on prend une durée d'exposition de plusieurs millions d'années, les engagements de dose correspondants seront élevés, bien que pour que la population mondiale, supposée stabilisée à 10 milliards d'individus, accumule une dose de 1 homme x rem/MWe, il faudrait intégrer les doses provenant de l'uranium 238 rejeté sur 10 millions d'années et le faire sur 10 000 ans pour l'iode 129.

La période du carbonne 14 (5 730 ans) tout en étant beaucoup plus courte est encore très longue par rapport à la vie humaine. L'engagement de dose collective due aux rejets envisageables de ce nucléide est de 5 hommes x rem/MWe x an dans les tissus mous et de 14 hommes x rem/MWe x an dans les os et la moelle rouge. La moitié de cette dose collective serait délivrée en 5 000 ans environ et au cours de 500 premières années seulement le quart de cette dose serait délivré. Finalement, le rapport en question ne prend en compte que cette durée de 500 ans pour le carbone 14.

Ce même rapport souligne que le stockage intérimaire des déchets ne donnera qu'une dose insignifiante comparée aux autres stades du cycle de combustibles et qu'il a estimé actuellement impossible d'évaluer les engagements de dose collective due aux stockages définitifs de déchets, compte tenu du fait que les méthodes exactes de réalisation de ces stockages ne sont pas entièrement arrêtées.

Le tableau 5 tiré du rapport de l'UNSCEAR synthétise les résultats des études de ce comité. Ces chiffres peuvent paraître devoir être élevés quand l'énergie nucléaire atteindra un développement de 1 000 ou 2 000 GWe, main il ne faut pas oublier :

- que l'engagement de dose collective due à l'utilisation pendant un an des engrais phosphatés est de 3 x 10^4 homme x rem.
- qu'en utilisant des modèles de calculs analogues, l'engagement de dose collective due aux diverses sources de rayonnement naturel est de 300 millions hommes x rem/an (NUREG - 0116, 1976).

TABLEAU 5

ENGAGEMENT COLLECTIF DE DOSE DU AU DEVELOPPEMENT DE L'ENERGIE NUCLEAIRE EN HOMME x REM/MWe x an

ETAPE DU CYCLE	ENGAGEMENT DE DOSE	
	ENSEMBLE DU DEVELOPPEMENT DE L'ENERGIE NUCLEAIRE	GESTION DES DECHETS
- extraction, traitement des minerais et étape du cycle avant le réacteur	0,2 - 0,3	
- fonctionnement des réacteurs	1,2 - 1,4	0,2 - 0,4
- retraitement	2,4 - 5,1	1,2 - 3,9
- recherche et développement	1,4	
total	5,2 - 8,2	1,4 - 4,3

N.B. - Nous avons arrondi les chiffres du rapport. Ce qui nous semble important étaient les ordres de grandeur.

- Les deux valeurs indiquées correspondent à l'engagement de dose sur des organes ou tissus différents.

D'autres éléments de comparaison sont donnés dans le tableau 6 tiré du rapport de l'UNSCEAR.

TABLEAU 6

ENGAGEMENT GLOBAL DE DOSE CORRESPONDANT A DIVERSES SOURCES D'IRRADIATION

SOURCE D'EXPOSITION BASEE SUR UN AN	ENGAGEMENT GLOBAL DE DOSE EN JOURS (1)
Exposition aux rayonnements naturels	365
Transports aériens commerciaux	0,4
Utilisation d'engrais phosphatés	0,04
Production d'électricité par centrale à charbon (puissance installée 10^6 MWe)	0,02
Utilisation des radioisotopes	3
Energie électrique produite par centrales nucléaires (80 GWe)	0,6
Explosions nucléaires (basées sur la moyenne 1951-1976)	30
Radiodiagnostic médical	70

(1) L'engagement global de dose est exprimé en durée d'exposition aux rayonnements naturels qui donneront le même engagement de dose que celui dû à l'utilisation pendant 1 an de la source considérée.

Impact des traitements de déchets

Les paragraphes précédents traitant de l'impact des rejets d'effluents considéraient en fait l'impact de la prise en compte du premier terme de l'alternative fondamentale en matière de gestion des déchets : "diluer-rejeter" ou "concentrer-confiner". Il faut donc maintenant examiner l'impact de la prise en compte du second terme. Ce second terme comprendra toutes les opérations de traitement de conditionnement et de stockage.

Nous examinerons d'abord l'impact des opérations de traitement et de conditionnement des déchets. On doit d'abord remarquer que la limite est difficile à définir entre certaines opérations du cycle du combustible et certaines opérations de traitement des déchets. Par exemple, la concentration par évaporation des raffinats du premier cycle d'extraction dans les usines de retraitement est en général considérée comme faisant partie du retraitement et non du traitement des déchets, la même opération pour les effluents liquides de faible activité est considérée comme faisant partie des traitements des déchets.

Pour cette raison et aussi compte tenu du fait que les opérations de traitement de déchets soient d'importance faible par rapport aux opérations du cycle proprement dit en ce qui concerne l'exposition des travailleurs, on est souvent conduit à intégrer les premières dans l'ensemble des secondes. C'est ce que fait l'UNSCEAR dans le rapport déjà cité.

On peut essayer de voir quels sont les différents impacts possibles des principaux modes de traitement et conditionnement :

- Evaporation de liquide de faible activité

risque d'irradiation des travailleurs :	très faible
rejet d'effluents radioactifs ou chimiques :	très faible
consommation de combustible :	faible

- Incinération de déchets de faible activité

risque d'incident des travailleurs :	très faible
rejet d'effluents radioactifs :	très faible
rejet d'effluents chimiques :	rejet de NO_x, composés chlorés, etc.
consommation de combustibles :	assez faible

- Tri de déchets solides

risque d'irradiation des travailleurs :	relativement important

- Conditionnement des déchets F.A.- M.A. (bétonnage, bitumage)

risque d'irradiation des travailleurs :	faible
rejet d'effluents radioactifs ou chimiques :	faible
consommation combustible, électricité :	faible

- Vitrification des produits de fission

risque d'irradiation des travailleurs :	faible
rejet d'effluents radioactifs :	faible
rejet d'effluents chimiques :	NO_x
consommation combustible, électricité :	relativement élevée

On doit aussi envisager comme impact possible l'utilisation de terrain. Elle est faible comparée aux surfaces utilisées pour l'ensemble des opérations du cycle et aussi par celle nécessitée pour les stockages. Pour faibles que soient ces impacts, ils ont cependant été chiffrés en particulier dans le rapport de la N.R.C. (NUREG - 0116, 1976) duquel est tiré le tableau 7.

TABLEAU 7

PRINCIPAUX IMPACTS DU TRAITEMENT DES DECHETS DU CYCLE DE COMBUSTIBLE NUCLEAIRE (RAPPORTE EN FONCTIONNEMENT PENDANT UN AN D'UN REACTEUR DE 1 GWe DE LA FILIERE A EAU ORDINAIRE)

Terrain utilisé temporairement		0,1 ha
Effluents chimiques	NO_x	0,66 t
	SO_x	0,03 t
Energie électrique		140 MWh
Combustible fissile consommé		50 t (équivalent charbon)

En toute rigueur l'ensemble de ces opérations comporte quelques rejets d'effluents radioactifs. Mais les nucléides que l'on y retrouve sont ceux étudiés dans les paragraphes précédents dans lesquels ont été étudiés les effets globaux des rejets correspondants. Il n'y a donc pas lieu d'en tenir compte séparément.

Enfin le dernier impact possible du traitement des déchets serait les conséquences des accidents en cours de traitement. Il est clair que ces conséquences ne pourraient être importantes s'il s'agit d'installations ne traitant que des produits peu actifs et faiblement contaminants. Si on prend comme exemple une installation de traitement d'effluents liquides de faible activité par évaporation, les liquides traités pourront avoir une activité spécifique de l'ordre de 10^{-2} à 10^{-1} Ci/m3; un accident important entraînant un rejet dans l'environnement de 10 m3 ne se traduirait que par un relâchement occasionnel compris entre 100 mCi et 1 Ci.

Les seuls accidents à prendre en compte seraient ceux dans lesquels serait impliquée une installation de solidification de produits de fission ou une installation de conditionnement de déchets α. La première est une installation active conçue comme toutes les installations de ce type avec des barrières de confinement offrant un haut degré de garantie. Les accidents maximum imaginables paraîssent être soit une explosion à l'intérieur du four de calcination, soit la rupture du pot de coulée du verre. Le relâchement d'activité à l'intérieur de la cellule de solidification serait très élevé et pourrait atteindre plusieurs milliers ou plusieurs dizaines de milliers de curies. Mais compte tenu du volume de la cellule et des systèmes de filtration, il est peu vraisemblable qu'il y ait un accident de ce type et surtout plusieurs accidents de ce type d'ici à la fin du siècle, l'impact sur l'environnement ne pourra être que négligeable.

Le cas de l'accident survenant dans une installation de traitement de déchets α a été étudié dans le rapport de la NRC (NUREG - 0116, 1976). L'accident considéré est celui survenant pendant le compactage d'un filtre contaminé au plutonium sur lequel seraient déposées 60g de PuO_2; les hypothèses sont le relâchement dans l'enceinte secondaire de 1 % et que le confinement secondaire apporte une réduction du relâchement de 10^5. Il n'y aurait alors que 2,6 μCi de Pu de rejeté dans l'environnement et on n'atteindrait pas le 1/100 de la CMA aux limites du site.

Il peut sembler cependant que certaines hypothèses sont assez optimistes. En cas d'incendie des déchets, la fraction atteignant l'enceinte secondaire peut probablement atteindre 10 % du plutonium continu et l'efficacité du confinement secondaire peut être moins bonne. Mais même si un tel accident entraînait que l'on atteigne la CMA, voire 10 fois la CMA aux limites du site de l'installation, les conséquences radiologiques réelles seraient extrêmement faibles.

On peut donc finalement conclure qu'en condition normale ou accidentelle, l'impact sur l'environnement du traitement des déchets ne peut être que faible.

Impact des stockages

Il y a lieu d'étudier successivement les stockages des différentes catégories de déchets compte tenu du fait que les possibilités de stockage sont différentes et les réalisations probables le seront aussi.

Stockages des déchets de faible et moyenne activité sans émetteur . Trois modes principaux de stockage sont utilisés :

- stockage en surface ou au voisinage de la surface (shallow land burial)
- immersion dans les fosses océaniques
- stockage dans les mines

Stockage en surface. Le principal impact d'un tel mode de gestion est la stérilisation des surfaces utilisées. Les surfaces utilisées jusqu'à présent sont très limitées. En France le site géré par la Société Infratome (Bardet, 1976) a une superficie de 12 ha et la capacité possible de stockage correspond à un volume de l'ordre de 250 000 m3. Aux Etats-Unis (Holcomb, 1978) les six sites de stockage en service à la fin de 1977 totalisaient une superficie de 300 ha et la capacité de stockage correspondante était de 6,7 x 10^6 m3. Ces chiffres correspondent à une surface obérée de 0,48 m2/m3 dans le cas français et 0,45 m2/m3 dans le cas américain. La concordance est très bonne d'autant plus que la technologie de réalisation est assez différente : tranchées de 91 m x 12,2m et profondeur de 6,1 m dans le cas américain. Monolithes de béton de 40 m de long et de 3 m de profondeur dans le cas français, surmontés d'un stockage en surface de 5 m de haut.

En conclusion, les superficies de terrain nécessaire demeureront faibles. Si l'on estime qu'un réacteur de 1 GWe produit 500 m3/an de déchets de cette catégorie, la superficie correspondant aux chiffres précédents sera de 200 m2/an.

D'après l'étude de W.F. HOLCOMB précédemment citée, les 300 ha des six sites USA suffiront à assurer les besoins américains jusqu'en 1992. Et si de plus, on utilise des techniques de réduction de volume avant stockage, compaction en particulier, la saturation de ces sites n'interviendra pas avant l'an 2000. En se basant sur les chiffres précédents, et sur les prévisions de puissance nucléaire installée dans la Communauté européenne prévues dans le rapport NRPB (1975), les surfaces utilisées pour ce genre de stockage (à supposer que d'autres méthodes ne soient pas utilisées) ne dépasseraient pas 130 ha en l'an 2000.

Un autre point important à considérer dans l'étude de l'impact sur l'environnement de ce genre de stockage est la durée pendant laquelle les terrains seront stérilisés. Il y a, en fait, deux problèmes distincts pour la stérilisation de la surface. Compte tenu du fait que tous les modes de gestion prévoient que les stockages seront recouverts en dernier lieu de terre végétale et engazonnés, on peut considérer que les sols pourront, après la fin des opérations de stockage et probablement une période limitée de surveillance et d'entretien, faire l'objet de certaines utilisations : parc de loisirs, élevage. Cependant, dans ce dernier cas, on devra contrôler qu'aucune radioactivité ne soit transférée des déchets vers l'herbe, en quantités dépassant les limites admissibles, ce sera en principe le cas puisque les migrations éventuelles se feraient par l'intermédiaire des eaux souterraines vers des exutoires situés en contrebas. Par contre, la libre utilisation du terrain, avec possibilité de travaux de génie civil, de construction, etc. ne pourra intervenir qu'à beaucoup plus long terme. Si le stockage ne comprend pas de déchets contenant des émetteurs α (ou en comprend dans des limites correspondant à une teneur moyenne de quelques milliers de CMA, (1) soit la teneur de granit en radium) ce qui déterminera la durée sera la teneur en produits de fission à vie longue et spécialement en strontium 90 et césium 137.

---

(1) C'est-à-dire que l'activité volumique des déchets correspond à celle d'une eau ayant une radioactivité égale à quelques milliers de CMA.

D'après les références mentionnées, l'activité moyenne était de l'ordre de 8 Ci/m3 incluant une partie importante de radionucléides à période courte aux Etats-Unis et de 3 Ci/m3 en France. Au bout de 300 ans, la radioactivité résiduelle ne dépassera pas quelques centaines de CMA et l'utilisation des terrains occupés par ces stockages ne poserait pas plus de problèmes que l'exécution des travaux dans du granit. En définive, l'impact de l'utilisation de sols pour le stockage sera très limité sur le plan des surfaces obérées et limité également dans le temps.

Un autre mode d'impact sur l'environnement de ce mode de stockage peut être la suite des incidents survenant au cours des opérations. Au site de stockage de Savannah River Laboratory en 20 ans d'opérations au cours desquelles a été enfoui un volume de déchets équivalent à celui produit par 200 réacteurs de 1 GWe fonctionnant 1 an, il s'est produit 68 incidents (rupture d'emballage, feu, etc.) dont plus de 90 % n'ont entraîné aucun relâchement d'activité au-delà des limites du site, les centres n'ayant entraîné que de faibles contaminations d'aire de surface faible au-delà de ces limites. En 10 ans de fonctionnement du site français de stockage, aucun cas de contamination en dehors du site ne s'est produit.

On peut donc conclure là aussi que l'impact est très limité. On ne peut enfin craindre l'impact d'un découvrement des stockages par érosion éolienne ou entraînement par les eaux, compte tenu de l'épaisseur du recouvrement utilisé (1,50 m au minimum pour les sites des Etats-Unis) et du fait que les déchets stockés n'ont comme contaminant que des émetteurs de faible période radioactive. Pour les mêmes raisons, l'action des animaux fouisseurs ou le transfert par les racines des plantes n'est pas à craindre.

Immersion dans les fosses océaniques. Ce mode de gestion peut avoir deux sortes d'impacts différents sur l'environnement :

- pollution radioactive du milieu
- sujétion dans l'utilisation du fond des mers.

Théoriquement, le problème de la pollution radioactive du milieu pourrait se poser de la même façon que dans le cas des stockages en surface où l'analyse de sûreté est basée sur la résistance du conditionnement combiné à la valeur de l'activité spécifique. C'est d'ailleurs en fait probablement la même chose pour les immersions. Néanmoins, les analyses de sûreté de ce mode d'évacuation des déchets ne prennent pas en compte la résistance du conditionnement, mais sont basées sur l'étude de la diffusion des nucléides et des différents chemins possibles de migration. Ce sont de telles études et notamment les études de Webb et Morley (1973) et celle de Shepherd (1975) qui ont été à la base des recommandations de l'AIEA pour l'application de la Convention de Londres sur la prévention de la pollution des mers résultant de l'immersion des déchets et autres matières (IAEA, 1975). L'impact théorique des opérations d'immersion en ce qui concerne la pollution radioactive ne peut donc être que faible, à partir du moment où les recommandations de l'AIEA sont suivies.

En fait, il est encore beaucoup plus faible, compte tenu du fait que les quantités immergées sont bien inférieures à celles considérées dans le modèle (100 000 t/an pour un site donné) comme le montre le tableau 8 résumant les opérations effectuées de 1967 à 1977 dans l'Atlantique Nord sous l'égide de l'OCDE (OECD, 1977).

Le deuxième problème qui peut sembler à priori très mineur est en fait jusqu'à présent le seul qui ait risqué d'avoir un impact non rigoureusement nul. En effet, de nouveaux câbles transatlantiques ont été immergés en 1976 sur un trajet qui tangentait le site d'immersion choisi pour les opérations réalisées sous l'égide de l'OCDE. Le problème qui risquait de se poser était celui posé par les opérations de réparation du câble. Ces opérations sont faites avec remontée du câble par dragage et ces opérations de dragage auraient pu être gênées par la présence dans le voisinage de blocs de béton. Mais cet impact ne peut être que faible : en effet, d'abord si les déchets à immerger sont constitués de blocs lourds, ceux-ci s'enfonceront presque complètement dans les sédiments (c'est d'ailleurs un facteur qui n'a pas été pris en compte dans les études de sûreté, mais qui accroît les marges de sûreté) et ensuite les surfaces des fonds des mers effectivement susceptibles d'être recouverts de déchets lors des opérations d'immersion sont infimes.

Les opérations d'immersion organisées sous l'égide de l'OCDE l'ont été depuis 1971 dans une zone constituée par un cercle de 70 milles nautiques de diamètre (centré sur le point de coordonnées 46°15' N et 17°25' O). Cette zone représente, comme on l'a souligné, moins de 0,2/1000 de la surface du fond de l'Atlantique Nord, mais les 50 000 t immergées en supposant que les rejets n'aient été constitués que de fûts de 200 l qui reposeraient horizontalement sur le fond ne représenteraient que 0,02/1000 de la surface de la zone. Ce dernier point permet d'ailleurs de conclure que du point de vue occupation du fond, la même zone est susceptible de servir encore pendant plusieurs siècles, même en supposant des augmentations des volumes rejetés croissant parallèlement avec le développement de l'énergie nucléaire.

<u>Stockage dans les mines</u> : Le troisième mode de stockage utilisé pour les déchets de faible et moyenne activité sans alpha est l'utilisation de mines désaffectées. C'est le cas en République fédérale allemande où la mine de sel de Asse est utilisée depuis 1965 pour le stockage de tels déchets. Les impacts de ce mode de stockage sont certainement encore plus faibles que ceux du stockage en surface en ce qui concerne les surfaces stérilisées. Celles-ci sont pratiquement nulles pour les emprises au sol et le sous-sol n'a pas de valeur en lui-même.

On pourrait d'ailleurs tenter de conclure que puisqu'il s'agit de mines abandonnées, ce sous-sol ne recèle plus de ressources intéressantes. Cependant, il est connu que la notion de valorisation des ressources minières est sujette à variation et dépend du degré d'avancement des techniques de récupération, ainsi que des ressources disponibles. Il est donc possible que dans l'avenir des mines dont l'exploitation a été abandonnée puissent être reprises et que le fait de les avoir utilisées comme stockage de déchets radioactifs (ou chimiques car la tendance pour le stockage définitif de déchets contenant des toxiques chimiques non récupérables et non dégradables tels que mercure, arsenic, cadmium, etc. est d'utiliser des mines abandonnées) puisse être un obstacle. Cependant, cet impact ne pourrait être que faible, compte tenu du fait qu'il ne s'agit que de déchets à période radioactive courte. Néanmoins, l'utilisation à grande échelle de ce mode de stockage (ce qui n'est pas réalisé jusqu'à présent, ni envisagé actuellement) pourrait constituer une certaine gêne dans la récupération de certaines substances minérales par les générations futures.

TABLEAU 8

DECHETS IMMERGES PAR LES PAYS EUROPEENS DEPUIS 1967

ANNEES	MASSE BRUTE (tonnes)	ACTIVITE APPROXIMATIVE	
		ALPHA (ESSENTIELLEMENT DES ACTINIDES) (Ci)	BETA-GAMMA (H-3 INCLUS) (Ci)
1967	10 900	250	7 600
1969	9 180	500	22 000
1971	3 970	630	11 200
1972	4 130	680	21 600
1973	4 350	740	12 600
1974	2 270	420	100 000*
1975	4 460	780	60 500**
1976	6 770	880	53 500***
1977	5 600	950	68 200
	51 630	5 830	357 200

* Cette activité est due presque exclusivement au tritium.

** Dont environ 30 000 curies de tritium.

*** Dont environ 21 000 curies de tritium.

Ce tableau montre aussi que les activités spécifiques moyenne (~ 7 Ci/t) sont également bien inférieures à celles résultant de l'application des modèles (10^2 Ci/t) et d'autant plus inférieures qu'elles comprennent une part importante de tritium qui n'intervient pratiquement pas dans le modèle.

Le second impact réside dans les sujétions que risque d'imposer la présence de déchets à l'utilisation présente ou future des fonds marins. On peut considérer deux utilisations: utilisation des resources et cheminement des câbles sous-marins. Il est évidemment très difficile de prévoir quelles pourront être les ressources minérales que l'on extraira dans le futur. Cependant, les progrès réalisés dans les connaissances des fonds océaniques au cours de ces dernières années permettent d'estimer que la probabilité de trouver de nouvelles substances qu'il sera possible d'extraire est faible. En ce qui concerne les nodules de manganèse, on sait déjà que dans l'Atlantique Nord, il n'y a que deux ou trois zones bien répertoriées qui sont susceptibles d'exploitation et que la majeure partie des fonds océaniques en dehors des plateaux continentaux et de leur prolongement n'est pas susceptible de recéler du pétrole. Par ailleurs, les surfaces obérées sont extrêmement faibles. La zone utilisée pour les rejets mentionnés plus haut est un cercle de 70 milles nautiques. Cette surface correspond à moins de 0,2 millième de la surface des fonds de l'Atlantique Nord.

Stockage des déchets de haute activité. Ainsi qu'il a été défini au chapitre 1, il s'agit de la quasi totalité des produits de fission séparés dans les usines de retraitement au premier cycle d'extraction par solvant comme raffinat. Ces déchets sont stockés sous forme liquide pendant des durées variables, solidifiés et seront en général stockés sous cette forme en surface pour des durées plus ou moins longues avant d'être envoyés au stockage définitif.

Stockage intérimaire sous forme liquide : On retrouve comme impact possible les stérilisations de terrain, mais on trouve aussi l'impact possible d'accident. La stérilisation du terrain est limitée. Une usine de retraitement de combustibles irradiés venant de réacteurs de la filière à eau ordinaire produira, en utilisant le procédé connu sous le nom de Purex, 500 l de solutions concentrées de produits de fission par tonne d'uranium irradié retraité, si le retraitement a lieu 14 mois après le déchargement du réacteur et si ces cuves ont une capacité de refroidissement d'au moins 16 K Cal/h x l. Pour une usine de 1 500 t/an, cela correspond à 750 m3/an, soit en 30 ans d'exploitation - si la solidification n'avait pas été effectuée jusque là - 22 500 m3 soit environ 200 réservoirs de la gamme 120/140 m3, en tenant compte des réservoirs de secours, ce qui correspondait à 2,5 ha environ.

Le rapport Flowers (1976) qui indique qu'il existait à Windscale mi-1976, 12 cuves de 70 m3 et 4 cuves de 150 m3 pour le stockage de ces solutions concentrées de produits de fission, alors que les quantités réellement stockées à cette date s'élevaient à 610 m3, estimait qu'il y aurait,toujours à Windscale, 1 800 m3 à stocker en 1985 et 6 000 m3 en l'an 2000. Pour ces quantités, il serait nécessaire de disposer de 50 grandes cuves. La surface nécessaire ne s'élèverait cependant qu'à un hectare. De plus, ces surfaces seraient récupérables après solidification encore que la décontamination de telles cuves constituera très probablement une des opérations les plus délicates de ce type.

L'impact possible d'accident durant le stockage doit être examiné de près. En effet, c'est avec ce type de déchets et ce mode de stockage que se sont produites les fuites du centre de HANFORD aux Etat-Unis. Au total, la quantité de liquide qui a fui des 20 réservoirs en question a été estimée à 1 700 m3 (Nat. Acad. Sci., 1976). Ces chiffres peuvent paraître considérables, mais il faut les rapprocher, pas tellement du nombre total de réservoirs (152) ou du volume stocké (de l'ordre de 200 000 m3), mais plutôt de la conception de construction de ces réservoirs : tôle en acier noir, sans double enveloppe, soudures non radiographiées. Il est évident que ce type de réservoir ne serait plus admis aujourd'hui et n'est d'ailleurs plus construit maintenant. Les réservoirs en question avaient été construits pendant ou juste après la guerre.

Mais, en fait, malgré cette construction rustique et l'importance de ces fuites, l'impact sur l'environnement a été pratiquement nul. La radioactivité s'est fixée dans le sol, de nature basaltique, certains nucléides ont davantage migré que d'autres, mais aucun ne s'est approché de la nappe phréatique de moins de 30 mètres. Et il n'y a eu, ni il n'y aura à la suite de ces incidents d'irradiation détectable ou calculable du public. Aucun accident de ce type n'est à craindre avec les réservoirs construits selon les normes actuelles. On estime généralement que la durée pendant laquelle on peut garantir la tenue de l'intégrité de ces réservoirs est de l'ordre de la cinquantaine d'années (Gauvenet, 1977).

Si les politiques de gestion envisagent la solidification de déchets de ce type dans des délais beaucoup plus courts, de l'ordre de 5 à 10 ans, comme il a déjà été mentionné, c'est en raison d'autres critères : optimisation économique de la gestion, car les cuves de stockage liquide sont très coûteuses, nécessité d'un contrôle et d'une surveillance plus rigoureuse et enfin, diminution du risque de l'accident maximal théoriquement possible. On peut, en effet, envisager un accident résultant d'un arrêt pendant une longue durée des refroidissements. La probabilité d'un tel accident est excessivement faible, compte tenu de l'importance des mesures prises. Dans les nouvelles cuves construites, il y a trois circuits de refroidissement indépendants, l'alimentation électrique peut être assurée par deux dispositifs de secours indépendants. Les conséquences d'un tel accident, sans être catastrophiques, seraient néanmoins importantes. Une étude effectuée par le Centre d'Oak Ridge (Finneman, 1967) estimait qu'une perte de refroidissement, pendant une durée de 10 h - ce qui est extrêmement long, compte tenu des dispositifs de secours - sur une cuve de 1 million de gallons, avec les condenseurs et filtres hors de service entraînerait un relâchement de 720 000 Ci. Le débit d'exposition correspondant au passage du nuage serait inférieur à 70 m R/h à 5 miles en cas de temps pluvieux et inférieur à 7 m R/h en cas de temps sec. Malgré la valeur élevée des quantités rejetées, l'impact radiologique serait limité. On peut d'ailleurs se prémunir contre un tel accident (bien qu'en fait sa probabilité soit extrêmement faible) en ajoutant des dispositifs de condensation à refroidissement atmosphérique fonctionnant sans apport extérieur d'énergie et sans dispositif mécanique. Mais on augmenterait encore le coût de ces stockages.

Stockage intérimaire sous forme solide : Après vitrification (ou autre procédé de solidification qui pourrait être utilisé dans le futur), les produits obtenus seront généralement stockés dans un stockage intérimaire implanté, soit sur le site de production, soit dans un site central. La principale raison est le dégagement de chaleur important de ces blocs, dégagement susceptible de poser des difficultés dans le cas de certains stockages géologiques et augmentant de toute façon les problèmes technologiques de réalisation du stockage.

Le tableau 9 indique la décroissance de ce dégagement thermique et fait comprendre l'intérêt d'un stockage intérimaire d'une durée de l'ordre de 20 à 30 ans.

TABLEAU 9

DEGAGEMENT THERMIQUE DE PRODUITS DE FISSION VITRIFIES (1)

TEMPS DE REFROIDISSEMENT (2)	TAUX DE DEGAGEMENT DE CHALEUR W/l DE VERRE
1 an	108
10 ans	11
20 ans	5,6
30 ans	3,8
50 ans	2,3
100 ans	1,2

(1) Verres provenant du retraitement de combustibles irradiés à 33 000 MWJ/t. Volume du verre 95 l/t de combustibles.

(2) Refroidissement compté à partir du déchargement du réacteur.

Un stockage intérimaire de blocs de verre est déjà réalisé sur le site de l'usine française de retraitement de Marcoule. Ce stockage (Bonniaud, 1976) comprend 140 puits métalliques de 10 m de haut et de 60 cm de diamètre pouvant recevoir chacun 10 blocs de 150 l de verre. L'impact sur l'environnement d'un tel stockage ne saurait bien évidemment être encore inférieur à celui du stockage sous forme liquide. Les surfaces nécessaires sont très limitées. Le stockage de l'atelier AVM à Marcoule couvre une superficie de 400 m2. Le stockage étudié par les Suédois dans le rapport KBS (1978) s'étend sur une surface de 4 000 m2, alors qu'il correspond à l'ensemble des déchets produits par 13 réacteurs suédois pendant 30 ans. Il faut en plus souligner que les verres sont susceptibles de monter en température sans conséquence pour l'environnement, autre qu'un éventuel risque de modification de structure pouvant entraîner une dégradation relativement limitée de la valeur du taux de lixiviation (facteur probable entre 10 et 100) qui ne serait pas significative sur la sûreté à très long terme, surtout si elle ne porte que sur une quantité très limitée de blocs de verre.

<u>Stockage final</u> : Le stockage final envisagé est l'évacuation dans une formation géologique adéquate. Le rapport des experts de l'OCDE (1977) mentionne trois groupes de roches :

- les formations salines
- les formations argileuses
- les roches dures (granites, basalte)

De nombreuses variantes technologiques sont envisagées pour la réalisation de tels stockages. Mais elles possèdent, toutes, les éléments fondamentaux suivants :

- installations, en surface, de réception des déchets, comportant en général des possibilités de contrôle, décontamination et éventuel ment des installations de reconditionnement,
- puits assurant la descente du personnel et des matériels des déchets eux-mêmes et puits assurant la ventilation. Un minimum d leux puits est toujours envisagé,
- dans la formation choisie, galeries de dessertes et puits ou cavités de stockage, l'exploitation du stockage pouvant être réalisée sur plusieurs niveaux.

Dans l'étude des impacts sur l'environnement d'un tel stockage, on doit distinguer l'impact du stockage et de son exploitation et celui du comportement à long terme du stockage. Les surfaces obérées au sol sont faibles. D'après le rapport suédois KBS (1978) on peut évaluer à environ 1 ha la superficie nécessaire pour l'ensemble des installations de surface correspondant au stockage des 9 000 blocs de verre provenant du fonctionnement pendant 30 ans des 13 réacteurs suédois. Le rapport ERDA (1976) mentionnait pour une installation de stockage, une superficie au sol de 6 ha. Ces chiffres correspondent à des pourcentages négligeables des sols nationaux et même des pourcentages négligeables des surfaces nécessaires pour le stockage de déchets chimiques ou de résidus industriels. De plus, ces terrains seront récupérables un certain temps après la fin de la période pendant laquelle sera fait l'enfouissement des déchets. On sera amené à garder opérationnelles, pendant une certaine durée (de l'ordre probablement de 20 à 50 ans) ces installations, de façon d'être à même, si on le désirait, d'intervenir dans le stockage, d'y effectuer des contrôles, voire éventuellement des opérations d'entretien, voire même de récupération éventuelle de certains déchets si on l'estimait nécessaire.

L'importance des volumes obérés dans les formations n'est pas en soi une nuisance, compte tenu des profondeurs choisies, bien supérieures à celles où pourraient être réalisées d'autres installations souterraines (abris, autres stockages, etc.). La seule nuisance de cet ordre pourrait provenir de la stérilisation de ressources naturelles. Ainsi qu'il a été souligné plus haut, l'estimation précise de l'intérêt potentiel de toutes les roches comme source de matières premières pour les générations futures est très difficile à faire. On peut cependant affirmer que certains choix ne seront pas susceptibles d'avoir de conséquences de ce point de vue. Pour les formations granitiques, le fait que le granite ne renferme pas de substances minérales potentiellement valorisables et surtout le fait des profondeurs envisagées pour le stockage (de 500 m à 1 000 m ou même 1 500 m) laissant des volumes énormes disponibles au-dessus du stockage pour une exploitation, permet cette affirmation qui est également vraie pour les formations argileuses.

Le cas des formations salines et, en particulier, du sel gemme (NaCl) est un peu plus délicat. En effet, malgré les énormes possibilités qu'offrent les océans, le sel gemme est actuellement encore exploité à large échelle. En France, le sel gemme fournit 50 % de la production totale de sel. Et, bien évidemment, ce sont les gisements les plus puissants et les plus homogènes qui sont à la fois les plus intéressants et pour l'exploitation de sel et pour les stockages des déchets. On ne peut donc exclure un certain impact constitué par la gêne apportée à certaines exploitations dans le futur. Cependant, cette gêne est limitée : on estime en France à 400 milliards de m3 les principales formations salines susceptibles d'être exploitées. La réalisation de stockage de déchets radioactifs même se montant à quelques dizaines ou même quelques centaines de milliers de mètres cubes correspondrait à un pourcentage d'utilisation négligeable et même faible devant les stockages d'hydrocarbures réalisés actuellement (en France 10 millions de m3) et d'ailleurs susceptibles de rendre le sel impropre à une exploitation industrielle.

Le cas des conséquences du comportement à long terme des déchets est une question plus complexe. La sûreté de ce type de stockage est basée sur l'existence d'un certain nombre de "barrières" entre les déchets et la biosphère. La première barrière est constituée par la matrice du conditionnement : dans ce cas, ce sera le verre. La seconde barrière sera le conteneur (ou les conteneurs) dans lesquels seront ces blocs. La troisième pourra être constituée par l'interposition de substances douées de capacité de "sorption" conduisant à la rétention des radionucléides susceptibles de s'échapper des déchets.

Enfin, la quatrième sera la formation géologique elle-même ou éventuellement les formations situées entre le stockage et la surface. Ces barrières sont indépendantes les unes des autres, ce qui montre le degré élevé de redondance que l'on peut introduire dans le schéma de confinement des déchets dans les structures géologiques. Le confinement apporté par ces barrières sera de nature diverse et sa probabilité d'évolution au cours du temps le sera aussi. Un conteneur métallique, pris comme deuxième barrière, apportera un confinement absolu, tant que le conteneur demeurera intact, ou tout au moins, tant que les phénomènes de corrosion, etc. ne l'auront pas percé, mais après il n'apportera pratiquement plus rien dans le confinement. Une matière telle que le verre, très résistante dans les conditions de stockage, mais à travers laquelle pourront se produire des phénomènes de diffusion, n'apportera pas un confinement intégral, mais seulement un confinement relatif, mais qui pourra être très élevé, si les phénomènes de diffusion sont faibles et qui pourra subsister pendant une très longue durée.

Les études d'impact de tels stockages devront intégrer l'ensemble de ces processus. Ils devront d'abord tenir compte des possibilités accidentelles de rupture de l'ensemble du confinement. Les causes peuvent être des événements naturels :

- chute d'une météorite. Les études de Claiborne et Gera (Claiborne et Gera, 1974) concluent que pour un stockage réalisé à 600 mètres de profondeur, la couverture ne peut être brisée que par une météorite de 2 km de diamètre au moins. La possibilité de chute sur une surface de 1 km2 est évaluée dans ce rapport à 2×10^{-14}/an. On peut d'ailleurs remarquer que la chute d'une telle météorite pourrait briser l'intégrité du stockage, mais que cela n'entraînerait pas nécessairement un relâchement important de radioactivité (les blocs de verre pourraient rester intacts). Par contre, il est évident que les conséquences de la chute d'une telle météorite sur la région seraient considérables, les effets pouvant être comparés à ceux de l'explosion d'une bombe thermonucléaire en ce qui concerne l'onde de choc et en une moindre mesure, l'onde de chaleur.

- occurence d'un phénomène géologique telle que formation d'une faille. Les même rapport estime la probabilité correspondante, pour un stockage s'étendant sur 8 km2, à 4×10^{-11}/an.

- effet de l'érosion. Ce phénomène ne peut être négligé pour un stockage situé à faible profondeur, mais ne peut jouer que sur des périodes supérieures au million d'années quand il s'agit de stockage situé à plus de 500 m de profondeur.

Des actions humaines peuvent entraîner la rupture du confinement géologique :

- guerre nucléaire. L'étude de Claiborne et Gera (1974) conclut que l'explosion en surface d'une bombe de 50 MT n'entraînerait pas la rupture du confinement.

- forage d'exploration ou de recherches minières. Une étude de Battelle évalue ce risque à 4×10^{-4}/an.

C'est donc pratiquement le seul risque qui n'apparaisse pas comme négligeable. Mais en fait, la probabilité mentionnée est une probabilité moyenne et elle variera beaucoup avec la nature des formations. Comme il a été souligné plus haut, si le stockage est situé dans des formations dans lesquelles la probabilité de trouver des ressources minérales est faible (granite), elle sera beaucoup plus faible que s'il s'agit de formations où ces probabilités sont élevées (sel). Cependant, les conséquences d'une telle rupture ne seront pas très considérables; il pourra en résulter un certain transfert de radioactivité, très comparable à migration à long terme qui va être examiné plus loin. Il faut cependant considérer à part le cas d'une exploitation d'un gisement de sel, ayant contenu des déchets de haute activité, par dissolution. Si les deux premières barrières n'ont pas résisté, le sel extrait contiendrait une certaine quantité d'actinides (ou éventuellement d'iode 129), les autres produits de fission auraient complètement décru, car un tel événement n'est pas susceptible de se produire avant plusieurs milliers d'années. COHEN l'a analysé (1976) et a conclu que le nombre de décès susceptibles d'être entraînés par un tel événement était inférieur à 1 par million d'années et pour 4 000 GWe x an (c'est-à-dire pour 100 réacteurs de 1 GWe fonctionnant pendant 40 ans). Il s'agirait donc en fait plutôt d'une crainte psychologique, mais on peut cependant mettre en doute les hypothèses prises dans les calculs de COHEN (effet d'ingestion de faibles quantités de plutonium par exemple) et pour cela ne pas choisir comme stockage pour les déchets contenant des nucléides à très longue période radioactive, des formations salines.

Si la formation reste intacte et que l'homme n'y pénètre pas, directement ou indirectement, dans le futur, il peut cependant se produire un mécanisme par lequel une fraction de radionucléides contenus dans les déchets pourrait être lixiviée et transportée vers la surface par les eaux souterraines. Cela peut se produire même dans un milieu anhydre comme une formation de sel gemme dans lequel de faibles quantités d'eau peuvent arriver et former des poches de saumure qui peuvent migrer sous différentes causes (effet SORET, etc.). Par ailleurs, la plupart de ces gisements contiennent de faibles quantités d'eau (de l'ordre de 0,5 %). Cela se produira dans les formations comme les argiles qui renferment toujours de l'eau ou les granites. Le rapport suédois sur la sûreté du stockage de blocs de verre (1978) évalue, pour les sites envisagés pour le stockage en Suède de déchets de haute activité, la circulation d'eau à 0,2 l/m2 x an à la profondeur considérée de 500 m, la perméabilité étant de 10^{-9}m/s. Les conséquences sont multiples. D'abord, une résistance des conteneurs sur des périodes très longues. Le conteneur en titane resterait intact pendant au moins 1 000 ans et se décomposerait entre 1 000 et 6 000 ans. Le conteneur interne en plomb ne serait dissout que très graduellement et mettrait plus d'un million d'années à être totalement oxydé. En définitive, et compte tenu de la faible lixiviation du verre, le rapport conclut que

- la dose maximale intégrée pour les personnes les plus exposées n'excédera pas 0,4 rem en 30 ans et que cette irradiation ne se produira pas avant 200 000 ans.
- la dose collective pour l'ensemble de la population mondiale, dans la période de 500 ans la plus défavorable sera inférieure à 0,01 homme x rem/MWe x an.

Une autre étude des conséquences de ces migrations a été publiée récemment (de Marsily et al, 1978). A partir d'un modèle mathématique d'écoulement des eaux dans la formation considérée et en fonction des conditions d'adsorption des différents radionucléides pour les terrains traversés, elle a calculé la durée du transfert et la concentration maximale dans les flux arrivant en surface.

Les 5 formations géologiques, dans lesquelles le stockage est toujours supposé réalisé à une profondeur de 500 m ont des perméabilités variant de 10^{-6} à 10^{-10}m/s. Ce tableau qui n'est, rappelons-le, que le résultat de calculs mathématiques, montre que, dès que l'adsorption est forte, la rétention est totale. Si l'adsorption est faible et le terrain peu perméable, la rétention est encore très bonne. Il n'y a que dans le cas où il n'y a pratiquement pas d'adsorption et que le terrain est relativement perméable qu'on pourrait retrouver en surface, avant toute dilution, des concentrations assez sensiblement supérieures à la CMA. Ceci permet bien de conclure, comme les études précédentes, qu'un stockage géologique bien étudié et bien réalisé sera susceptible d'assurer, même à très long terme, un confinement entièrement satisfaisant. Car une concentration de l'ordre de la CMA à long terme et avant dilution ne peut être considérée comme un impact important sur l'environnement. Mais cela montre, à contrario, que l'on ne pourrait choisir n'importe quelle formation géologique pour y stocker ces déchets.

Le tableau 10 résume les résultats obtenus

TABLEAU 10

MIGRATION DE RADIONUCLEIDES A TRAVERS DES FORMATIONS GEOLOGIQUES

FORMATION GEOLOGIQUE	VITESSE MOYENNE DE PORE (m/s)	TAUX DE TRANSMISSION DE LA FORMATION %	DUREE DE TRANSFERT (années)	RAPPORT CONCENTRATION/CMA	DATE DU MAXIMUM DE CONCENTRATION (années)
	Iode 129 (période $1,6.10^{7}$ années) (a)				
1	5.10^{-6}	100	6	0,58	10 000
2	10^{-6}	100	29	2,9	10 000
3	4.10^{-8}	100	725	28,0	10 700
4	2.10^{-9}	99	14 500	250,0	20 000
5	10^{-11}	93	2 840 000	170,0	$1,45.10^{6}$
	Neptunium 237 (période $2,13.10^{6}$ années (b)				
1	5.10^{-6}	99,7	10 500	0,67	18 000
2	10^{-6}	99	52 500	1,13	40 000
3	4.10^{-8}	91	505 000	1,13	275 000
4	2.10^{-9}	43	$14,9.10^{6}$	0,57	2 400 000
5	10^{-11}	10^{-16}	$3,3.10^{7}$	$6,2.10^{-18}$	83.10^{6}
	Plutonium 239 adsorbé (période $2,44.10^{4}$ années) (c)				
1	5.10^{-6}	8.10^{-6}	$1,4.10^{6}$	3.10^{-9}	460 000
2	10^{-6}	3.10^{-21}	6.10^{6}	$1,4.10^{-24}$	1 150 000
3	4.10^{-8}	0	-	0	-
4	2.10^{-9}	0	-	0	-
5	10^{-11}	0	-	0	-
	Plutonium 239 non adsorbé (période $2,44.10^{4}$ années) (a)				
1	5.10^{-6}	100	6	1,3	10 000
2	10^{-6}	100	29	6,0	10 000
3	4.10^{-8}	99	725	66,0	10 700
4	2.10^{-9}	80	14 500	470,0	20 000
5	10^{-11}	6.10^{-11}	2 840 000	$8,5.10^{-10}$	700 000

(a) coefficient de distribution 0 : pas d'adsorption.

(b) coefficient de distribution 15 ml/g : adsorption faible.

(c) coefficient de distribution 2 000 ml/g : adsorption forte.

Stockage des déchets alpha. En première analyse, le stockage de cette catégorie de déchets peut être assimilé à celui des déchets de haute activité. En effet, après quelques siècles, compte tenu de la décroissance des produits de fission, les déchets de haute activité seront des déchets α si l'on fait exception du TC-99 et de I-129. Les études d'impact à long terme des stockages géologiques étudiées dans le paragraphe précédent sont valables.

Il y aura néanmoins, quelques particularités. D'abord, l'absence pratiquement complète des problèmes dus au dégagement de chaleur. Pour les déchets de gaine, le flux thermique est environ 100 fois plus faible et pour les autres déchets alpha 10^5 fois plus faible que pour les déchets de haute activité (NUREG - 0116, 1976). Il en résultera qu'il sera possible de stocker ces déchets dans des chambres que l'on pourra remplir entièrement (du type de celles que l'on obtient en exploitation minière par la méthode chambres et piliers) au lieu d'être obligé de les mettre dans des puits régulièrement espacés. Ensuite la variété d'origine et de conditionnement et aussi de composition des déchets. Ces caractéristiques évolueront d'ailleurs car dans les diverses usines de retraitement et de fabrication de combustible au plutonium, la tendance est de pousser aussi loin que possible le recyclage des effluents contenant du plutonium et récupération du plutonium dans les déchets solides. Mais de toute façon, l'impact sur les surfaces de volumes utilisées sera faible. Le rapport de la Nuclear Regulatory Commission des Etats-Unis déjà cité (NUREG - 0116, 1976), évalue à 40 ha les surfaces nécessaires au fond de la mine pour stocker les déchets de gaine correspondant à la marche pendant 25 ans de 160 réacteurs de 1 GWe et à 80 ha les surfaces correspondantes pour les autres déchets alpha. Par contre, le risque d'impact des déchets sur la formation géologique va être beaucoup moins important avec les déchets alpha. En effet, c'est surtout le dégagement thermique qui risque d'apporter des perturbations. Une récente note de l'US. Geological Survey (1978) mentionnait que l'effet thermique peut se traduire par les ruptures de liaison dans les minéraux hydratés entraînant un relâchement d'eau. Cet effet peut apparaître à des températures inférieures à 100°C dans des minéraux, comme la montmorillonite ou l'illite. C'est la raison pour laquelle on est amené à prévoir d'espacer autant les puits de stockage des déchets de haute activité.

En toute rigueur, un autre point est à examiner dans les études d'impact des stockages de déchets alpha : l'éventualité du risque de criticité. Un certain nombre d'éléments se présente de façon différente dans les deux catégories de stockage :

- la densité de déchets dans le stockage (rapport volume déchets/volume du dépôt) est beaucoup plus grande dans le cas d'un stockage de déchets alpha.
- la présence de matériaux modérateurs comme certaines matrices de conditionnement (bitume par exemple).

Cependant, il ne semble pas qu'en pratique, il puisse y avoir de réels problèmes. En effet, le rapport H/X (nombre de noyaux d'hydrogène/nombres de noyaux fissiles) dans les déchets est bien supérieur aux valeurs du domaine dans lequel la criticité est possible et, par ailleurs, un certain nombre d'absorbeurs de neutrons sera toujours présent. Cependant, un dernier point doit être examiné : la fixation des atomes fissiles lixiviés des déchets. Ces atomes seront fixés, soit par une barrière artificielle située autour du stockage, soit par le terrain lui-même au cours de la migration des ions actifs. L'étude des conditions de fixation sur une barrière artificielle peut être menée relativement facilement; en prenant les hypothèses les plus pessimistes (circulation de l'eau dans une seule direction, fixation de la totalité des ions fissiles), on peut voir si un problème risque de se poser ou non et si la réponse est affirmative, prévoir l'addition de matériaux jouant le rôle de poison neutronique. Le cas de la fixation sur le terrain lui-même est plus difficile à étudier, compte tenu de la complexité des cheminements et de celle des différentes formations. On peut, cependant, bien entendu, faire une étude de sûreté basée sur des hypothèses pessimistes et qui permettront de conclure à l'absence réelle de risque. Il faut noter deux facteurs importants :

- L'intérêt d'avoir un stockage placé à une profondeur suffisante et possédant au moins une formation douée de propriété de sorption des ions radioactifs suffisamment profonde. En effet, une excursion critique qui se produirait dans une telle formation et qui ne pourrait être que de très faible puissance (l'apport des noyaux fissiles étant progressif) n'aurait en fait aucun impact sur l'environnement.
- L'importance de n'avoir qu'une quantité limitée de noyaux fissiles dans les déchets. Si l'on admet, ce qui est vrai en première approximation, que les risques de criticité sont proportionnels à la quantité de matière fissile présente dans les déchets, on peut conclure que le risque serait de 50 à 100 fois plus important dans le cas de stockage de combustibles irradiés non retraités que dans le cas de stockage de déchets.

Stockage des déchets de démantèlement des installations nucléaires

Le consensus actuel concernant les installations nucléaires est qu'il n'y a pas d'obligation de démanteler immédiatement après l'arrêt, mais qu'il faut se mettre dans une situation d'attente sûre. Cependant, il est très probable que ces installations seront, après éventuellement des périodes intérimaires pouvant s'étendre sur plusieurs décennies, totalement démantelées. Il en résultera une production de déchets dont il faut examiner les impacts sur l'environnement. Mais il est possible qu'en certains pays ou en certaines occasions - par exemple lors de la période de fin d'utilisation de l'énergie nucléaire - un certain nombre d'installations nucléaires ne soient pas démantelées. Ces installations elles-mêmes constitueront alors des déchets. Leur impact proviendra des surfaces utilisées et des risques de migration à long terme des radionucléides. Les surfaces utilisées sont faibles, l'ensemble des bâtiments actifs d'une centrale de 900 MWe de la filière à eau sous pression couvre une superficie de 1 300 m2 environ. Les risques de migration à long terme des radionucléides se présentent de façon assez différente pour les réacteurs et pour les usines de retraitement de combustibles irradiés. Les autres installations nucléaires ne peuvent intervenir à ce point de vue, soit qu'elles ne contiennent après déclassement que des quantités très faibles de radioactivité, soit que leur nombre soit très faible.

Déchets de démantèlement des réacteurs nucléaires : Tous les matériaux soumis au flux neutronique issu du coeur vont s'activer durant le fonctionnement du réacteur et en particulier les pièces métalliques (cuves, grilles, plaques support, peaux de recouvrement de béton) mais aussi dans une moindre mesure le béton et, pour les filières qui en comportent, les empilements de graphite. Les nucléides formés ont des périodes très différentes : comme le montre le tableau 11.

TABLEAU 11

CARACTERISTIQUES DES PRODUITS D'ACTIVATION

activation de l'acier	période	rayonnement (1)
^{60}Co	5,27 ans	γ 1,3 Mev
^{55}Fe	2,6 ans	K
^{59}Ni	8 x 10^4 ans	K 100 Kev
^{63}Ni	12,5 ans	γ 100 Kev
activation du béton (2)		
^{45}Ca	165 j	β 100 Kev
^{152}Eu	12 ans	β 1,8 Mev
		(γ 340 Kev)
^{154}Eu	16 ans	γ 1,2 Mev

(1) ne figurent dans ce tableau que les rayonnements principaux pour chaque nucléide.

(2) en fait les bétons sont armés avec de l'acier doux et on y retrouve donc aussi du ^{55}Fe.

Ce tableau montre que pour les pièces métalliques, il y aura deux sortes de problèmes. A court terme le problème principal est l'irradiation due au Co-60. A long terme, c'est la teneur en nickel actif. Mais les teneurs en nickel sont faibles et les activités résultantes aussi, comme le montre le tableau 12, et il s'agit en outre de rayonnement β .

En fait, en plus de l'activation, une deuxième source de production de déchets pour le démantèlement va être la contamination du circuit primaire. Cette contamination peut être constituée par des dépôts de produits de corrosion provenant de ce circuit primaire et activés par passage dans le coeur. Elle peut aussi être due aux produits de fission et éventuellement aux actinides provenant du combustible et résultant des ruptures de gaines. Le premier cas ne se présente pas de manière différente de celui de l'activation, sauf que l'on peut effectuer une décontamination. Pour le second cas on estime les valeurs de la contamination, un an après l'arrêt du réacteur, (CEE, in preparation) à :

- 100 μCi/cm2 pour les produits de fission
- 5 μCi/cm2 pour les actinides

mais elles dépendent essentiellement du nombre et de la localisation des ruptures de gaines. Si, comme on peut l'admettre, la qualité des gaines va en s'améliorant avec le développement de l'énergie nucléaire, ces valeurs devraient en fait, dans le futur être sensiblement plus faibles.

TABLEAU 12

ACTIVITES SPECIFIQUES (Ci/g) DUES A L'ACTIVATION DES COMPOSANTS D'UN REACTEUR PWR APRES 25 ANS DE FONCTIONNEMENT EN FONCTION DU REFROIDISSEMENT

	1 an	5 ans	25 ans	100 ans	1000 ans
cuve de réacteur(1)	1×10^{-4}	$4,5 \times 10^{-5}$	$2,1 \times 10^{-6}$	$1,8 \times 10^{-7}$	2×10^{-9}
	1×10^{-10}	$4,5 \times 10^{-11}$	$2,1 \times 10^{-12}$	$1,8 \times 10^{-13}$	2×10^{-15}
peau interne	$1,3 \times 10^{-4}$	6×10^{-5}	9×10^{-6}	$4,4 \times 10^{-6}$	5×10^{-8}
	$1,3 \times 10^{-10}$	6×10^{-11}	9×10^{-12}	$4,4 \times 10^{-12}$	5×10^{-14}
couvercle	2×10^{-12}	9×10^{-13}	$4,3 \times 10^{-14}$	$3,7 \times 10^{-5}$	4×10^{-18}
grille supérieure	$9,8 \times 10^{-3}$	$4,4 \times 10^{-3}$	$7,4 \times 10^{-4}$	$4,1 \times 10^{-4}$	$4,5 \times 10^{-6}$
structure du coeur	$1,3 \times 10^{-2}$	$5,8 \times 10^{-3}$	9×10^{-4}	$4,6 \times 10^{-4}$	5×10^{-6}
écran thermique	$1,5 \times 10^{-1}$	$6,7 \times 10^{-2}$	$1,1 \times 10^{-2}$	$5,5 \times 10^{-3}$	6×10^{-5}
grille inférieure	$1,1 \times 10^{-2}$	5×10^{-3}	8×10^{-4}	$4,1 \times 10^{-4}$	$4,5 \times 10^{-7}$
béton (zone interne) (2)	$1,5 \times 10^{-4}$	$5,7 \times 10^{-5}$	$4,7 \times 10^{-7}$	$2,3 \times 10^{-9}$	-
béton (zone externe)	$1,6 \times 10^{-9}$	$5,4 \times 10^{-10}$	$1,9 \times 10^{-10}$	$3,8 \times 10^{-12}$	-

(1) les deux valeurs indiquées correspondent, de même que pour la peau intérieure, aux valeurs extrêmes de l'activation des différentes zones de la pièce considérée.

(2) pour les premières années en décroissance, l'activation de la zone interne du béton est due à celle du fer 55.

Sinon il sera nécessaire d'effectuer une décontamination d'ensemble de tout le circuit primaire avant démontage. On peut admettre que l'efficacité moyenne de cette décontamination sera de 99 %. Les ordres de grandeur précédents ne seront pas alors fondamentalement changés. Une des principales questions que va poser la gestion des déchets de démantèlement va être celle de la limite en dessous de laquelle les matériaux pourront être réutilisés sans restriction, pour des usages non nucléaires, et de celle en dessous de laquelle les matériaux pourront être réutilisés pour certains usages ou évacués comme déchets non radioactifs.

Dans les règlements de transport de matières radioactives, la limite en dessous de laquelle les matériaux ne sont plus considérés comme radioactifs est de 0,002 μCi/g soit 2×10^{-9} Ci/g. La prise en compte d'une telle valeur conduit à la possibilité de rejeter ou de réutiliser sans précautions particulières, un grand nombre des pièces ou parties démantelées. L'impact potentiel sur l'environnement - dans ce cas l'irradiation potentielle des invidivus - est-il négligeable ou non? Il faut bien voir que les 2×10^{-9} Ci/g correspondent à la radioactivité d'un minerai à 0,25 % d'uranium, ce qui est un minerai relativement courant. Il faut voir aussi quels sont les débits de dose correspondants dans la plupart des cas très faibles. Après quelques décennies, l'irradiation due au Co-60 est négligeable et l'activité résiduelle des pièces métalliques est due aux isotopes du nickel qui sont des émetteurs de faible énergie et l'irradiation correspondante est très faible. C'est ainsi que pour la cuve elle-même du réacteur, l'irradiation après 100 ans pour la zone la plus active n'est que de 0,2 mrad/h à 1 m et pour les autres zones, elle est inférieure à 0,05 mrad/h à 1 m. Le béton restera irradiant plus longtemps, du fait de l'émission γ de Eu-154. Au bout de 100 ans, la zone la plus active correspondra à une valeur de l'irradiation de 1,4 mrad/h à 1 m, mais pour le béton des autres zones, l'irradiation sera aussi inférieure à 0,05 mrad/h.

Il semble donc que l'on puisse conclure qu'en pratique la réutilisation, après une période adéquate de décroissance des matériaux provenant du démantèlement d'un réacteur n'aurait pas d'impact important sur l'environnement. Cependant, il est possible que l'on préfère dans un certain nombre de pays stocker définitivement ces matériaux comme déchets pour éviter les irradiations potentielles, si faibles soient-elles. L'impact sur l'environnement serait la surface correspondante nécessaire pour le stockage. Les volumes à stocker seraient les suivants pour une centrale PWR de 1 200 MWe :

- pièces contaminées	500 à 750 m3
- pièces faiblement activées	40 à 60 m3
- pièces de moyenne activité	150 à 200 m3
- bétons	150 à 200 m3

Même en admettant que tout cet ensemble soit envoyé aux déchets sans traitement, le total se monterait à environ 1 300 m3, soit un volume de déchets de l'ordre de celui produit en trois ans d'exploitation du réacteur, soit en définitive 10 % du total des déchets produits en 30 ans de durée d'exploitation du réacteur.

<u>Déchets de démantèlement des usines de retraitement</u> : Les déchets obtenus lors des opérations de démantèlement poseront des problèmes différents de ceux venant des réacteurs. Ils ne contiendront pas de produits d'activation et seront tous plus ou moins contaminés par des nucléides émetteurs α , plutonium en particulier. En fait, les déchets obtenus seront comparables à ceux provenant de l'exploitation des usines de retraitement et en particulier des déchets de procédés et des déchets technologiques. Le seul point à examiner concerne donc les volumes qu'il y aura à comparer à ceux de l'exploitation. Peu d'études ont été publiées sur le démantèlement des usines de retraitement et peu d'installations de ce type ont été démantelées. L'étude la plus complète concerne l'usine Eurochemic pour laquelle des démontages partiels ont été effectués et l'étude du démantèlement total effectuée. (Detilleux, 1976), le démantèlement de 3 des 29 cellules constituant la partie active du bâtiment principal a été effectué et a entraîné la production de moins de 100 m3 de déchets. Mais il s'agissait de cellules peu actives et peu contaminées (stockage de nitrate de plutonium par exemple), le démantèlement de l'ensemble des cellules correspondrait à un volume de l'ordre de 4 000 m3 de déchets. On ne peut que difficilement transposer ces chiffres à une grande usine et on peut juger que l'on a sous-estimé les difficultés de décontamination d'installations, comme les évaporateurs de produits de fission. Cependant, si l'on se rappelle que la production de déchets d'une usine de retraitement de 800 t/an serait d'environ 3 000 m3/an pour l'ensemble des déchets de procédés et des déchets technologiques, on ne peut que conclure que les déchets venant du démantèlement des usines de retraitement ne seront guère susceptibles de faire changer l'ordre de grandeur des volumes à

stocker, alors que ces opérations n'apportent pas non plus de déchets de nature nouvelle. On ne peut donc que conclure que l'impact de la gestion des déchets de démantèlement d'une usine de retraitement ne pourra être que très faible.

Stockage des stériles de minerais

Des différents stockages de déchets radioactifs, ceux des stériles de minerais peuvent à priori paraître comme ceux qui ont l'impact minimal sur l'environnement, tout au moins en ce qui concerne l'impact radiologique. En effet, l'extraction de minerai et son traitement dans les usines de concentration n'ont modifié en rien la radioactivité contenue dans ce qui est laissé sur place (stériles de mines et d'usines), l'uranium enlevé, correspond même, du moins théoriquement, à une diminution de la nuisance à très long terme. Cependant le radium va se trouver au cours de ces opérations dans des formes très différentes; au lieu d'être piégé dans une formation géologique dont les caractéristiques permettaient naturellement ce piégeage (sinon il n'y aurait pas eu de gisement d'uranium) il va se trouver dispersé en surface dans un matériau finement broyé : ses possibilités de transfert vont donc être très différentes de celles qui existaient dans l'état initial. Une seconde nuisance potentielle découlera de l'éventuelle utilisation des stériles dans les constructions d'habitation enfin, et comme dans le cas des autres stockages, la stérilisation des terrains utilisés pour ces stockages pourra constituer une nuisance.

Rayonnement et transfert de radioactivité des stockages de stériles : On doit théoriquement analyser le cas des deux catégories de stockages de stériles : les stériles de mine, c'est-à-dire la partie du minerai qui est extraite de la mine, mais dont la teneur est trop faible pour justifier l'envoi à l'usine de traitement et qui est stocké le plus souvent sur le carreau de la mine et les stériles d'usine constitués par la partie du minerai resté insoluble lors de l'attaque chimique. En pratique on peut se borner à examiner cette seconde catégorie, car pour la première la teneur en uranium (et donc en radium) ne peut être que très faible, souvent inférieure à 500 ppm et de plus l'état physico-chimique et minéralogique initial n'est pas modifié.

La contamination de l'environnement et l'exposition de l'homme à partir des stockages de stériles de minerais peut se faire de trois façons différentes : libération du radon, transport par le vent de particules de petit diamètre, enfin lixiviation de ces stockages par les précipitations atmosphériques ou les eaux de ruissellement. La quantité de radon produite dans les stériles est estimée (APS, 1977) à 63 Ci/GWe x an si on effectue le recyclage de l'uranium et du plutonium et à 90 Ci/GWe x an sans recyclage. Ces chiffres correspondraient à une émission de radon de 480 x 10^{-12} Ci/m2 x sec, alors que l'émission moyenne à la surface du globe est de 0,42 x 10^{-12} Ci/m2 x sec soit environ 1000 fois plus faible. Mais en fait seulement une partie de ce radon va diffuser dans l'environnement. Cette fraction va dépendre de l'épaisseur de la couche de stérile elle-même et de l'épaisseur de sable humide, se trouvant en partie supérieure des stériles.

Une épaisseur de 15 cm de sable à 4 % d'humidité suffit à réduire l'émission d'un facteur de l'ordre de 2, alors que l'on admet également que seulement 20 % du radon produit s'échappera de la masse du stockage. Ceci montre bien que l'incidence sur l'environnement de l'émission de radon des stériles de minerais ne pourra être que très faible et on pourrait, si on l'estimait nécessaire, la réduire encore en augmentant l'épaisseur du recouvrement. A ce type d'impact, on doit rattacher ceux qui découlent du fait que de grands volumes de terre auront été déplacés et surtout remis, au lieu de stockage, après broyage, c'est-à-dire sous une forme susceptible d'être transférée (El-Hinnawi, 1978). Les conséquences pourront être le transport de poussières vers des zones non contrôlées, la pollution de cours d'eau par ces poussières, ces poussières pouvant être ou non contaminées par le radium.

On sera également amené à mettre un recouvrement suffisant pour pallier ce risque de transfert des fines particules dont sont en partie constitués ces stériles. Les stériles sont, suivant la pratique courante, envoyés au champ d'épandage par transfert hydraulique. Après décantation, l'eau surnageante est rejetée et après la fin des opérations, l'ensemble du champ d'épandage sèche, du moins pendant les périodes sans pluie. Ce risque de transfert sera encore diminué, si après recouvrement on plante une végétation, destinée à retenir l'humidité plantée.

La lixiviation des stockages peut se traduire par un transfert de radioactivité. Dans la pratique actuelle environ 1 % du radium est entraîné vers les cours d'eau récepteurs (EWV, 1978) en l'absence de station d'épuration. La concentration en radium soluble dans les eaux de rejet des bassins de sédimentation varie de 200 à 600 pCi/l. La quantité de radium soluble rejetée par une usine produisant de 500 à 700 t d'uranium par an sous forme de concentrés, sera de l'ordre de 1 Ci/an. La concentration en radium dans les cours d'eau récepteurs variera bien évidemment avec le débit de ceux-ci, mais n'atteint en général qu'une faible fraction de la CMA, sauf si le débit est très faible.

La rivière réceptrice joue un rôle épurateur, le radium se retrouvant dans les sédiments dans les premiers kilomètres en aval du point de rejet. Le transfert de radium vers les végétaux et animaux se fait avec des facteurs de concentration pouvant ne pas être négligeables, mais il ne saurait en résulter de conséquences sanitaires sensibles pour les populations locales.

Utilisation de stériles pour la construction de bâtiments : Une telle utilisation pourrait se traduire par une augmentation très sensible de radon à l'intérieur de ces bâtiments, du fait de l'atmosphère relativement confinée qui peut s'y trouver. En fait ceci s'est réellement produit aux Etats-Unis où des constructions (maisons, écoles) ont été faites en utilisant comme sable certains stériles des usines de concentration de minerai de Grand Junction (Colorado). Les concentrations maximales mesurées se sont montées jusqu'à 100 pCi/l, alors que la teneur naturelle est de 0,1 pCi/l. L'on ne serait plus alors dans le domaine des faibles doses et on ne pourrait conclure que l'impact serait nul. Cependant il est très aisé d'éviter tout risque à court ou moyen terme, en interdisant l'utilisation de ces stériles dans la construction. On peut considérer qu'une telle interdiction sera respectée sur des durées pouvant aller jusqu'à un ou même plusieurs siècles, mais au-delà, l'efficacité d'une telle mesure est plus douteuse. Toutefois, le recouvrement des stériles par une couche suffisante d'argile ou d'un matériau similaire devrait permettre de diminuer ce risque dans une proportion importante.

Stérilisation des terrains : De même que dans le cas des sites de stockage de déchets de faible activité en surface, les terrains utilisés pour l'épandage des stériles de minerais ne sont pas, théoriquement du moins, stérilisés, car après la fin de l'exploitation et une période de stabilisation des sols, on pourra utiliser ces terrains. Ils seront cependant impropres à l'agriculture, compte tenu de leur composition (sables très chargés en sels minéraux). Les surfaces correspondantes seront cependant relativement limitées comme l'indique le tableau 13 (APS, 1977).

TABLEAU 13

STERILES CORRESPONDANT A LA PRODUCTION DE 1 GWe x an DANS DES REACTEURS DE LA FILIERE A EAU ORDINAIRE

	avec recyclage U-Pu	sans recyclage
minerai	1,5 x 10^5 t	2,1 x 10^5 t
surface bassin stérile	16 ha	22 ha
volume bassin stérile	2,1 x 10^6 m3	2,9 x 10^9 m3

Ces surfaces ne correspondent quand même pas à des valeurs négligeables. Si l'on admet qu'en l'an 2000, la puissance électronucléaire dans le monde sera de l'ordre de 2 000 GWe, on aboutit à une surface annuelle stérilisée de 32 000 ha. Ce chiffre en fait n'a pas une grande signification - et nous le citons uniquement pour donner l'ordre de grandeur du problème - car les terrains sont loin d'avoir des utilités ou des potentialités comparables dans des régions semi-désertiques ou dans des régions peuplées et cultivées. D'ailleurs, on peut envisager de le réduire dans des proportions relativement importantes en utilisant ces stériles pour remblayer les galeries et chambres des exploitations minières. Il s'agit d'une technique, mise en oeuvre par transfert hydraulique, bien développée, mais qui ne peut en pratique être utilisée pour la totalité des stériles. On peut admettre comme ordre de grandeur qu'environ la moitié des stériles pourrait ainsi être renvoyée dans la mine, diminuant ainsi d'autant l'impact sur l'environnement, le diminuant en fait encore davantage car on diminue par cela même le risque de dégâts en surface, consécutifs au maintien de cavités souterraines sur de longues périodes.

On peut conclure de tout cela que, si, comme pour les autres stockages de déchets, le stockage des stériles de minerais n'est pas susceptible d'exercer des impacts importants sur l'environnement, il n'en demeure pas moins que des contraintes strictes devront s'exercer sur ces dépôts, pour éviter une utilisation abusive dans la construction. L'efficacité de ces mesures à très long terme peut devenir moins grande et il pourrait alors en résulter un certain risque pour l'environnement. Néanmoins, ce risque demeurerait limité et le sera encore plus par la mise en application des mesures mentionnées et qui sont résumées dans la publication de l'AIEA traitant du sujet (IAEA, 1977).

Impact à court terme et à long terme

Un certain nombre d'actions humaines peuvent engendrer un impact sur l'environnement à long terme (déboisement entraînant un changement de climat, construction d'un barrage constituant un risque potentiel augmentant avec le temps). Nous avons vu que le rejet d'effluent de période radioactive longue entraînait un engagement de dose pour des durées très longues. Le stockage de déchets radioactifs comportant des émetteurs de longue période peut entraîner - comme d'ailleurs tout stockage de déchets chimiques comportant des composés toxiques non biodégradables ou non chronodégradables - à long terme, voire à très long terme, une légère contamination radioactive de l'environnement. Il est certain que le seul domaine d'activité humaine qui ait essayé de prendre en compte l'impact à long terme est l'énergie nucléaire. Mais il est normal que cette prise en compte ne puisse revêtir le même degré de certitude que pour l'impact à court terme. On doit d'ailleurs distinguer le long terme du très long terme. La prise en compte du long terme - de l'ordre de quelques millénaires ou quelques dizaines de millénaires - peut être faite dans des hypothèses raisonnables : non changement important de l'espèce humaine, ni de son cadre général de vie. Mais on ne peut prévoir quelle sera dans quelques milliers d'années la population du globe, ni sa répartition géographique. Mais l'environnement pourra déjà avoir radicalement changé. On peut se rappeler que :

- il y a 2 000 ans le désert du Sahara était fertile.
- il y a 7 000 ans, les îles anglo-normandes étaient reliées à la Normandie.
- il y a 10 000 ans, les volcans du Massif Central français étaient encore actifs.
- il y a 10 000 ans que s'est produite la dernière glaciation.

Par contre la prise en compte du très long terme (millions d'années) nous paraît être dénuée de sens. On ne peut savoir quelle aura été l'évolution de l'espèce humaine, de son mode de vie, si même elle subsistera. On ne peut savoir les bouleversements géographiques qui auront pu survenir.

Les nuisances à long terme. Cependant toutes les études récentes poussent les études des nuisances à long terme jusqu'à 10 millions d'années, voire jusqu'à 100 millions d'années, comme par exemple l'étude de Cohen (1977) qui évalue le nombre de doses susceptibles de causer des cancers mortels. De nombreux indices ont été proposés pour tenir compte de ces nuisances à long terme. Celui qui nous paraît le plus approprié est celui proposé dans le rapport KBS (1978) : cet indice de risque ("hazard index") est la somme des activités des différents nucléides contenus dans 1 t de déchets multipliée par la toxicité relative de chacun d'eux. La toxicité relative était déterminée par l'engagement de dose entraîné par l'ingestion de 1 Ci des radionucléides, en pondérant l'action sur les différents organes.

La figure 8 donne la variation de l'indice de risques des différents nucléides contenus dans les déchets de haute activité (PF) venant du retraitement de 1 t de combustibles. On y voit qu'après 5 à 600 ans, l'indice de risques dû aux seuls produits de fissions est inférieur à celui du minerai dont on est parti, et que même en tenant compte des actinides contenus dans ces déchets (0,5 % de plutonium et la totalité de l'américium et du curium) il devient inférieur à celui du minerai initial après 1 100 ans environ. D'autres études comme celle de Hamstra (1975) donnent des résultats un peu plus optimistes ou comme celle de Culler (1977) des résultats un peu plus pessimistes. Mais l'ordre de grandeur demeure le même : au bout d'une durée de l'ordre de 1 000 ans, la nuisance potentielle des produits de fission solidifiés est inférieure à celle du minerai dont on est parti ce qui ne veut pas dire que l'on pourrait stocker, après ce temps, ces déchets sans aucune précaution, car les conditions de transfert de la radioactivité ne sont pas forcément les mêmes.

Les impacts possibles à long terme. Si l'on se place à l'échelle de quelques dizaines de milliers d'années, quel pourra être l'impact possible résultant de la gestion des déchets suivant les événements actuels ou réalisables prochainement? Des nucléides gazeux ne restera que le carbone 14, des produits de fission à vie longue, l'iode 129 qui devra être stocké en formation géologique dans des verres ou d'autres matrices, comme les déchets de haute activité, dont il ne restera que les actinides. Il restera enfin les déchets α . Mais si la formation géologique a été bien choisie et les barrières bien réalisées, l'impact ne pourra être que faible : faible contamination des eaux de surface qui n'aura probablement aucune signification radiologique. Il est évident que le stockage en surface de déchets contenant des quantités relativement importantes d'émetteurs α pourrait entraîner des conclusions différentes.

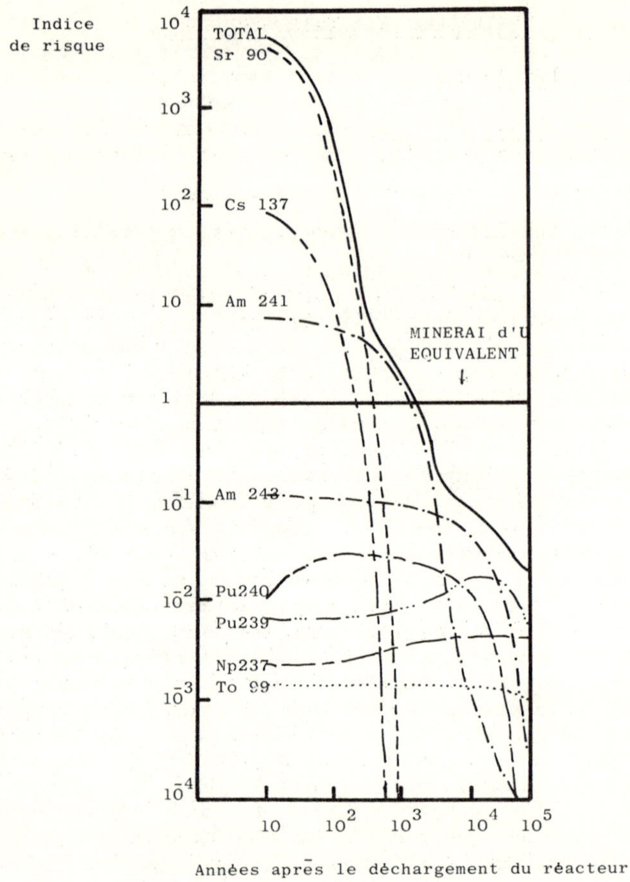

Fig. 8 Indice de risque des déchets de haute activité.

Quant à l'iode 129, dont la décroissance radioactive n'aura été que faible, on peut estimer que sur de telles durées, la garantie apportée par les deux premières barrières ne sera plus aussi grande. Si c'était le cas, on pourrait supposer qu'en définitive tout l'iode 129 se retrouverait dans les océans, compte tenu du fait que ce nucléide est peu adsorbé par les terrains naturels. Dans une telle hypothèse, on bénéficierait de l'énorme dilution apportée par le volume des océans et de la dilution isotopique apportée par l'iode inactif contenu dans l'eau de mer à une concentration moyenne de 60 mg/l. Si la quantité cumulée d'iode 129 produit dans le monde en l'an 2000 se retrouvait dans le milieu marin, la concentration moyenne dans l'eau de mer serait de 2 à 3 x 10^{-14} Ci/m3 (Sousselier, 1976) et le risque d'irradiation correspondant serait très faible. Cependant avec une utilisation continue pendant plusieurs siècles de l'énergie nucléaire, l'impact ne serait probablement pas négligeable. Ceci veut simplement dire que l'on devra choisir, pour le stockage des déchets contenant l'iode 129, les formations donnant les garanties de confinement à terme les plus élevées et étudier la réalisation de barrières géochimiques, efficaces aussi pour ce nucléide.

INCIDENCES D'AUTRES HYPOTHESES

Incidences d'autres cycles de combustibles

Incidences de non retraitement. Les conséquences sur la production de déchets ont été examinées précédemment. On peut schématiser leurs conclusions:

- doublement des quantités de minerais nécessaires
- suppression des déchets α et remplacement des déchets de haute activité par les combustibles eux-mêmes.

On a vu que les stériles de minerais étaient un déchet dont la gestion entraînait actuellement un impact potentiel non négligeable.

La suppression des déchets α peut, à première vue, paraître être un facteur favorable au non retraitement. Il faut cependant voir que cette méthode de gestion conduirait à ce que la totalité du plutonium se retrouve dans les déchets et non seulement un faible fraction, de l'ordre du pourcent. Cependant si la nuisance potentielle définie par exemple par l'indice de risque est de l'ordre de 100 fois plus élevée, s'il n'y pas retraitement, il ne faudrait pas conclure trop rapidement que le risque réel serait aussi 100 fois plus élevé. D'autres facteurs interviennent aussi. Du fait que les volumes de déchets sont moins importants dans l'option du non retraitement, on pourra choisir des conditionnements plus performants; on pourra peut-être aussi choisir comme lieu de stockage définitif des formations géologiques donnant des garanties encore meilleures. En fait le risque potentiel le plus important est le transfert de radioactivité par les eaux souterraines. La quantité transférée dépendra du débit d'eau susceptible de lixivier les déchets, de la valeur du taux de lixiviation, de la quantité de radionucléides dans les déchets et de l'efficacité des barrières, géochimiques en particulier. Le volume des déchets et surtout leur surface contribuent donc à augmenter ce risque potentiel.

D'autres facteurs peuvent par contre intervenir dans l'autre sens. Si la lixiviation des déchets à long terme est relativement élevée et que les produits lixiviés se trouvent piégés dans une seule barrière, on peut se trouver plus proche des conditions de criticité. En ce qui concerne ce dernier point, on a d'ailleurs avancé que le stockage des combustibles irradiés pourrait constituer un risque. En fait (NUREG - 0116, 1976) pour des raisons de dégagement de chaleur, on sera amené à disposer les conteneurs dans le stockage selon un réseau, avec une maille de l'ordre de 5 m, alors que la maille du réseau nécessaire pour une réaction en chaîne est de l'ordre de 0,50 m. Le stockage direct des éléments combustibles comme déchets semble supprimer l'impact entraîné par les rejets d'effluents gazeux. En fait, on ne connaît pas à terme comment pourra s'effectuer le relâchement des isotopes gazeux. Si l'on peut penser que pour le tritium, il n'y aura aucune possibilité d'impact : ce radionucléide serait piégé au cours de sa migration dans les formations géologiques, il est plus difficile de conclure pour le Krypton 85 (s'il diffuse à travers le conditionnement, il peut atteindre rapidement la surface, néanmoins, sa période est relativement courte) et surtout pour l'iode 129, radionucléide difficilement fixable dans le sol et les critères de choix de formations géologiques ne seront peut-être pas les mêmes pour des déchets contenant ce nucléide et pour les combustibles irradiés. Enfin, il faut remarquer que l'on ne connaît pas encore les méthodes qui seront utilisées pour conditionner les combustibles irradiés avant stockage définitif. Il est donc difficile de définir quel serait l'impact global de ce mode de gestion. Mais on peut d'ores et déjà conclure que le fait d'utiliser deux fois plus de minerais et celui de laisser la totalité du plutonium dans les déchets entraînent des impacts non entièrement négligeables de ce mode de gestion et probablement sensiblement supérieurs à ceux entraînés par le mode de gestion comportant le retraitement.

Incidence du devenir du plutonium : Si l'on retraite les combustibles irradiés, le plutonium peut être utilisé comme combustible des réacteurs de la filière surrégénérateur ou recyclé dans les réacteurs de la filière à eau ordinaire. Son envoi aux déchets après séparation semble une hypothèse peu vraisemblable et nous ne la prendrons pas en compte. Les deux autres hypothèses entraînent des conséquences différentes. Le recyclage du plutonium dans les réacteurs à neutrons thermiques entraîne un enrichissement du plutonium en isotopes supérieurs et le recyclage n'est possible que deux ou trois fois. Il y aurait en définitive, diminution des quantités d'uranium nécessaire - et aussi des besoins d'enrichissement - mais finalement la quantité de plutonium restant dans les déchets sera relativement importante. Il ne semble pas que ce point ait fait jusqu'à présent l'objet d'étude quantitative, ce le sera certainement dans le cadre des études de l'INFCE. Si par contre le plutonium est utilisé comme combustible dans la filière surrégénérateur, il peut complètement "brûler" et l'impact est considérable sur les besoins en uranium naturel et donc sur les productions des stériles de minerais, qui ne seraient plus que de 1 000 t/GWe x an, contre 45 000 t/GWe x an dans le cas du recyclage de l'uranium et du plutonium et de 80 000 t/GEe x an avec stockage définitif du combustible irradié. Le développement de la filière à neutrons rapides permettrait aussi la mise en application d'autres schémas de gestion des déchets, comme la "transmutation" des déchets contenant des émetteurs α et en particulier de l'américium et du curium.

Incidence de l'utilisation d'autres filières de réacteurs : Si jusqu'à la fin de ce siècle, le développement de l'énergie nucléaire sera essentiellement basé sur la filière de réacteur à eau ordinaire, avec vraisemblablement une importance croissante de la filière à neutrons rapides dans la dernière décennie, il est plus difficile de prévoir ensuite quelle pourra être la part des différentes filières. On peut estimer que la filière des réacteurs à haute température a une grande probabilité de développement compte tenu de ses nombreux avantages potentiels. Le développement d'autres filières, comme celles des réacteurs à sels fondus ou à refroidissement par fluide organique, semble plus douteux. Les réacteurs à haute température utiliseront certainement le thorium comme matière fertile et l'uranium 233 comme matière fissile. Le cycle correspondant comprendra donc un retraitement.

La production de produits de fission sera pratiquement la même que dans les réacteurs utilisant l'uranium 235 ou le plutonium comme matière fissile. Mais la grande différence sera qu'il n'y aura pas de production de plutonium, ni de transplutoniens. Cela ne veut pas dire forcément que le choix de cette filière n'entraînera la production d'aucun déchet contenant des émetteurs de longue période. En effet, la filière devra d'abord démarrer avec de l'uranium enrichi et ensuite on peut avoir la production de certaines quantités de radium 226 par décroissance radioactive du thorium 230 venant de la décroissance de l'uranium 234 obtenu par capture, mais les quantités ne peuvent être que faibles et, en ce qui concerne les déchets, la filière des réacteurs utilisant le couple Th-U-233, conduirait à des impacts encore plus faibles et à une gestion, plus simple que les filières actuellement développées ou en voie de l'être.

Incidences d'autres sources d'uranium : Il a toujours été supposé que l'uranium serait extrait de roches terrestres à une teneur moyenne de l'ordre de 0,2 %. Ceci correspond à la production actuelle et à ce qui est prévu au cours des années et probablement des décennies à venir. Mais une autre source potentielle qui fait l'objet de nombreuses études est l'eau de mer. Au cas où, dans l'avenir, l'uranium serait tiré de cette source, il en résulterait la suppression d'un impact sur l'environnement que sont les stockages de stériles de minerais avec la stérilisation de terrain, les risques de transfert de poussières, l'irradiation due au radon et les risques de transfert au radium. Du point de vue protection de l'environnement, il serait donc certainement intéressant d'explorer à fond les possibilités de cette source d'uranium, bien que les coûts semblent pour l'installation trop élevés, ainsi que les besoins en énergie.

Incidences d'autres schémas de gestion des déchets

Incidence du confinement total : Pour le krypton 85 et le tritium, il s'agirait seulement d'effectuer la séparation de ces nucléides avec une efficacité suffisante. Ensuite, et compte tenu de leur période, le stockage pendant la durée nécessaire à une décroissance suffisante, ne poserait pas de problèmes et l'intégrité des barrières de confinement pourrait être garantie sur de telles périodes. Les quantités à stocker ne poseraient pas non plus de problèmes aigus. Nous avons évalué à 60 000 bouteilles de gaz comprimé de 50 l, la quantité cumulée en l'an 2020 pour le krypton 85 : le stockage en puits de mine ou le stockage au fond des fosses océaniques (Beaujean et al., 1973) pour lequel la sûreté serait assurée par la solubilité du krypton dans l'eau de mer à la pression considérée et par la durée nécessaire à la diffusion du krypton vers les couches d'eau supérieures, pourrait être réalisé sans impact pratique pour l'environnement. La question se présente de manière similaire pour le tritium.

Le cas est un peu différent pour le carbone 14, compte tenu de sa période. On a vu plus haut que l'impact du rejet dans l'environnement de ce radionucléide ne serait peut-être pas entièrement négligeable, si l'on prend en compte l'engagement de dose sur de très longues durées et si les quantités estimées ne sont pas supérieures à ce qui sera effectivement produit. On sera peut-être amené à l'avenir à effectuer le piègeage de ce nucléide. L'impact sur l'environnement de son stockage ne devrait pas poser plus de problèmes que celui des déchets α. Enfin les déchets dans lesquels serait confiné l'iode 129 seront probablement ceux pour lesquels on devra chercher le maximum de garantie.

Incidence du recyclage intégral des déchets : Il s'agit du recyclage des émetteurs α dans les réacteurs, de façon à les transformer en nucléides à période courte. Ceci semble théoriquement possible dans les réacteurs à neutrons rapides. En effet le rapport section de fission/section de capture est plus favorable que dans les réacteurs à neutrons thermiques. Il n'est actuellement pas facile d'évaluer l'impact sur l'environnement qu'aurait la prise en compte d'un tel mode de gestion, qui suppose, d'abord, une séparation, dans les différentes catégories de déchets, des émetteurs α. Les études, actuellement menées dans divers pays, ont justement pour but, de permettre une étude coût bénéfice comparée. Mais à priori, en ce qui concerne le stockage, l'impact d'une telle méthode serait favorable : en effet elle permet d'éviter le besoin de stockage de très longue durée si la transmutation est effectuée. Il faut d'ailleurs remarquer, que la séparation des émetteurs α dans les déchets, même si elle n'est pas suivie de transmutation, permet de "séparer les facteurs" dans les stockages et de réaliser d'une part des stockages pour les déchets pour lesquels se posent les problèmes dus au dégagement thermique et d'autre part les stockages pour les déchets α : il sera alors probablement plus facile de réaliser des barrières géochimiques plus performantes.

Méthodes avancées de stockage ou d'évacuation : D'autres méthodes de stockage ont été proposées, qui sont en fait des variantes dans les méthodes de stockage dans les formations géologiques. Certaines, comme l'utilisation des calottes glacières ou celle faisant appel à la théorie de la tectonique des plaques ne semblent pas devoir être retenues. Par contre, celle préconisant les stockages dans les formations géologiques subocéaniques pourra, peut-être, être utilisée dans l'avenir. Il s'agit en fait d'une méthode combinant le stockage en couche géologique stable et l'immersion dans les fosses profondes. Cette méthode offrirait des avantages sensibles, diminuant le risque d'intrusion involontaire à long terme dans les stockages et aussi celui de stérilisation d'espace ou de ressources pour les générations futures. Si les études en cours prouvant la faisabilité de ce mode d'évacuation, montrent que sa sûreté est au moins égale à celui du stockage en couche géologique et que le coût est acceptable, sa mise en application diminuerait encore l'impact sur l'environnement du stockage des déchets.

Enfin un mode d'évacuation définitif des déchets qui a été proposé est l'évacuation spatiale. Qu'il s'agisse de mise en orbite solaire ou d'envoi vers le solaire, il est certain que cette méthode conduirait à l'élimination définitive des déchets pour lesquels elle serait appliquée. Les études (ERDA, 1976) ne permettent pas de conclure, dans leur état actuel, à la faisabilité et encore moins à l'économie de la méthode. Une objection que l'on fait souvent à cette méthode est les conséquences d'accident lors du lancement. En fait, un tel accident aurait pour conséquence l'immersion des déchets contenus dans la tête de la fusée correspondante. Le tout est l'évaluation de la probabilité d'un tel accident. S'il ne concerne qu'un lancement sur mille, l'immersion, dans certaines conditions de confinement du millième des déchets, aurait probablement moins d'impact potentiel que le stockage, même en couche géologique, de la totalité des déchets. Mais cette méthode pourrait se heurter à d'autres impacts, du moins si on voulait l'appliquer à la totalité des déchets : besoins en énergie, besoins en matériaux spéciaux pour enrobage, coût total. Il ne semblerait pas raisonnable d'utiliser une telle méthode pour tous les déchets, ni même probablement pour tous les déchets de longue période. Mais la question devrait être examinée de près pour l'iode 129, pour lequel il sera peut-être le plus difficile de garantir que l'impact sur l'environnement à long terme sera entièrement négligeable.

Incidence de changements dans l'environnement

Il est bien évident, surtout en ce qui concerne les déchets contenant des nucléides de longue période radioactive, que des changements peuvent intervenir dans l'environnement, indépendamment de la présence des déchets, changements dont il faut aussi examiner les conséquences sur le comportement de ces déchets. Ces changements peuvent venir de l'homme ou de la nature.

Changements apportés par l'homme : Les activités humaines sont susceptibles, de nombreuses façons, de modifier l'environnement d'un stockage. Il n'est pas nécessaire de prendre en compte le court terme, car on peut raisonnablement supposer que l'existence du stockage continuera à être connue pendant au moins plusieurs siècles. Il sera facile en effet de "marquer" le site de façon sûre et de transmettre les informations pendant de telles durées, même en cas de bouleversements sociaux, guerres, etc. Cela semble pouvoir s'appliquer au cas des actions susceptibles d'avoir une action directe (urbanisation s'étendant sur un stockage) ou à celles qui pourraient avoir des actions indirectes (construction de barrages pouvant entraîner la modification du régime des eaux souterraines). Mais là aussi la question se présenterait différemment pour des échéances beaucoup plus lointaines. On ne peut plus être certain que sur des durées de l'ordre de plusieurs millénaires, l'existence d'un stockage de déchets radioactifs continuera à être connue. Des activités humaines telles que des travaux publics (urbanisation, construction de barrages, de routes) ou travaux miniers pourraient se répercuter sur les stockages. Mais, compte tenu de ce qui vient d'être dit, aucune conséquence n'est à prévoir pour des stockages de déchets de courte période. Les conséquences à long terme seront d'autant plus faibles que les stockages seront réalisés à plus grande profondeur : c'est certainement vrai pour les actions directes (travaux miniers) et aussi pour les actions indirectes (modification des circulations d'eaux souterraines). Ceci est une raison pour ne pas stocker en surface ou même à faible profondeur des déchets contenant des émetteurs de longue période, même si l'étude des conditions hydrogéologiques et des barrières géochimiques naturelles montre qu'un stockage serait sûr, dans les conditions actuelles. On peut assimiler aux changements apportés par l'homme les changements qui surviendraient dans les conditions de vie, mode d'alimentation de l'humanité. On peut par exemple supposer que dans quelques millénaires, l'homme tirera des mers et même du fond des océans une part importante de son alimentation. Ceci est une raison pour être très prudent dans l'immersion des déchets contenant des émetteurs de longue période, et c'est ce qui a conduit l'AIEA à prendre dans ses nouvelles recommandations (IAEA, 1978) comme limite d'activité des déchets dont l'immersion est interdite 1 Ci/t pour les déchets α (et 10^{-1} Ci/t pour le radium 226) au lieu de 10 Ci/t.

Changement de l'environnement naturel : Les changements dans les conditions naturelles ne paraîssent pas susceptibles d'être tels à court terme qu'ils puissent changer les éléments pris en compte dans les analyses de sûreté. Par contre, en ce qui concerne les stockages de déchets de longue période, ces analyses prennent en compte ces changements. C'est en particulier le cas pour l'étude de sûreté suédoise (KBS, 1978) qui étudie par exemple, les conséquences d'une nouvelle glaciation. Là aussi, on peut conclure que s'il s'agit d'un stockage réalisé à une profondeur suffisante (de l'ordre de 3 à 400 m au moins) les conséquences de changement de l'environnement ne mettraient pas en cause la sûreté du stockage.

SYNTHESE ET CONCLUSIONS

Synthèse des résultats de l'étude

L'étude détaillée des différents stades de la gestion des déchets radioactifs telle qu'elle est actuellement réalisée et des techniques qui sont développées et seront mises en oeuvre au fur et à mesure du développement de l'énergie nucléaire a clairement montré que les impacts potentiels sur l'environnement sont actuellement et resteront dans le futur, faibles voire négligeables. Les principales conclusions par lesquelles se sont terminés les différents paragraphes peuvent être synthétisées ainsi :

- Opérations de traitement et de conditionnement : Impact sur l'environnement très faible, en particulier si on le compare à celui de l'ensemble des opérations du cycle. Pas de risque d'accident pouvant avoir des conséquences importantes.

- Rejets d'effluents :

  Kr-85 : Pas d'impact radiologique. Faible risque dû à l'augmentation de l'ionisation de l'air et faible impact potentiel sur les conditions de travail dans les usines d'air liquide, si aucune rétention n'est effectuée. Méthodes de rétention en cours de développement, pas de problème de stockage.

  Tritium : Impact radiologique très faible, surtout si le tritium est rejeté dans la mer et à fortiori dans les couches océaniques profondes.

C-14 : Impact radiologique négligeable, sauf si l'on estime nécessaire d'intégrer les doses à la population sur de très longues durées (dizaine de milliers d'années).

I-129 : Pas d'impact si l'on piège ce nucléide. Eventuel problème du très long terme (millions d'années) mais quantités faibles à stocker dans les formations géologiques offrant le maximum de garanties.

- Stockage des déchets :

Faible activité sans alpha - Impact négligeable sur le plan des risques radiologiques et très faible dans les autres domaines (stérilisation de terrain, impact sur les ressources naturelles, etc.) tant en ce qui concerne le stockage en surface que l'immersion dans les fosses océaniques.

Haute activité (produits de fission) - Le stockage sous forme liquide est le seul stade de gestion au cours duquel peut théoriquement se produire un accident entraînant le rejet dans l'environnement de quantités relativement importantes de radioactivité mais la probabilité d'un tel accident est très faible. Stockage intérimaire sous forme solide : aucun impact. Stockage définitif en formations géologiques : possibilité à long terme de migration de radionucléides vers les eaux de surface mais seulement à de très faibles concentrations et sans réelle signification biologique. Risque un peu plus élevé d'intrusion involontaire de l'homme qui devient très faible si la formation géologique choisie est suffisamment profonde (plusieurs centaines de mètres) et si elle est réalisée dans une roche ne possédant pas de ressources minérales potentielles.

Déchets alpha - Conclusions analogues à celles des déchets de haute activité, si le stockage est fait en formation géologique à une profondeur suffisante.

Déchets de démantèlement : impacts potentiels négligeables.

Stériles de minerais : impact faible si des précautions adéquates sont prises dans la stabilisation des aires d'épandage et si une couverture d'épaisseur suffisante est réalisée.

L'étude de l'incidence possible d'autres formes de développement de l'énergie nucléaire ne fait pas apparaître de changement notable dans ces conclusions, tandis que des méthodes plus sophistiquées de gestion des déchets sont en cours d'étude, qui permettraient, si on l'estimait justifié, de réduire encore les impacts potentiels.

Impacts réels et hypothétiques de la gestion des déchets

Mais on peut aussi mettre en doute certaines des hypothèses qui ont permis d'aboutir aux conclusions précédentes. En fait, deux questions importantes à prendre en compte sont la signification des faibles doses et la durée d'intégration des doses dans le cas de rejet de radionucléides de période longue. Pour la première, la Commission internationale de protection radiologique a apporté les éléments de réponse dans sa publication 26 du moins en principe. Mais l'étude n'est pas facile à faire en pratique, surtout quand il s'agit d'irradiations potentielles de l'ordre du mrad/an ou même moins. S'agit-il d'un impact réel? ou plutôt d'un impact hypothétique mais dont on ne peut, du moins actuellement, démontrer qu'il est entièrement nul? Il est bien évident qu'une réponse définitive à cette question permettrait de simplifier éventuellement la gestion des déchets. Quant à la deuxième question, on a été amené, plusieurs fois, au cours de ce rapport, à en souligner l'importance. L'intégration des doses sur de très longues périodes ne semble pas plus raisonnable que la prise en compte du très long terme. Qu'il s'agisse de population régionale ou mondiale, on ne peut savoir ce qu'elle sera dans 10 000 ans, à fortiori dans 100 000 ans. Les calculs faits sur de telles périodes ont-ils une signification? Il semblerait raisonnable de ne prendre en compte, du moins pour les analyses du type coût-bénéfice, qu'une durée de l'ordre de 1 000 ans ou de quelques milliers d'années, étant entendu - et ce rapport l'a bien montré - que l'impact sur l'environnement, même à beaucoup plus long terme, ne pourra être que mineur.

C'est pour cela que la prise en compte de très long terme relève plus d'une option philosophique que de données scientifiques. Peut-on accepter de faire courir un risque même s'il est extrêmement faible à des générations futures, suite à des actions menées actuellement et profitant à l'humanité à court terme? Il semble que la réponse ne puisse être que oui à partir du moment où l'on a la garantie que même à très long terme, le risque reste faible. D'ailleurs, si l'on veut se placer à extrêmement long terme, on constate que l'utilisation de l'énergie nucléaire diminuera le risque radiologique. L'utilisation de la fission de l'uranium (directe ou indirecte via celle du plutonium) aboutira dans un avenir lointain à une diminution des teneurs naturelles en radium et l'on sait que ce radionucléide est plus toxique que le plutonium, surtout en ce qui concerne le risque d'ingestion. D'une façon plus générale, l'homme, en utilisant l'énergie nucléaire, ne crée pas la radioactivité, il augmente pendant une certaine durée la radioactivité ayant transféré des nucléides à période longue (uranium 235) en nucléides à période plus courte (produits de fission). De nombreuses études ont montré qu'au bout de quelques centaines d'années, l'indice de nuisance des produits obtenus était inférieur à celui de l'uranium dont on est parti et en particulier les études de Claiborne (1972), Hamstra (1975), le rapport KBS (1978) qui concluent que cette durée est comprise entre 300 et 1 000 ans, la différence venant des différentes façons possibles d'évaluer cet indice de nuisance. Mais ce qui compte est l'ordre de grandeur. Ceci ne veut pas dire que les déchets radioactifs ne représenteront plus aucune nuisance au bout d'un millier d'années ni que l'on pourrait considérer qu'il serait alors possible de les abandonner dans n'importe quel environnement. Si l'homme n'a pas créé de nuisances radio-actives nouvelles, il a modifié l'environnement naturel, les conditions de lixi-viation et les risques de transfert. Mais les études faites ont amplement montré qu'il était possible de prendre en compte ces nouvelles conditions de façon à garantir que l'impact soit, au plus, très faible.

Mise en oeuvre de la politique de gestion et impacts potentiels

L'impact de la gestion des déchets telle qu'elle a été menée jusqu'ici ne peut donc conduire qu'à des impacts négligeables et on dispose de méthodes et de procédés dont la mise en oeuvre conduira à des résultats identiques. Mais en fait, le développement de l'énergie nucléaire n'en est encore qu'à sa phase initiale. Tous ces modes de gestion ne sont pas encore - et ne peuvent d'ailleurs pas avoir été - mis en oeuvre à grande échelle. On peut mettre en doute la simultanéité de cette mise en oeuvre et du développement de l'énergie nucléaire. Cependant, jusqu'à présent, au fur et à mesure du développement de l'énergie nucléaire, on a mis en place les mesures de sûreté, les procédures de contrôle adaptées à ce développement. Les exigences en matière de sûreté ont plutôt été de plus en plus strictes avec ce développement.

Les délais de mise en application ne semblent pas non plus devoir être mis en doute en ce qui concerne les traitements, conditionnements et mise en stockage intérimaire. Mais la question se pose un peu différemment en ce qui concerne l'envoi au stockage définitif spécialement pour les déchets de haute activité et les déchets alpha. On a vu, en effet, que les méthodes actuellement réalisées et prévues pour les stockages intérimaires offrent un degré très élevé de sûreté pour plusieurs décennies. Et il sera certainement utile de disposer de ce délai pour terminer les études des stockages définitifs et les réaliser. Mais il ne faudrait pas que, compte tenu de la sûreté des stockages intérimaires, on attende 50 ans voire 100 ans pour la mise en oeuvre de ces stockages définitifs. Les quantités en cause entraîneraient que les opérations de transfert constitueraient alors un poids non négligeable pour les générations futures qui n'utiliseraient peut-être plus l'énergie nucléaire. Il faut être conscient qu'il sera d'autant plus difficile de faire procéder à des opérations de conditionnement et de mise en stockage que l'utilisation de l'énergie nucléaire serait sur une pente descendante. On pourrait alors aboutir, si des stockages intérimaires étaient utilisés, sans avoir été prévu pour, comme stockage définitif, à des situations entraînant un impact non négligeable sur les générations futures. Il est donc indispensable que dans un délai raisonnable, les stockages définitifs soient réalisés et que les déchets en relevant y soient transférés.

Conclusion

Les déchets radioactifs sont souvent considérés dans l'opinion publique comme constituant un problème non bien maîtrisé et risquant d'entraîner des impacts importants sur l'environnement et la santé de l'homme. Ce jugement découle probablement beaucoup du fait que les méthodes de traitement et de conditionnement n'ont pas toutes entièrement atteint le stade industriel et que, pour plusieurs catégories de déchets, les installations destinées au stockage définitif n'ont pas été réalisées. Mais comment pourrait-t-il en être autrement alors que l'on en est encore qu'au début du développement de l'énergie nucléaire, alors que certains déchets n'ont pas encore été produits en quantités importantes? Au contraire, on ne peut que conclure que les développements ont atteint un stade très avancé. Il est sûr que certaines méthodes de gestion devront être infléchies dans l'avenir, comme on l'a vu, comme par exemple la gestion des stériles de minerais qui devront être soigneusement recouverts, que d'autres devront en temps opportun être mises en oeuvre comme la rétention du krypton 85 et de l'iode 129. Des stockages définitifs devront être réalisés, dans un délai raisonnable, dans des formations géologiques stables à des profondeurs adéquates, pour les déchets contenant des émetteurs radioactifs de longue période. Mais tout cela est prévu et on ne voit pas s'obstacles qui pourraient s'opposer, sur le plan technique, à une mise en oeuvre. On peut donc, en fait, conclure raisonnablement que dans l'avenir l'impact sur l'environnement de la gestion des déchets continuera à être minime comme il l'est à présent.

REFERENCES

AEN, Pratique de gestion des déchets radioactifs en Europe Occidentale - Agence Européenne pour l'Energie Nucléaire, Septembre (1971) 146 pp.

AKA, Spent nuclear fuel and radioactive wastes - Rapport AKA Committee au gouvernement suédois - MALMO (1976) 94 pp.

APS, Report to the American Physical Society by the Study Group on Nuclear fuel cycle and waste management. American Physical Society, Review of Modern Physics - 50, No. 1 (1978) 184 pp.

Bardet, G., Expérience de 7 années de stockage de déchets solides de faible et moyenne activité en surface ou en tranchées bétonnées. Symposium AIEA - SM 207/39 - Vienne (1976).

Battelle Laboratories, High Level Radioactive waste management alternatives - Battelle Pacific Northwest Laboratories BNWL 1900 (1974).

Beaujean et al., Gaseous radioactive emissions from reprocessing plant and their possible reduction. Symposium AIEA - SM 172 - Aix en provence - on environment behavior of radionuclides released in the nuclear industry. Conf. 730-503.

Bendixsen, C.L., Separation and Storage of krypton-International Symposium on the management of waste from the LWR fuel cycle, Denver (1976).

Blomeke, J.O., Projections of wastes to be generated - Symposium on the management of wastes from the LWR fuel cycle - Denver (1976).

Boeck, W.L., Krypton 85, a global contaminant - Electrical process in atmosphere - Congrès de Garmisch-Partenkirchen (1974).

Bonniaud, R., La vitrification continue des produits de fission. Revue Générale Nucléaire, 490 (1976).

Claiborne, H.C., Neuron Induced Transmutation of High Level Radioactive Waste - Rapport ORNL-TM 3964 (1972) 85 pp.

Claiborne, F., Potential Containment failure mechanisms and their consequences at a radioactive waste repository in bedded salt in New-Mexico - ORNL - TM 4639 (1974).

Cohen, B.L., Environmental hazards from high level wastes. Proceedings of the Conference on the management of radioactive wastes: waste partitioning as an alternative. Seattle (1976) PB 254.

Cohen, B.L., The disposal of radioactive wastes from fission reactors, Scientific American - Vol. 236-6, 21, (1977).

Commission des Communautés Européennes - Rapport sur la gestion des déchets radioactifs provenant des centrales de puissance (à paraître prochainement).

Culler, F.L. et al., Management of high level and alpha bearing wastes. IAEA - CM-36/15 - International Conference on Nuclear Power - Salzbourg (1977).

de Marsily, G., et al., Les déchets de l'industrie nucléaire et le sous-sol - Annales des Mines 184, 35-56 (1978).

Detilieux, E.J., Status of the decommissioning programme of the Eurochemic reprocessing plant. Symposium on the management of wastes from the LWR fuel cycle. Denver (1976). Conf. 760-701.

El-Hinnawi, E., Les effets écologuques de l'énergie d'origine nucléaire - Bulletin de l'AIEA - Vol. 20 - 2 (1978).

Finneman, P., Progress in waste disposal research and development. Power reactor technology and reactor fuel processing, Vol. 10, No. 1 (1966-1967).

Gauvenet, A., Les résidus radioactifs, Revue de l'Energie, No. 290, 292-293 (1977).

Hamstra, J., Radiotoxic Hazard Measure for buried solid radioactive wastes - Nuclear Safety - Vol. 16, No. 2, 180 (1975).

Holcomb, W.F., A summary of shallow land burial of radioactive wastes at commercial sites between 1962 and 1976 with projection - Nuclear Safety, 19, No. 1, 50 (1978).

IAEA, Document INF.CIRC/205/Add.1 - Janvier (1975) 22 p.

IAEA, Gestion des déchets des mines et des usines de traitement des minerais d'uranium et de thorium - AIEA - Collection Sécurité No. 44. Vienne (1977).

IAEA, Guide to the safe handling of radioactive wastes at nuclear power plant Safety Series AIEA (encours de publication).

IAEA, Groupe de consultants sur la mise en application des responsabilités de l'Agence vis-à-vis de la Convention de Londres. (Discussions non publiées).

IAEA, International Symposium ou transuranics nuclides in the environment. IAEA - SM - 199 - San Francisco (1975).

IAEA, L'énergie d'origine nucléaire dans les pays en développement - Bulletin de l'AIEA - Vol. 19 - No. 3 (1977).

IAEA, The Agency's responsibilities in connection with the convention on the prevention of marine pollution by dumping of wastes AIEA - Doc. GCV/1889 (1978) 29 pp.

ICRP, Recommendations of the International Commission on Radiological Protection - ICRP Publ. 26, Pergamon Press (1977) 56 pp.

Internationale Kommission zum Schutze des Rheins Gegen Verunnreinigung - Albeits gruppe R. über der Folge der Ableitung Radioaktive Stoffe in Einzugsgebiet des Rheins. Reichs Institut für Volksgesundheit - Octobre (1974).

KBS, Handling of spent nuclear fuel and final storage of vitrified high level reprocessing waste - Kärn Bränsle Säkerhet report - Stockholm (1978).

Kirk, W.P. et al., In vivo solubility of Kr-85 in guinea pig tissues. Health Physics - Vol. 28.

National Academy of Sciences, Radioactive wastes at the Hanford Reservation - National Academy of Sciences (1976) 45 pp.

OECD, Mécanisme de surveillance internationale pour l'évacuation des déchets radioactifs en mer. L8observateur de l'OCDE - No. 88, 19, (1977).

OECD, Objectifs, concepts et stratégie en matière de gestion de déchets radioactifs provenant des programmes nucléaires de puissance. Agence pour l'Energie Nucléaire de l'OCDE (1977) 156 pp.

ORNL, Siting of fuel reprocessing plants and waste management facilities Report ORNL 4451 (1970) 405 pp.

Pautrot, P. et Arnault, J.P. - Tritium extraction plant of Laue-Langevin Institute - Trans Ans, Vol. 20, 202 (1975).

Royal Commission Environmental Protection, Nuclear Power and the Environment - Sixth Report - Royal Commission on Environmental Protection - Londres (1976) 237 pp.

Shepherd, J.C., A simple model for the dispersion of radioactive wastes dumped on the deep sea bed. Fisheries Research Technical Report No. 29, 19 pp.

Sousselier, Y., Les déchets radioactifs et leur gestion - Revue Générale Nucléaire No. 3, 160-165 (1978).

Sousselier, Y. et al., La stockage à très long terme des produits de fission - Symposium AIEA sur la gestion des déchets - Vienne (1976) 37-48.

Thomas, T.R., Separation and retention of iodine - International Symposium on the management of wastes from the LWR fuel cycle. Denver (1976).

UK National Radiological Protection Board, The predicted radiation exposure of the population of the European Community resulting from discharges of Kr-85, H-3, C-14 and I-129 from the nuclear power industry to the year 2000 - Rapport du National Radiological Protection Board U.K. publié par la C.C.E. - Luxembourg (1975) 87 pp.

UNSCEAR, Sources and Effects of Ionizing Radiation, U.N.S.C.E.A.R., United Nations, (1977) 725 pp.

USERDA, Alternatives for managing wastes from reactors and post fission operations in the LWR fuel cycle by Pacific Northwest Laboratory - Rapport ERDA 76-43, W Washington, D.C. (1976).

US Geological Survey, Geologic disposal of high level radioactive waste: earth science perspectives - US Geological Survey Circulaire 779 (1978) 15 pp.

US Nuclear Regulatory Commission, Environmental Survey of the reprocessing and waste management portions of the LWR fuel cycle - Rapport NRC - NUREG 0116 (1976) 256 pp.

WAES, Workshop on Alternative Energy Strategics "Energy supply demand integration to the year 2000". MIT Press-Cambridge (1977).

Webb, G.A.M. and Morley, F., A model for the evaluation of the deep ocean disposal of radioactive waste. National Radiological Protection Board - Harwell (1973) 26 pp.

9

L'IMPACT SUR L'ENVIRONNEMENT DES TRANSPORTS DE MATIERES RADIOACTIVES DU CYCLE DE COMBUSTIBLES NUCLEAIRES

Y. Sousselier
Institut de Protection et Sûreté Nucléaire
Commissariat à l'Energie Atomique
92260 Fontenay-Aux-Roses, France

INTRODUCTION

Tout transport de marchandises entraîne un certain impact sur l'environnement: utilisation de matières premières et d'énergie pour la construction et la traction des véhicules, bruits occasionnés par le transport, mauvaises odeurs et/ou produits nocifs rejetés par le véhicule, nombre de transports entraînant un encombrement des voies de transport ou nécessitant la création de nouvelles voies, risques d'accidents pouvant entraîner des dégâts matériels sur d'autres véhicules ou des morts ou blessures des usagers. Dans le cas de transport de matières dangereuses il y a en plus les risques de relâchement dans l'environnement de ces matières en cas d'accidents.

Les transports de matières radioactives pourront entraîner les mêmes impacts. Cependant il est à priori évident que le nombre de transport de matières radioactives nécessité par le développement de l'énergie électronucléaire est très faible, par rapport, par example, à celui du transport des carburants et autres matières dangereuses - ceci étant une conséquence du contenu énergétique très élevé de l'uranium - et que donc tous les modes d'impact énumérés en premier plus haut seront faibles, voire négligeables, et qu'il faudra surtout examiner le problème des accidents. Il faudra, aussi examiner celui du risque d'irradiation en conditions normales de transport. Les impacts des transports sur l'environnement pourront être différents - aussi bien pour les transports de matières radioactives que pour les autres transports - suivant le mode de transport utilisé: route, voie ferrée, transport aérien ou maritime. On devra examiner l'incidence du choix de tel ou tel mode de transport.

Cette étude d'impact sera limitée au transport proprement dit, c'est-à-dire qu'elle n'incluera pas les opérations de chargement et déchargement des matières radioactives dans les emballages, mais elle prendra en compte les opérations de chargement des colis sur les véhicules. Elle n'étudiera pas le problème des actions malveillantes, sabotages, qui dépendent beaucoup trop des conditions locales et qui sont très variables d'ailleurs suivant les circonstances. On peut cependant remarquer que nombre de transports radioactifs, et en particulier ceux de combustibles irradiés, sont "auto-protégés" contre cette nature de risque par la masse de l'emballage et la résistance de ses différents composants, tandis que d'autres (transport de matières fissiles) feront l'objet de mesures de protection, en raison du risque de détournement de cas matériaux, qui les garantiront aussi contre le risque de sabotage.

LES TRANSPORTS DU CYCLE DE COMBUSTIBLE

Le développement de l'énergie électronucléaire jusqu'à l'an 2000 sera essentiellement basé sur la filière de réacteurs à eau ordinaire avec probablement une part, commençant à devenir importante au cours de la dernière décennie du siécle, de réacteurs surrégénérateurs au plutonium.

Transports de Cycle de Combustibles d'un Réacteur de 1 GWe

Le tableau 1 donne, pour chaque stade du cycle de combustible associé au fonctionnement pendant un an d'un réacteur de la filière à eau ordinaire de 1 GWe, les tonnages à transporter et le nombre de véhicules nécessaires.

Tableau 1

TRANSPORTS ANNUELS LIES AU CYCLE DE COMBUSTIBLE D'UN REACTEUR A EAU ORDINAIRE DE 1 GWe

	TONNAGE(1)	ACTIVITE(3) Ci	NOMBRE DE VEHICULES(4)
minerai	80 000 t/an	60	
concentrés	180 t/an	-	9
UF_6 naturel	240 t/an	-	20
UF_6 enrichi	50 t/an	-	4
combustibles neufs	30 t/an	-	5
combustibles irradiés	500 t/an	150×10^6	6
plutonium	300 kg/an	-	3
déchets de haute activité	130 t/an	150×10^6	2
déchets de faible et moyenne activité	730 m3/an(2)	3 000	25

(1) Il s'agit des tonnages nets des matières à transporter, sauf pour les combustibles irradiés et les déchets de haute activité où il s'agit du tonnage total, conteneur compris.

(2) Pour les déchets de faible activité, le chiffre important est

le volume.

(3) L'activité des concentrés UF_6 et combustibles neufs est faible, celle indiquée pour le minerai est celle du radium contenu, celle indiquée pour les déchets de haute activité ne tient pas compte de la décroissance radioactive.

(4) Tous les transports sont supposés faits par la route, à l'exception des combustibles irradiés et des déchets de haute activité faits par voie ferrée et il s'agit alors de wagons.

Il est évident que pour les matières qui sont transportées en vrac (minerais) ou en emballage dont le poids est faible devant celui de la matière elle-même (concentrés, UF_6, combustibles neufs) le poids à prendre en compte est celui de la matière. Pour les combustibles irradiés ou les déchets de haute activité pour lesquels le rapport charge utile/poids total de l'emballage varie de 1/15 à 1/30, c'est le poids de l'emballage qui compte. Tous les transports du cycle n'ont pas la même importance en ce qui concerne l'impact potentiel. Les transports de minerais en particulier peuvent être laissées de côté dans l'étude: les usines de traitement de minerais sont situées près des gisements. La plupart de ceux-ci - tout au moins pour les districts uranifères mondiaux correspondant aux réserves les plus importantes - sont situés dans les régions peu peuplées, voire semi-désertiques. Enfin ces regions ont un niveau d'irradiation naturel élevé, dû justement aux gisements. Aussi pour une étude faite à un niveau mondial, on peut négliger l'impact de ces transports. Par contre pour certaines études faites au niveau national ou régional, il faudrait, en toute rigeur, en tenir compte.

Le tableau 1 ne comporte pas de données sur les distances sur lesquelles ces tranports sont effectués. C'est cependant un élément important. Mais il diffère trop d'un pays à l'autre. On peut cependant souligner que les transports de concentrés (yellow-cake) se font sur de grandes distances, que ceux de combustibles irradiés également, tout au moins pour la période actuelle pendant laquelle il n'y a que peu d'usines de retraitement en fonctionnement. Mais on peut prévoir que ce facteur changera d'ici 10 à 20 ans, le développement de l'énergie nucléaire entraînant la construction d'installations nouvelles.

Expérience Actuelle de Transport

L'énergie nucléaire en est encore à un stade de développement, en ce qui concerne le production d'électricité. Mais l'expérience en matière de transport de matières radioactives est déjà étendue, ce qui permet de conclure, que cette étude n'est pas seulement une étude prospective, mais qu'elle repose sur des éléments réels. C'est ainsi que plus de 750 t de combustibles irradiés ont été transportées en Europe à partir de réacteurs de la filière à eau ordinaire de 1966 à 1977 (KBS, 1978) et que 4 200 t de combustibles irradiés de la filière gaz graphite l'ont été en France de 1966 à 1977. Aux Etats-Unis on estimait en 1973 (Brobst, 1973) le nombre d' expéditions de matière radioactives à 800 000, mais 95% concernaient des radioisotopes à usage médical ou industriel qui n'entrent pas dans le cadre de cette étude. La même source estimait pour le cycle de combustible à 40 000 environ le nombre d'expéditions ayant eu lieu la même année. Ce chiffre ne comprenait pas les transports de minerais. En France, le nombre de transports radioactifs effectués en 1975 - à l'exclusion des radioisotopes - et des transports de minerais s'est élevé à 6 500.

Prévisions du Nombre Total de Transport

Le nombre de transports va croître dans l'avenir, de manière similaire à l'accroissement de la puissance nucléaire installée. Peu d'études publiées sur les prévisions globales de transport en dehors d'analysis faites pour les Etats-Unis. On peut mentionner le récent rapport suédois sur le retraitement et le stockage des déchets (KBS, 1978) qui évalue, pour 13 réacteurs installés avec une puissance totale de 10 700 MWe, à 300 t/an le tonnage de combustibles irradiés à transporter soit environ 100 transports par an. Il est prévu d'expédier ces combustibles irradiés en France par voie maritime: le nombre de ces transports sera de 16 par an. Le rapport étudie aussi les transports en retour de déchets de haute activité (blocs de verre) et les estime à 300 conteneurs par an, soit 3 à 4 transports par an. L'étude américaine la plus complète, couvrant les prévisions de transport de matières radioactives aux Etats-Unis est un peu ancienne (Blomeke et al., 1975) mais elle semble toujours suffisamment exacte pour le sujet qui nous occupe. Le tableau 2 synthétise les principaux résultats.

Tableau 2

PREVISION DE TRANSPORTS DU CYCLE DE COMBUSTIBLES AUX USA

		1980	1990	2000
puissance installée GWe		85	410	850
combustibles neufs	tonnes	3 600	11 800	21 100
	nombre de transport	670	2 500	5 300
combustibles irradiés	tonnes	2 200	7 800	17 700
	nombre de transport	2 100	6 450	12 350
plutonium	tonnes fissiles	7.3	51	158
	nombre de transport	20	143	438
déchets de haute activité	nombre de transports	58	215	475

Une étude plus rècente a été publiée par le U.S. Department of Energy (DOE, 1977). Elle porte sur les transports correspondant à 200 réacteurs en fonctionnement, ce qui correspond à peu près neufs à 1 200 par an, celles des combustibles irradiés à 2 000 ou à 8 000 par an suivant qu'elles seront faites par voie ferrée ou par voie routière. L'étude précedente admettrait une certaine répartition entre les deux modes. Enfin l'étude D.O.E. évalue le nombre d'expédition de déchets de faible activité produits par le réacteur à 4 400 par an. Ces chiffres recoupent bien ceux de l'étude précedente , mais ont aussi l'intérêt de montrer, pour le nombre de transports de combustibles irradiés, l'influence considérable du mode de transport: ce nombre est divisé par 4 si les expéditions sont faites par rail (wagons groupés mais pas nécèssairement trains complets) au lieu de la route. Enfin l'étude de la NRC (NUREG-0170, 1977) aboutit pour 1985 à 6 500 expéditions pour le combustible neuf à 18 000 expéditions pour l'UF_6 naturel et 2 400 pour l'UF_6 enrichi. Nous pouvons tenter, à partir des chiffres précèdents, d'évaluer le nombre total de transports de matières radioactives dans le monde. Nous estimerons que la puissance totale installée sera de 2 000 GW_e (tableau 3).

Tableau 3

PREVISION DE TRANSPORTS RADIOACTIFS DANS LE MONDE EN L'AN 2000

	TONNAGE A TRANSPORTER
concentrés	400 000 t
UF_6 naturel	500 000 t
UF_6 enrichi	100 000 t
combustible neuf	60 000 t
combustible irradié	45 000 t
plutonium	450 t
déchets de faible et moyenne activité	1 250 000 m3
déchets de haute activité	3 000 m3

- Tous les chiffres indiqués correspondent aux tonnages nets à transporter excluant donc, contrairement au tableau 1, le poids des conteneurs, même pour les combustibles irradiés.
- les déchets de haute activité correspondent aux produits de fission vitrifiés.

Ces chiffres ne sont, bien entendu, à être considérés que comme des ordres de grandeur.

REGLES APPLICABLES POUR LES TRANSPORTS RADIOACTIFS

Le nombre des transports, examiné au paragraphe précèdent, et les chiffres élevés des radioactivités transportées montrent clairement que les transports radioactifs constituent un risque potentiel certain. Ce risque a été neutralisé par l'élaboration de règles de transport très strictes dont l'application s'est traduite par le fait qu'aucun transport de matières radioactives effectué jusqu'à présent n'a jamais donné lieu à un accident ayant entraîné une irradiation notable du public ou des agents d'exécution. Il est important pour juger de l'impact potentiel futur d'examiner de degré des risques que laisse subsister la règlementation et de voir comment celle-ci est appliquée.

Reglementation

Le texte à la base de toute le règlementation actuelle est le règlement de transport des matières radioactives de l'AIEA. Toutes les règlementations nationales et internationales sont pratiquement alignées sur ce texte qu'il suffit donc d'examiner. Cette règlementation est principalement axée sur deux principes:

- la sûreté doit être apportée par l'emballage
- l'emballage doit être adapté à la nature du risque

Du point du vue de la radioactivité, les emballages sont divisés en trois catégories:

- type B
- type A
- emballage industriel

Dès que la radioactivité transportée dépasse un certain seuil - qui dépend d'une part des radiosotopes transportés et d'autre part de leur aptitude à se disperser - l'emballage doit être du type B. Ces emballages sont de mature à résister à tous les accidents dans lesquels ils sont susceptibles d'être impliqués. En dessous de la limite du type B et jusqu'au millième de cette limite, l'emballage doit être du type A. Ce type d'emballage garantit le confinement de la matière radioactive dans les conditions normales de transport les incidents de manutention étant compris. En conditions accidentelles par contre, l'emballage peut être détruit, mais la limite d'activité admissible a justement été déterminée de façon que, même alors, aucune personne du public ou du personnel chargée des transports ne puisse recevoir une dose supérieure aux normes. En dessous du millième de la limite du type B, l'emballage requis est dit du type industriel et ne garantit le confinement de la matière que dans des limites moins rigoureuses. Quel que soit le type d'emballage à utiliser, la règlementation prévoit que cet emballage doit être tel qu'il limite le débit de dose de rayonnement émis par la matière transportée à des valeurs bien définies et qui sont:

- 200 mrem/h à la surface externe de l'emballage
- 10 mrem/h à 2 m du véhicule de transport
- dans le cas de transport routier, 2 mrem/h en tout point de véhicule normalement occupé

L'emballage doit aussi, pour le transport des matières fissiles, garantir contre le risque de criticité. On distingue de ce point de vue, trois classes de sécurité nucléaire suivant que la guarantie est apportée:

- quel que soit le nombre et la disposition des colis
- quel que soit la disposition des colis mais pour un nombre limité
- moyennant des précautions spéciales et pour un nombre limité

La règlementation peut donc paraître moins stricte vis-à-vis du risque de criticité que vis-à-vis de celui de l'irradiation , mais en fait les hypothèses à prendre en compte sont très sévères (empilement de colis après accident de façon à avoir le réactivité maximale, c'est-à-dire, en général, réseau régulier inscrit dans une sphère, ce qui est en pratique inconcevable) et le degré de sûreté est le même, en fait, vis-à-vis des deux natures de risque.

La réglementation prévoit aussi des limites très strictes pour la contamination radioactive des surfaces extérieures du colis. Les valeurs principales de ces limites sont de:

10^{-4} μCi/cm2 pour les émetteurs
10^{-5} μCi/cm2 pour les émetteurs

Limites et Applications de la Règlementation

A strictement parler, la réglementation n'élimine donc pas intégralement tous les risques d'irradiation du public ou de contamination du milieu, mais elle maintient ces risques à un niveau extrêmement bas, tans dans les conditions normales de transport que dans les conditions accidentelles. Mais une autre question se pose: quelle est la façon dont sont respectées toutes les prescriptions règlementaires? Il y a en fait deux aspects à cette question: la façon dont sont respectées en pratique ces prescriptions et le risque d'erreur involontaire dans leur exécution. En fait la règlementation prévoit de façon détaillée un certain nombre de procèdures administratives: approbation des modèles d'emballages par l'autorité compétente sur présentation d'un dossier comprenant, en particulier, le détail des essais règlementaires auxquels ont été soumis les modèles et qui constitue une véritable analyse de sûreté de l'emballage, contrôle de qualité obligatoire dans la fabrication et l'entretien des emballages, rédaction, avant chaque expédition d'un document: l'attestation d'expédition, engageant la responsibilité de l' expéditeur sur le respect des prescriptions règlementaires.

On peut cependant remarquer plusieurs choses:

a) l'activité de la matière transportée, quand il ne s'agit pas de substances pures est souvent difficile à déterminer. C'est en particulier le cas pour les déchets et dans une moindre mesure, pour le combustibles irradiés. Les calculs de protection ont été faits pour des activités données et un spectre de rayonnement donné; il pourrait se produire que l'activité réelle du chargement soit supérieure aux prévisions et que par voie de conséquences, les limites règlementaires de débit de dose soient dépassées. En toute rigueur, une partie du chargement devrait alors être déchargée, mais on peut se demander si cela - qui est certainement à l'heure actuelle effectivement réalisé - continuera à être fait dans une période où le nombre des transports exigera des cadences d'expéditions très strictes. Cependant l'impact sur l'environnement ne pourra être que très faible. On peut difficilement supposer que les erreurs faites sur les estimations de l'activité conduisent à des valeurs des débits de dose excédant plus de 20% les valeurs règlementaires et en particulier les 10 mrem/h à 2 m qui interviennent le plus dans l'estimation de l'irradiation potentielle du public. Et on peut surtout difficilement suppose que cela se produire pour plus de 10% des expéditions (car une chose est de décharger un emballage chargé et une autre de prévoir l'activité des futurs chargements et de la faire accepter au contrôle d'expédition) sur lesquelles on est susceptible de commettre des erreurs et qui, dans des catégories de matière ne représentent probablement pas plus du 1/3 du total des expéditions de matière radioactive. L'erreur que l'on pourrait faire sur le calculs d'irradiation potentielles ne dépasserait pas - de ce fait - 1%.

b) Les modèles d'emballages lourds (de masse supérieure à 10 t) ne subissent pas eux-mêmes les épreuves règlementaires qui sont faites sur des modèles réduits. On pouvait à priori mettre en doute la validité de cette méthode, admise dans la règlementation. En fait tous les essais faits sur des emballages réels ont confirmé la validité de cette méthode (Robinson et al., 1976).

c) La détermination des valeurs de contamination des surfaces extérieures des emballages est délicate. Les valeurs mesurées font souvent l'objet de controverse entre expéditeur et destinataire, ce qui conduit d'ailleurs certains à admettre un phénomène de "ressuage" entraînant des valeurs pouvant être pour cette contamination 10 fois, ou même quelques dizaines de fois, plus élevées à l'arrivée qu'au départ. Bien que ce point fasse encore l'objet d'études, on peut d'ores et déjà conclure que l'impact, en ce qui concerne l'irradiation potentielle est très faible (Sousselier, 1977). Il apparaît donc que la règlementation est suffisament détaillée et qu'elle doit permettre d'assurer ce pourquoi elle a été faite, c'est-à-dire la sûreté des transports. Cependant ou peut aussi craindre des erreurs involontaires lors des opérations de chargement, fermeture des emballages etc. Par example, absence de joint d'étanchéité ou joint mal posé, boulons serrés inégalement, pour les petits emballages, couvercle non disposé correctement. Cela s'est d'ailleurs produit un certain nombre de fois. Il est certain que l'augmentation du nombre des transports est une cause potentielle de relâchement dans l'observation des prescriptions règlementaires et qu'il sera nécessaire dans l'avenir de renforcer les procèdures de contrôle des opérations de chargement et d'expédition. L'adoption de méthodes transposées de celle des check lists utilisée dans l'aviation (et qui ont certainement contribué de façon non négligeable à l'amélioration de la sûreté dans cette branche de l'activité humaine au cours des dernières décennies) devrait permettre de garantir dans l'avenir une observation complète des prescriptions et un respect total de la règlementation.

IMPACT DES TRANSPORTS SUR L'ENVIRONNEMENT

Une règlementation stricte, complète, précise existe et est capable (elle l'a d'ailleurs montré jusqu'à présent) d'assurer la sûreté des transports. Moyennant un certain nombre de procédure, de mise en oeuvre relativement aisée, on peut estimer qu'elle pourra dans l'avenir, même avec l'augmentation importante des transports, continuer à assurer ce rôle. Cependant la rèlgementation admet certaines valeurs pour l'irradiation externe ou la contamination, dont il faut examiner l'impact potentiel. De même qu'il faut examiner celui de ces transports indépendamment de leur contenu radioactif. Enfin il faut examiner si les prescriptions règlementaires sont à même d'assurer l'intégrité des emballages du type B dans tous les accidents possibles et de voir les conséquences possibles.

Impact des Transports en Conditions Normales

Impact radioactif. Le premier impact possible est l'irradiation du public et du personnel chargé du transport. L'étude la plus complète publiée sur ce sujet est celle de la NRC (NUREG-0170, 1977) et nous nous y référerons, sauf indication contraire, dans tout ce paragraphe. La question ne se pose pas exactement de la même façon pour tous les modes de transport qu'il faut donc examiner successivement.

Transport routier. Il a été indiqué au début de ce rapport que l'on n'y inclueraît pas les opérations de chargement et déchargement des matières radioactives. Les principaux impacts radiologiques sont ceux de l'irradiation des manutentionnaires, des chauffeurs (et éventuellement des accompagnateurs) et du public. La NRC estime la dose annuelle maximale reçue par un chauffeur à 2,2 rem, en se basant sur une dose maximale par voyage de 73 mrem (3mrem/h x 20 h + 33 mrem/h correspondant aux operations diverses) et à 30 voyages par an. Il semble que dans des conditions autres qu'américaines, européennes, par example, il faille compter sur un nombre plus important de voyage pouvant atteindre 80, mais de durée plus courte (10 h en moyenne), la dose reçue au cours des opérations divers semble surévaluée, les chauffeurs n'ayant pas, en général, les responsibilités de chargement sur le véhicule et d'arrimage. On arrive cependant à un total d'irradiation potentiel compris entre 2,4 et 2,8 rem/an. Mais un certain nombre de réglementations nationales limite le débit de dose dans la cabine du chauffeur à 0,75 mrem/h. Le rapport de la

NRC souligne d'ailleurs que la valeur de 2 mrem/h n'est que très rarement atteinte en pratique et estime que la moyenne est de l'ordre de 0.2 mrem/h, ce qui correspond à une dose maximale individuelle pour les chauffeurs de 870 mrem. En fait les doses effectivement reçues sont bien inférieures. Le D.O.E. estime que les chauffeurs de camion ne reçoivent pas plus de 0,3 mrem par transport de combustibles neufs et que même pour les transports de combustibles irradiés, la dose qu'ils reçoivent ne dépassent pas 30 mrem par transport. (1977). Si l'on admet qu'un chauffeur fait 80 transports par an répartis moitié-moitié entre irradiant et peu irradiant, on aboutit à un total de l'ordre de 1,2 mrem/an, chiffre un peu supérieur à l'estimation de la N.R.C., mais du même ordre de grandeur. L'évaluation de la dose reçue par le public est beaucoup plus délicate. On doit distinguer la dose reçue par le public circulant sur la même route que le véhicule transportant l'emballage radioactif, la dose reçue par le public se trouvant au voisinage de cette route et enfin celle reçue pendant que le véhicule transportant les matières radioactives est à l'arrêt. C'est le premier mode qui est susceptible de causer l'irradiation maximale qui peut se produire pour un chauffeur d'un véhicule suivant le camion transportant des matières radioactives. Le N.R.C. estime invraisemblable qu'un même conducteur reçoive une dose supérieure à celle qu'il recevrait en étant au cours de l'année pendant 20 h à une distance de 30 m derrère un camion transportant des matières radioactives correspondant à l'irradiation maximale autorisée. Dans ce cas l'irradiation reçue serait de 1,9 mrem/an. Bien entendu cette valeur n'est susceptible de s'appliquer qu'à nombre très restreint de personnes du public. Le second mode correspond au cas d'une personne vivant au voisinage d'une route sur laquelle circule des transports radioactifs. Si l'on suppose une distance de 30 m et un trafic annuel de 250 camions transportant des châteaux de combustibles irradiés à une vitesse de 50 km/h, on aboutit à une irradiation de 0,009 mrem/an. Cette valeur est négligeable, d'autant plus que le calcul ne fait pas intervenir la protection apportée par les murs des bâtiments. Enfin le dernier mode correspond à l'irradiation des personnes circulant ou séjournant près du véhicule lors d'un arrêt. La NRC évalue à 1,3 mrem/an la dose maximale que peut recevoir un individu: c'est celle résultant d'un séjour de 3 minutes à 1 m d'un château de combustible irradié. Il est en fait peu vraisemblable, compte tenu des dimensions des véhicules que l'on soit à 1 m du château; par contre, un séjour de 3 minutes n'est certainement pas le maximum envisageable. Néamoins, au total la valeur globale semble correcte. Une autre étude (Yadigaroglu, 1974) examine le cas de l'irradiation reçue par le chauffeur d'un autre véhicule doublant un véhicule transportant des château de combustibles irradiés et aboutit à une dose de $2,1 \times 10^{-3}$ mrem, ce qui conduit à des valeurs d'irradiations encore plus faibles que les hypothèse précédentes. Il reste bien entendu le fait que cette irradiation pourra toucher un nombre important de personnes. Cette analyse de l'impact global sera faite plus loin.

Transport ferroviaire. Des évaluations similaires à celles du transport routier peuvent être faites. Compte tenu de la longueur de trains, la dose maximale pouvant être reçue par le mécanicien - et le cas échéant par ses aides et les convoyeurs - ne peut être, selon la NRC, que de 1,2 mrem/an. La valeur de l' irradiation potentielle pour les usagers du rail est estimé négligeable par le rapport de la NRC. On peut cependant remarquer qu'en Europe, le nombre de trains de voyageurs est beaucoup plus élevé, sur un parcours donné, qu'aux Etats-Unis. Par contre, les vitesses à prendre en compte y sont plus élevées que celles indiqués dans le rapport de la NRC et l'effet de protection par les structures des wagons plus important que pour la route. Des employés de chemin de fer peuvent, d'après l'étude de la NRC, recevoir une dose annuelle pouvant se monter jusqu'à 1,65 mrem, reçu pendant des stationnements près de wagons à l'arrêt. Cette estimation semble faible, compte tenu de la durée effective des stationnements. Par contre les manutentions dans les gares de triage étant de plus en plus automatisées, le nombre de personnes susceptibles de recevoir cette dose, même si elle peut se monter à 5 ou 10 mrem/an, est relativement limité.

Transport aérien. D'après de la NRC, c'est ce mode de transport qui conduit aux doses d'irradiation les plus élevées. Mais cette étude envisage l'ensemble des transports de matières radioactives, alors que nous ne considérons dans notre étude que les transports du cycle. La grande majorité des transports de matière radioactive faits par avion concerne les radioisotopes à usage industriel ou médical (radiopharmaceuticals) et ceux-ci sont souvent transportées dans les avions-passagers. Les seules matières du cycle de combustible qui soient partiellement transportées par avion sont des concentrés d'uranium et l'UF_6 enrichi et ces transports sont faits par avion cargo. Ces matières sont peu irradiantes. Des mesures faites à bord d'avion cargo transportant des concentrés ont donné une valeur de 1 mrem/h pour le débit de dose dans la soute et de 0,1 à 0,2 mrem/h pour celui de la cabine de pilotage. Ces chiffres correspondraient à une irradiation maximale potentiale pour les équipages de 160 mrem/an.

Impact non radioactif. Il s'agit des impacts mentionnés dans l'introduction de ce rapport: consommation de carburante, encombrement des routes, conséquences d'accidents ne comportant pas de relâchement de produits radioactifs. Pour juger de ces impacts, il faut replacer les transports radioactifs dans l'ensemble des transports de matières dangereuses, voire dans l'ensemble des transports. Le rapport de la NRC évalue à 2 millions le nombre de colis de matières radioactives transportées aux Etat-Unis en 1975 pour un total de 100 millions d'expéditions pour le total des matières dangereuses et 500 milliards d'expéditions de l'ensemble des matières transportées. En plus, environ 85% des matières radioactives transportées correspondraient à des radiosotopes et il n'y avait pour l'ensemble du cycle de combustibles, déchets inclus, que 300 000 colis, ce qui représenterait 0,3 % des matières dangereuses et moins d'un millionième de l'ensemble des transports de marchandises. Il est bien évident que les impacts de consommation et carburants et d'encombrement dans les routes ne peuvent être que négligeables.

Conditions Accidentelles

Impact des accidents entraînant un relâchement de produits radioactifs. On doit examiner séparément le cas des accidents concernant des matières transportées en emballage du type B ou en emballage d'autres types puisque dans le second cas, la réglementation admet qu'en cas d'accident, l'enveloppe de confinement peut ne pas résister et que les matières radioactives peuvent être dispersées dans l'environnement, ce que la réglementation n'admet pas pour les transports en emballage du type B. Dans ce dernier cas, ce ne pourrait être que des conditions extrêmes d'accident qui entraîneraient des relâchements d'activité.

Accidents "ordinaires". Par opposition aux accidents "extrêmes" dont on vient de parler, nous désignerons par accidents ordinaires, les accidents se produisant de façon assez fréquent lors des transports de toutes marchandises. La réglementation admet donc pour le transport en emballage de type A ou industriel qu'il puisse y avoir alors dispersion du contenu. Les élements sur lesquels se fonde la réglementation sont que, dans ces accidents, l'experience montre qu'il n'y a pas plus des 1/1000 du contenu qui est dispersé dans l'environnement et que seulement 1/1000 au plus de ce qui s'échappe est susceptible d'être inhalé par une personne du public se trouvant sur les lieux de l'accident. La valeur de A_2 (limite de la valeur de l'activité transportable en emballage de type A) pour un radionucléide donné est alors 10^6 fois, la quantité maximale inhalable. En fait ce raisonnement ne concerne que les solides dispersables. En ce qui concerne les solides non dispersables, le phénomène à considérer est l'irradiation potentielle et c'est à partir de là que sont calculées les valeurs de A_1 correspondantes, A_1 et A_2 différant justement, car la première valeur se rapporte aux activités de matière non aisément dispersable et la seconde, au cas contraire. Pour les liquides et les gaz transportés en emballage du Type A, il est prévu épreuve de chute de 9 m et on se rapproche du type B. Le point que nous avons réellement à examiner est celui des solides dispersables. Dans ce cas, l'impact maximal semble être l'inhalation d'une quantité égale à la valeur maximale admissible du ou des radionucléides en question. En fait, en cas d'accident, on peut envisager que plusieurs emballages du type A se trouvent sur le véhicule et si l'on admettrait que dix emballages se trouvent impliqués, on pourrait conclure qu'une personne présente pourrait recevoir 10 fois la dose admissible. En fait les choses ne se passeraient pas tout à fait ainsi, les conditions de mélange dans l'atmosphere et de dispersion, ne sont pas exactement les mêmes. Cependant on peut admettre que l'impact maximal théorique serait l'absorption par un ou quelques individus, d'une quantité de radioactivité égale à quelques fois la dose maximale admissible. Néanmoins, la prise en compte de cet impact a conduit quelques réglementations à limiter le nombre de colis du type A transportés sur un même véhicule (Sousselier et Cohendy, 1976). Mais en fait l'expérience montre que les impacts réels sont toujours restés très faibles. Aux Etats-Unis de 1968 à 1973, on a enregistré (Sisler, 1974) 53 incidents ou accidents dans lesquels étaient impliquées des matières radioactives. Dans 25 cas, il n'y a eu aucun relâchement d'activité, dans 16 cas, la contamination a été limitée au véhicule et dans 10 au voisinage immédiat. Seulement dans 2 cas, il y a eu dispersion proprement dite de la radioactivité. Au Canada, sur 300 000 expéditions de matières radioactives effectuées de 1957 à 1973, il n'est survenu que 40 incidents. Aucun d'entre eux n'a entraîné un relâchement d'activité dans l'atmosphère ou dans les cours d'eau et seulement 6 on entraîné une contamination limitée du terrain autour du lieu de l'accident (Smythe et al., 1974). Le dernier cas à examiner est celui du transport des concentrées d'uranium. La réglementation ne demande que des emballages dits industriels, ce qui est normal, compte tenu du fait que la radiotoxicité de l'uranium est très faible, plus faible, en tout cas que la toxicité chimique. Mais compte tenu des quantités transportées, un impact possible, en cas d'accident, est la contamination de l'environnement. Ceci s'est d'ailleurs produit à deux reprises (Sousselier et Cohendy, 1976; Hornsby et al., 1978). Il n'y a eu, dans ces deux cas, aucune contamination de personnes, mais la décontamination de la chaussée et des parties avoisinantes a constitué un certain impact. On peut cependant estimer que dans l'avenir, cet impact aura plûtôt tendance à diminuer, car plusieurs pays adoptent pour ces transports - et aussi pour le transport des déchets - le transconteneur (freight container) du type utilisé dans les transports maritimes avec les bateaux porte conteneurs. Le risque de dispersion de matière en cas d'accident est extrêmement réduit.

Accidents "extrêmes". De nombreuses études ont été menées pour examiner le comportement d'emballage du type B dans les accidents dont les conditions paraîssent plus sévères que celles qui correspondent aux tests réglementaires et déterminer les conséquences extrêmes qui pourraient en résulter. En premier lieu, un certain nombre d'essais réels ont été faits. On peut mentionner la série de tests faits par la Sandia avec un château de transport de combustibles irradiés projeté d'une part à 120 km/h sur un mur de béton, d'autre part recevant l'impact d'une locomotive lancée sur lui à la même vitesse, tests à la suite desquels l'intégrité du confinement avait été conservée (Sandia Lab., 1978). On peut mentionner les tests réalisés avec des emballages de transport d'oxyde de plutonium lancés à des vitesses de 232 m/s sur un sol rocheux et où l'intégrité du confinement a aussi été conservée (Bonzon, 1976). Les résultats de ces tests amènent à conclure que les hypothèses prises en compte dans l'étude de la NRC (NUREG-0170, 1977) sont exagérément pessimistes. Néanmoins, les conclusions de cette étude montrent que les conséquences seraient faibles. Pour les transports de plutonium, l'engagement collectif de dose pour un relâchement de 1,23 x 10^6 Ci en zone urbanisée serait au maximum de 3,15 x 10^6 hommexrem, mais la probablité d'un tel accident serait de 10^{-11} (soit 1000 fois inférieur à celle de la chute d'une météorite géante). La même étude estime qu'en 1985, le nombre de cancers susceptibles de découler des suites des accidents de transport de matières radioactives en emballage du type B serait 0,002, chiffre qui n'a évidemment aucune signification en soi et qui ne fait que démontrer à nouveau l'extrême petitesse du risque. Le cas des transports maritimes a fait l'objet d'études spéciales, compte tenu des conditions particulières: risques de feu de plus longue durée que pour des accident terrestres, conséquences des pressions externes élevées en cas d'immersion. L'étude suédois K.B.S. (1978) étudie le cas d'un navire transportant des châteaux de combustibles irradiés. La probabilité d'une collision, selon les statistiques maritimes, est de 10^{-7}/km, l'étude estime que la probabilité qu'un château soit endommagé lors d'une collision est de 10^{-1}, ce qui semble extrêmement pessimiste. Le tableau 4 donne les résultats de cette étude.

Tableau 4

CONSEQUENCES D'ACCIDENT SUR UN NAVIRE TRANSPORTANT DES COMBUSTIBLES IRRADIES

	ACTIVITE RELACHEE Ci	DOSE INDIVIDUELLE MAXIMALE REM	DOSE COLLECTIVE HOMME x REM	PROBABILITE/KM
Collision	2 000	0,59	2 800	5 x 10^{-9}
Collision et conteneurs coulés	3 900	0,03	11 300	5 x 10^{-9}
Feu de longue durée	6 050	1,8	8 200	3 x 10^{-10}
Feu suivi de conteneurs coulés	11 600	0,08	33 000	2,7 x 10^{-9}
Collision suite de feu	2 200	5,8	28 000	2,7 x 10^{-10}
Collision suite de feu et d'immersion	3 900	0,03	11 300	3 x 10^{-11}

Une étude plus détaillée avait été faite par la Battelle pour le compte du Department of Energy (1977). En estimant qu'un conteneur chargé de 2,5 kg de plutonium à haute teneur en isotopes supérieures laisse échapper la totalité de son contenu et que le plutonium se retrouve sous forme soluble, (ce qui semble une hypothèse très pessimiste) la dose moyenne reçue serait de 0,004 mrem et l'engagement de dose serait de 240 homme x rem si l'accident avait lieu sur le plateau continental. Au cas où l'accident se traduirait par une immersion dans les couches profondes de l'océan, ces valeurs seraient divisées par 1000 tant pour la dose individuelle que pour l'engagement de dose, c'est-à-dire sans aucune signification. Le même rapport étudie le cas de perte d'un château de combustibles irradiés contenant 3 t d'uranium. Les engagements de dose sont faibles dans le cas où le château coule en eau profonde et inférieures à 150 hommes x rem. Ils sont beaucoup plus élevé si cela se produit sur le plateau continental où ils se montent à 1 000 hommes x rem et même à 150 000 hommes x rem au cas où un feu sévère aurait précédé la coulée. Mais les hypothèses prises en compte semblent très péssimistes. En particulier la résistance au feu semble très sous-estimée. De même la rapport ne tient pas compte de l'enfoncement du château dans les sédiments meubles qui constituent une fraction importante du plateau continental. La vitesse limite de chute suivant l'orientation du château pendant sa chute varie entre 75 km/h et 95 km/h. Ces valeurs ne sauraient conduire à des dommages importants sur l'emballage, mais par contre, entraineraient leur pénétration presque complète dans les sédiments. Les phénomènes de corrosion, susceptibles de jouer pour rompre les enceintes de confinement en seraient retardés, Si l'on tient enfin compte que les sédiments marins ont une capacité de sorption non négligeable vis-à-vis de nucléides comme le césium 137 ou les actinides, on peut conclure que les chiffres de l'étude Battelle sont beaucoup trop élevés et qu'en fait un tel accident n'aurait que des conséquences relativement peu importantes.

Il est enfin intéressant de noter que les pertes de navire, qui sont relativement peu nombreuses, affectent surtout les navires de petit tonnage. En 1974 sur 311 pertes de navires enregistrées dans le monde (dont 51 à la suite d'incendies) il n'y en avait que 40 pour les bateaux jaugeant plus de 4 500 tonneaux (dont 9 à la suite d'incendie). La probabilité à prendre en compte pour ces études ne devrait donc pas être celle concernant l'ensemble des transports maritimes, mais une valeur plus faible. Enfin un certain nombre de mesures complémentaires sont en général prises pour les navires transportant de telles cargaisons. Les japonais ont ainsi récemment annoncé que les bateaux qui transporteraient les combustibles irradiés en Europe seraient construits avec des doubles coques et auraient les équipements principaux doublés.

Impact des accidents sans relâchement de produits radioactifs. La majeure partie des accidents dans lesquels se trouvent impliqués des matières radioactives ne se traduiront pas par des relâchements de produits radioactifs. Sur 144 accidents recensés par le D.O.T., il n'y en a eu que 36 dans lesquels il y a eu contamination de l'emballage et/ou des véhicules et un seul dans lequel il y a eu a proprement parler relâchement d'activité dans l'environnement. Le Ministère Français des Transports dans ses statistiques pour l'année 1977 relevait, pour un total de 227 accidents concernant des matières dangereuses, seulement 3 incidents concernant des matières radioactives dont aucun avec relâchement de radioactivité. L'impact de ces accidents ne peut être que négligeable par rapport à celui de l'ensemble des accidents entraînés par le transport des marchandises.

On peut cependant essayer de le chiffrer. C'est ce qu'a fait le rapport du D.O.E. (1977). Il estime que pour 200 réacteurs de 1 GWe en fonctionnement, il y aura 100 000 véhicules x miles de transport routier, ce qui, rapporté aux chiffres connus de transport routier, représente 13 accidents par an - de caractère non radioactif - entraînant 9 blessés par an et 1 mort tous les 2 ans. Il y aura en plus 15 000 wagons x miles de transport ferroviaire, ce qui correspondrait à 2 accidents par an, à 5 blessés et à mort tous les 2 ans. Ces 3 derniers chiffres semblent très surestimés, tout au moins dans le contexte européen pour lequel aussi bien le nombre d'accidents de chemin de fer/wagon x km que celui du nombre de tués et blessés par accident semble beaucoup plus faible que ceux pris en compte dans le rapport mentionné. D'ailleurs le rapport NRC estime à la même valeur le taux d'accident routier et ferroviaire exprimé en véhicule x km. Il faut cependant examiner de plus près un point particulier qui est l'impact potentiel dû aux masses unitaires de certains transports. Les "châteaux" (conteneurs) transportant les combustibles irradiés et les déchets de haute activité ont des masses très élevées: de 32 à 104 t (IAEA, 1977). Si les châteaux correspondant aux masses les plus élevées sont practiquement toujours transportés par voie ferrée, ceux de la gamme 30-50 t sont souvent transportés par voie routière. Si ces châteaux sont impliqués dans des accidents, en particulier collision avec d'autres véhicules, aucune conséquence radiologique n'en résulera dans la presque totalité des cas, mais par contre les conséquences sur les autres véhicules pourront être importantes. Un "château" de 35 t se renversant sur une voiture de tourisme l'aplatirait comme le ferait une presse. Une semi remorque, transportant un tel château, déportée dans un virage et quittant la route pourrait aisément traverser le mur d'une maison Et, ce qui serait certainement plus grave, une collision entre un tel château et une citerne transportant d'autres matières dangereuses pourrait se traduire par la rupture de cette dernière, non prévue pour résister à un tel impact, impact, entraînant éventuellement le relâchement de produits nocifs. La probabilité d'un tel événement est évidement actuellement très faible, compte tenu du petit nombre de transports de tels châteaux. En France en 1974, il y a eu par exemple 60 transports de tels châteaux. Néanmoins il serait prudent à l'avenir de limiter le plus possible les transports routiers d'emballages de masse supérieure à 20 ou 30 t. Les statistiques montrant que les transports ferroviaires sont plus sûrs, les wagons étant plus résistants que les semi-remorques et enfin les conséquences de l'accident maximal envisagé (rupture d'une citerne d'une matière dangereuse non radioactive) étant en général moins graves sur voie ferrée que sur route.

Impact des Transports en Conditions Anormales

Ainsi qu'il a été mentionné plus haut on peut envisager en plus des conditions normales et des conditions accidentelles des conditions anormales, c'est-à-dire des transports pour lesquelles toutes les conditions prévues par la réglementation ne seraient pas respectées ou des transports pour lesquels se seraient produites des erreurs lors des opérations de chargement, fermeture, arrimage etc.

Le non respect de certains prescriptions réglementaires ne paraît pas de nature à pouvoir provoquer des impacts autres que négligeables. En effet, si l'on exclut le non respect important et délibéré (ce qui releverait des actions malveillantes que nous ne prenons pas en compte dans ce rapport) il ne peut s'agir que d'erreurs ou de déviations minimes. Par examplе valeur maximale du débit de dose au contact ou à 1 m dépassant les valeurs réglementaires. Il est bien évident que même si le dépassement était important et allait jusqu'au double des valeurs autorisées, les conséquences pour l'irradiation du public seraient négligeables, compte tenu du fait que ceci ne pourra n'arriver que très rarement - disons dans 1 ou 2 % des cas - compte tenu des précautions prises. L'impact sur l'irradiation des chauffeurs, pour le voyage dans lequel se serait produits cette erreur, serait plus important (un chauffeur pourrait peut-être recevoir une dose de 50 à 60 mrem au cours de ce voyage au lieu de 30 mrem) mais sans signification biologique pour lui et à fortiori, sans incidence sur l'irradiation globale dûe aux transports.

Des erreurs au départ semblent susceptibles à priori d'entraîner des conséquences plus importantes. Il semble que l'on puisse distinguer les erreurs dans les opérations de fermeture et les erreurs de chargement ou d'emballage. Les premières se sont déjà produits comme on l'a mentionné. Ce sont aussi des erreurs de ce genre qui ont été à l'origine des contamination survenues aux Etats-Unis en 1971 lors de transport aérien de MO-99 (NRC, 1971). La dépression régnant dans les soutes de l'avion a causé la fuite du liquide contenu dans le récipient primaire mal fermé au départ. C'est à la fois un bon exemple de ce qui semble être le risque maximal (en l'occurence aucune irradiation significative au plan biologique, mais un coût élevé en opérations de contrôle et de décontamination) et un accident ne concernant pas les transports du cycle de combustibles. Pour ces derniers, il semble que des conséquences importantes soient moins à craindre, en particulier par ce que l'on transporte peu de liquides. D'ailleurs les erreurs relevées dans les transports de plutonium n'ont entraîne aucune conséquence. On objectera que les conséquences auraient été plus importantes en cas d'accident, mais la probabilité de ceux-ci est encore plus faible puisqu'il faut multiplier les probabilités d'accidents par celles du risque d'erreurs.

Les erreurs de chargement semblent possible quand il y a chargement à distance (combustibles irradiés par exemple) mais paraîssent difficilement envisageables autrement. Des erreurs plausibles semblent pouvoir être le chargement d'un élément combustible peu refroidi au lieu d'un élément très refroidi ou d'un élément très irradié, au lieu d'un élément peu irradié. Les conséquences n'en seraient que minimes sur les valeurs des débits de dose et, comme on l'a déjà fait observer, sans grande signification. On pourrait craindre par contre deux autres types de conséquences. Premièrement, une augmentation des risques de criticité, mais le réglementation demande que les calculs correspondants soient faits avec les éléments correspondants à la réactivité maximale. Deuxièmement une détérioration des éléments transportés si la chaleur dégagée par les éléments chargés excède les capacités du château. En fait, en ce qui concerne le transport proprement dit, l'impact serait nul, sauf peut-être en cas d'accidents en conditions extrêmes, mais dans ce cas, on ne tient pas compte dans le calculs d'impacts, de l'intégrité des éléments. L'impact réel qui pourrait subsister surviendrait lors des opérations de déchargement, mais les contrôles faits à l'arrivée du château permettraient de s'apercevoir du problème avant l'ouverture de celui-ci et donc d'empêcher un impact sérieux.

Enfin l'erreur dans le choix de l'emballage semble surtout envisageable, dans les transports du cycle de combustible, pour les déchets, car l'on connaît souvent mal leur activité. Il serait alors possible que le transport soit effectué en emballage du type A, voire en emballage industriel (si l'on croit à tort être dans la catégorie des matières solides de faible activité) au lieu de l'être en emballage du type B. Cette erreur n'entraînerait aucun impact sur l'environnement si aucun accident ne survenait pendant le transport. En cas d'accident entraînent la destruction de l'emballage, l'impact pourrait ne pas être négligeable. Cependant les matières solides de faible activité font l'objet de prescriptions strictes sur la non dispersion et il faudrait que la destruction de l'emballage soit suivie d'un incendie pour entraîner une contamination sensible de l'environnement. Cette contamination resterait en tout cas bien inférieure à celle envisagée dans le paragraphe consacré aux conditions extrêmes, et la probabilité que cet événement se produise est extrêmement faible. Néanmoins un certain nombre de réglementations prescrivent que de tels transports soient effectués dans des transconteneurs de type maritime, ce qui diminue encore de façon importante ce risque.

Impact Global

Les paragraphes précédents ont donné un grand nombre d'éléments permettant d'évaluer l'impact sur l'environnement des transports de matières radioactives. Il n'est pas très aisé d'en déduire un impact global, car les hypothèses faites dans les différentes études ne sont pas les mêmes. Cependant des études partielles d'impact global ont été faites.

L'ERDA dans son rapport 77-43 (1977) a étudié l'irradiation entraînée par le transport de châteaux de combustibles irradiés. Ce rapport prend également en compte l'impact entraîné par les accidents, en faisant l'hypothèse que, dans ce cas, il n'y aura aucun relachêment de produits radioactifs (ce qui semble réaliste, compte tenu de ce qui a été dit sur le degré extrêmement élevé de résistance des châteaux de transport) mais en observant qu'il en résultera un accroissement de la dose au public, compte tenu des retards que l'accident entraînera et du nombre de personnes qui seront amenées à séjourner près du château pour les opérations de réparation, éventuellement de lutte contre l'incendie etc. Les tableaux 5 et 6, tirés de ce rapport, donnent les résultats de cette étude.

Tableau 5

IRRADIATION DECOULANT DU TRANSPORT DE COMBUSTIBLES IRRADIES

	TRANSPORT ROUTIER		TRANSPORT FERROVIAIRE	
	Nombre de personnes exposés	Doses reçues hommexrem/1000 véhicule x km	Nombre de personnes exposés	Doses reçues hommexrem/1000 wagon x km
conducteur	2	5×10^{-2}	3	$6{,}09 \times 10^{-4}$
autres travailleurs	12	$1{,}5 \times 10^{-2}$	17	$7{,}71 \times 10^{-3}$
public:				
usager des transports	6 600	$4{,}1 \times 10^{-2}$	780	$1{,}06 \times 10^{-3}$
autre public	85 000	$1{,}3 \times 10^{-3}$	85 000	$5{,}8 \times 10^{-3}$

On peut noter que la dose moyenne reçue par les personnes de la catégorie "autre public", c'est-à-dire par les personnes vivant ou séjournant près de la route ou de la voie ferrée est inférieure à 1 µrem. On reviendra sur ce point dans la conclusion.

Tableau 6

IRRADIATION DECOULANT D'ACCIDENT SURVENU LORS DE TRANSPORTS DE COMBUSTIBLES IRRADIES

ACCIDENT ROUTIER		ACCIDENT FERROVIAIRE	
Nombre de Personnes exposées	Doses reçues homme x rem/accident	Nombre de Personnes exposées	Doses reçues homme x rem/accident
115	$5{,}19 \times 10^{-1}$	65	$2{,}65 \times 10^{-1}$

A partir de ces chiffres et de ceux des tableaux 1 et 3, on peut tenter une estimation globale et d'impact des transports de combustibles irradiés en l'an 2 000. Il y aura dans le monde 45 000 t à transporter annuellement, soit 9 000 châteaux, en supposant qu'à cette date tous les transports se feront avec des châteaux de grande capacité. Si on admet que les distances moyennes par transport seront de 1000 km, on arrive aux chiffres du tableau 7, en estimant, avec le rapport de l'ERDA, une probabilité d'accident de 2×10^{-4}/1000 véhicule x km, soit un nombre moyen d'accident de 1,8 par an.

Tableau 7

IRRADIATION DECOULANT DU TRANSPORT DES COMBUSTIBLES IRRADIES EN L'AN 2000 (en homme x rem)

CAS DU TRANSPORT ROUTIER		CAS DU TRANSPORT FERROVIARE	
conditions normales	accident	conditions normales	accident
972	1	137	0,5

On peut conclure que le transport par voie ferrée conduirait à une diminution sensible de l'irradiation potentielle et que l'incidence des accidents est négligeable. Une autre étude globale d'impact a été faite dans le rapport déjà plusieurs fois mentionnée de la NRC (NUREG-0170, 1977). Comme on l'a déjà indiqué, elle ne concerne que 200 réacteurs en fontionnement. On a déjà aussi souligné que les hypothèses prises en compte dans cette étude paraissent systématiquement majorantes et exagérément pessimistes. Il est intéressant cependant de noter les principaux résultats (tableau 8).

Tableau 8

IRRADIATION ENTRAINEE PAR LES TRANSPORTS DU CYCLE DU COMBUSTIBLE DE 200 REACTEURS DE 1 GWe

en homme x rem

		CONDUCTEURS	AUTRE TRA-VAILLEURS	PUBLIC	TOTAL
combustible irradié	route	188	306	71	565
	rail	3	261	35	299
U_3O_8		467	710	832	2 009
UF_6 nat		71	27	28	126
UF_6 enrichi		13	1	2	16
UO_2		131	14	65	210
déchets		647	43	5 490	6 180

Le rapport n'explicite pas les hypothèses prises en compte dans le calcul des doses occasionnées par le transport des concentrés d'uranium (U_3O_8) et des déchets. Mais il semble que les valeurs respectives de l'irradiation du chauffeur et du public ne seraient pas cohérentes. Il y a un rapport 3 pour les irradiations des chauffeurs entre le transport des concentrés et de combustibles irradiés (ce qui semble acceptable, compte tenu du fait que la quantité des concentrés à transporter sera de 10 fois environ celle de combustibles irradiés) et un rapport 12 pour l'irradiation du public(ce qui semble beaucoup trop élevé, compte tenu du fait que la réglementation AIEA admet que l'irradiation maximale en tout point à 1 m de la surface extérieure du chargement peut être évaluée à 2 mrem/h pour les concentrés d'uranium). De même pour le déchets, compte tenu de la forte proportion de déchets de faible activité, le chiffre de l'irradiation du public semble très surestimé.

SYNTHESE ET CONCLUSIONS

Synthese des Résultats de l'Etude

Les transports de matières radioactives découlant du fonctionnement du cycle de combustible nucléaire ne représentent actuellement qu'un nombre limité de transport. En l'an 2000, même si l'on admet qu'il y aura alors en fonctionnement des réacteurs electronucléaires d'une puissance totale de 2000 GWe, le nombre de transport restera limité, les 400 000 t de concentrés d'uranium à transporter ne représenteront que de l'ordre de 20 000 transports, les 45 000 t de combustibles irradiés correspondront au transport annuel de 9000 châteaux, soit 30 châteaux par jour pour l'ensemble du monde. Ces chiffres sont négligeables par rapport à ceux du transport des autres matières dangereuses/explosifs, matières corrosives, toxiques, inflammables etc.

Ces transports sont soumis depuis de nombreuses années déja à une réglementation précise, complète, stricte. Réglementation qui découle de celle édictée par l'AIEA et qui est en vigueur dans tous les pays du monde, ce qui garantit partout une sûreté également élevée y compris pour les transports internationaux. Cette réglementation ne supprimera pas intégralement tout risque. En particulier elle admet certaines valeurs pour le débit de dose ou les contaminations externes. Elle admet aussi que pour les contenus peu radioactifs, il puisse y avoir, après accident, une certaine dispersion de la matière. Mais cette réglementation permet de guarantir que tout risque grave est exclu, en particulier grâce à la sévérité des épreuves auxquelles doivent être capables de résister les emballages dits au type B.

L'impact radioactif que seront susceptibles d'avoir les transports de matières radioactives restera très limité. En conditions normales, ce sont les chauffeurs de camion qui recevront les doses les plus élevées: elles ne semblent pas pouvoir dépasser la valeur de 1 rem/an, ce qui correspondra au cinquième des valeurs admises pour les travailleurs de l'énergie nucléaire par la Commission Internationale de Protection Radiologique. Les personnes du public vivant ou se trouvant à proximité de transports de matières radioactives ne recevront pas de dose supérieure à 2 mrem/an et ne recevront en moyenne qu'une dose de l'ordre de quelques μ rem/an, voire inférieure à 1 μrem/an. L'impact global, en conditions normales, se traduira en l'an 2000 par une dose totale reçue par les chauffeurs de 1000 à quelques milliers d'homme x rem et pour le public par des doses qui seront probablement également de l'ordre de quelques milliers d'homme x rem et ne dépasseront pas quelques dizaines de milliers d'homme x rem, même en adoptant des hypothèses résolument pessimistes, Ces valeurs ne peuvent correspondre à un impact réel important si on se rappelle par exemple que l'irradiation naturelle de l'ensemble de la population du globe est de 300 millions d'homme x rem/an ou que l'usage seul d'en engrais phosphatés conduit à une irradiation mondiale de 30 000 homme x rem/an (UNSCEAR, 1977).

L'impact des accidents sera également très limité. Il ne se produira vraisemblablement, chaque année, vers l'an 2000 que quelques dizaines d'accidents dans lesquels seront impliquées des matières radioactives et la quasi totalité de ces incidents n'entraînera aucun relâchement de matières radioactives pouvant avoir une signification biologique. Les conséquences les plus graves envisageables semblent être un accident maritime survenant sur un plateau continental et au cours duquel un château transportant des combustibles irraidés serait pris dans un incendie de longue durée puis coulerait sur des fonds rocheux. Ces conséquences seraient l'engagement d'une dose collective de l'ordre de quelques dizaines de milliers d'homme x rem, ce qui est probablement sans conséquence biologique réelle ou au maximum pourrait se traduire par l'apparition au cours des quelques centaines d'années à venir, de quelques cancers supplémentaires.

L'inobservation des réglements ou des erreurs faites lors des opérations de chargement, de fermeture et de contrôle ne sont pas susceptibles de changer les conclusions précédentes, sauf évidemment si ces erreurs devenaient fréquentes ou si ces inobservations devenaient très importantes.

Conclusion

L'expérience acquise jusqu'à maintenant dans le transport des matières radioactives a montré que ces transports étaient très sûrs et pouvaient être effectués sans impact pour l'environnement. Certes avec le développement important que va connaître, au cours des deux prochaines décennies, l'énergie électronucléaire, les transports du cycle de combustibles vont énormément se développer. Mais l'expérience actuelle des transports est déjà très importante, grâce aux transports des radioisotopes, à ceux entraînés par les programmes de recherches et aux programmes militaires. Si l'étude détaillée des impacts montrent qu'il peut y avoir un certain impact en ce qui concerne l'irradiation ou un certain risque en ce qui concerne les accidents, il faut voir que dans le premier cas le problème qui se pose est celui des "microdoses" et dans le second celui des "microrisques".

On discute beaucoup de l'effet des faibles doses, mais ce qu'on envisage en général sont les doses de l'ordre de quelques centaines de mrem par an comme peuvent en recevoir des travailleurs de l'énergie nucléaire ou de quelques millirems par an, comme peuvent en recevoir quelques membres du public vivant au voisinage des installations nucléaires. Dans le cas des transports, la question qui se pose est de savoir s'il faut prendre en compte des doses de quelques microrads par an. Il semble bien que cette prise en compte n'aurait aucune signification. De même en ce qui concerne les probabilités d'accident, comme on l'a mentionné, on a pu estimer à 10^{-11}, la probabilité d'un accident "extrême" auquel ne résisterait pas un conteneur de transport de combustibles irradiés et souligné que cette probabilité était de l'ordre du millième de celle de la chute d'une météorite géante. En fait tous ces impacts semblent négligeables.

Les seuls impacts potentiels qui restent sont alors:

- l'irradiation des chauffeurs - très diminué par l'utilisation systématique des transports ferroviares,

- le risque des conséquences sur l'environnement d'un accident impliquant un conteneur de masse élevée, susceptible d'entraîner des conséquences à cause de sa masse. Là aussi, l'utilisation des transports ferroviaires diminuerait ce risque.

- les conséquences d'erreurs humaines lors des opérations préparatoires au transport. L'utilisation de procédure, analogue à celles des "chek-lists" utilisée dans l'aviation, doit aussi permettre de maintenir très bas ce risque

Les transports de matières radioactives dans le monde n'ont jamais entraîné jusqu'à présent d'impact significatif pour l'environnement. Les techniques très élaborées de construction de conteneurs, l'existence d'une réglementation universellement appliquée, stricte, précise, complète, permettront de garantie que l'impact de ces transports dans l'environnement continuera dans l'avenir d'être aussi négligeable et que ces transports continueront à être une des branches les plus sûres de l'activité humaine.

References

Blomeke, J.O. et al., Shipments in the nuclear fuel cycle projected to the year 2000. Nuclear News, 18, 62 (1975).

Bonzon, L.L., Special tests of plutonium shipping containers - AIEA - SR 10/22 - Vienne (1976) p. 195.

Brobst, W.A., Transportation accidents : how probable. Nuclear News, 16, 48 (1973).

Department of Energy, Shipments of nuclear fuel and waste. Rapport D.O.E./EV 004, Washington D.C. (1977).

Hornsby, R.T. et al., A highway accident which involved a spill of natural uranium oxide concentration, 5ème symposium international sur l'emballage et le transport des matières radioactives - LAS VEGAS - mai (1978).

IAEA, Etude sur le projet de centres régionaux de cycle de combustible - Transport des matières radioactives - janvier (1977) 25 pp.

IAEA, Reglement de transport des matières radioactives - Collection Sécurité No. 6. 159 p.

KBS, Handling of spent nuclear fuel and final storage of vitrified high level reprocessing waste, Kärn Bränsle Säkerh Report 4 Vol.; Stockholm (1978).

Ministere des transports, Statistiques 1977 des accidents et incidents de la circulation routière concernant le transport des matières dangereuses, Paris (1978).

PNL, Consequence of postulated losses of LWR spent fuel and plutonium shipping packages at sea - Rapport PNL 2093, (1977), 52 pp.

Robinson, R.A. et al., use of scale models to assess structure response of nuclear shipping containers under accident condition, Symposium AIEA SR 10/9, Vienne (1976).

Sandia Laboratories, Crash testing of Nuclear Fuel Shipping containers - Rapport 77-1462, (1978), 42 pp.

Sisler, J.A., A transportation research and development program, 4ème symposium international - Miami, (1974).

Smythe, W.D. et al., The transportation of radioactive materials on Canada, 4ème symposium international - Miami, (1974).

Sousselier, Y., Experience de mise en oeuvre et d'application du réglement dans les transports de combustibles irradiés et de plutonium. Conference internationale IAEA - CN-36/227 - Salzbourg (1977).

Sousselier, Y. et G. Cohendy, Considération sur les différentes classes d'emballages. Symposium AIEA sur les emballages de transfert de matières radioactives SR 10/33, Vienne (1976).

UNERDA, Alternatives for long terme management of defense HLW, ERDA rapport 77-43, Washington D.C. (1977).

UNSCEAR, Sources and effects of ionizing radiation, United Nations (1977) 725 pp.

U.S. Nuclear Regulatory Commission, Report of aircraft radioactive contamination incident. Appendix A - Vol. II - NCR Nureg 170 (1971).

U.S. Nuclear Regulatory Commission, Final Environmental Statement in the transportation of radioactive material by air and other modes, NUREG-0170 (1977).

Yadigaroglu, G., spent fuel transportation on high ways : the radioactive does to the traffic - 4ème symposium international sur l'emballage et le transport des matières radioactives - Miami - Septembre (1974), p. 798.

10

THE FUTURE OF NUCLEAR ENERGY

Essam E. El-Hinnawi
Chairman of Energy Task Force
United Nations Environment Programme
P.O. Box 47074, Nairobi, Kenya

"If a problem is too difficult to solve, one cannot claim that it is solved by pointing to all the efforts made to solve it."

H. Alfvén; Energy and Environment
Bull. Atomic Scientists, May 1972.

INTRODUCTION

It has been stated in the Introduction (El-Hinnawi; this volume) that the total installed nuclear generating capacity in the world in 1978 was 110.8 GW(e) from 227 power reactors operating in 20 countries, constituting about 6% of the world electrical power capacity, and that estimates of future nuclear energy production are subject to large margins of uncertainty. Estimates of the world nuclear power growth by the turn of the century vary between 900 to 2000 GW(e). This wide variation is due to the complexity of the factors involved in energy forecasts, which include: socio-economic, environmental, geo-political and technical ones. In the present paper, the main elements affecting the future development of nuclear power are reviewed.

* Permanent Address:
Research Professor of Energy and Natural Resources
National Research Centre
Dokki, Cairo
Egypt

NUCLEAR POWER AS A PUBLIC ISSUE

The nuclear debate began, essentially in the U.S.A., in the late 1960s when citizens in a number of communities became sensitive to the potential environmental impacts of nuclear power plants, and opposed specific siting proposals (see, for example, Surrey and Hugget, 1976; Nelkin and Fallows, 1978). At that time the debate was mainly "technical" and the scientific community joined the debate over problems that indicate basic technical uncertainty, for example, the significance of low-level radiation hazards, the possibility of accidents, the measures necessary to cope with potential hazards, the ways of storage and management of radioactive wastes, etc. Soon, however, the debate extended to cover socio-economic problems, for example, problems of human fallibility, diversion of nuclear material for non-peaceful purposes, the possibility of sabotage and terrorism, the economics of nuclear power and other issues which have politicized the debate. While some nuclear opponents are motivated primarily by the inadequacy of the technical solutions of some problems pertaining to the nuclear fuel cycle, others join the debate because they are disenchanted with centralization, corporate dominance, or with government and its institutions.

Using all this complex of issues, critics of nuclear power development have raised a series of value questions: What kind of society is implied by a nuclear economy? What are possible costs of a nuclear programme to future generations? What level of risk is society willing to accept? How can an equitable distribution of risks and benefits be guaranteed? Does the government have the institutional capacity to manage long-term risks? Ultimately, who should be making such decisions? Essentially, then, the centre of conflict has shifted from technical uncertainties about environmental impacts, to a set of broad, political questions about the credibility of decision-making authority (Nelkin and Fallows, 1978).

The development of nuclear power should, therefore, be based upon an understanding of the ways in which people think about risk and uncertainty. Although detailed risk evaluations have been made for the nuclear fuel cycle, showing generally consistent low overall risks, risk aversion is a factor in public and political reluctance to accept nuclear power in some regions and countries (Døderlein, 1977, Otway and Pahner, 1976; see also Liverman and Thorne, 1977; Davies et al., 1977; Slovic et al., 1979 a, b). Risk analyses show that the perceived risks from nuclear power are judged to be extremely high (Slovic et al., 1979 b). Døderlein (1977) in discussing the protection of the public interest concluded that "All human action will entail unforeseen consequences. Plans for action cannot be based solely on factual proofs and logical deductions: they must finally rest on political decisions made in the face of uncertainty, a point which gives a perspective on the place of scientific and technological fact in a world of individual and societal values. Clearly, a number of irrational, emotional, and ethical factors and demands may be of importance in a choice of power plants for electricity production. But an evaluation of rational and quantifiable factors tells us how many lives, which environmental improvements, what economic and social advantages we have to sacrifice in order to satisfy such emotional demands. In protecting the public interest, the importance of the irrational factors in the nuclear issue must be recognized, but as many problems as possible should be decided on a reasoned, factual and rational basis."

ISSUES OF MAJOR CONCERN

The environmental aspects of production and use of energy have become an increasingly important factor in the development of national energy policies and strategies and this has been conspicuously so in the case of nuclear energy. Public concern has focused on a number of issues, the most important of which are:

(a) the effects of radiation on man, both somatic and genetic, which may be associated with some stages of the nuclear fuel cycle; (b) the safety of nuclear installations; (c) the environmental impacts associated with radioactive waste management; (d) the availability of plutonium in case of reprocessing of spent fuel and fear of its theft and misuse and (e) the socio-economic aspects of nuclear power development. (See, Edsall, 1974, 1975; Royal Commission on Environmental Pollution, 1976; El-Hinnawi, 1977; NAS, 1979; SIPRI, 1979, for reviews of these issues).

Effects of Radiation on Man:

Man has always been exposed to ionizing radiation from various natural sources: external, such as cosmic rays and radioactive substances in the ground and in building materials, and internal sources in the form of naturally occurring radioactive substances in the human body. In addition, man has been also exposed to radiation from man-made ionizing radiation. On the average, a member of the world population receives a whole body dose of about 100 mrem/y from natural radiation and about 50-80 mrem/y from man-made sources, of which less than 1.0 mrem/y is due to radiation from coal and nuclear power industry. Although our knowledge about the health effects of ionizing radiation has greatly advanced over the past years, the effects of low levels of radiation have not yet been well-defined and have recently been the subject of public concern and controversy.

Some recent studies (Bross and Natarajan, 1972; Mancuso et al., 1977; Najarian and Colton, 1978) have been interpreted by some to indicate that the commonly used risk estimates, which are based on the UNSCEAR and BEIR Committee Reports, underestimate the effect of low-dose radiation, i.e., that the linear theory is not conservative in estimating risk at low doses but rather underestimates it. Bross and Natarajan believe that they have identified subgroups in the population which are especially sensitive to radiation damage. Their belief derives from finding an association between some "indicators of susceptibility" (viral infections, bacterial infections, and allergy) shown by leukemic children from birth until diagnosis of leukemia. They concluded "the apparently harmful effects of antenatal irradiation are greatly increased in certain susceptible subgroups of children possessing the indicators associated with a slightly higher intrinsic risk of leukemia." However, Smith et al. (1973) pointed out that the incidence of these indicator diseases before the clinical onset of leukemia is the same in children who had received no irradiation in utero as in those who had (see also, Rothman, 1977).

Mancuso et al. (1977) and Kneale et al. (1978, 1979) claim to have demonstrated a radiation-induced excess of cancers in workers at the atomic energy plant at Hanford. Najarian and Colton (1978) claim a similar excess of leukemia and cancer in workers at the Portsmouth Naval Shipyard in New England. The data on which some of these claims have been based, and the analyses involved have been criticized by a number of authors (Mole, 1978; Reissland and Dolphin, 1978; Marks et al., 1978). In the Hanford workers, there is the possibility of an association with the work experience for the cancer of the pancreas and multiple myeloma (Marks et al., 1978), and if the cancer-doubling estimates of Mancuso et al., were correct, something other than radiation was the cause of the observed cancers. Similarly, the absence of any apparent latent period effect casts doubt on conclusions about the contribution of radiation to the high numbers of cancer deaths among the Portsmouth Naval Shipyard workers. Evans et al., (1979) recently found an increase in chromosome damage (chromosome aberrations in peripheral blood lymphocytes) in dockyard workers in the U.K. with increasing exposure to low levels of radiations. However, they pointed out that the exposed population studied is small, subject to very low levels of radiation exposure for periods up to 10 years and is therefore unlikely to provide useful data on the incidence of malignant disease. Thus, although these claims of higher risks from low levels of radiation described by these authors have become the subject of considerable public debate, examination to date of the results does not support the findings of Bross and Natarajan; Mancuso et al. and Najarian and Colton.

The difficulties in establishing the effects of very low dose levels of a few millirem or less per year to the general population from the nuclear fuel cycle are two-fold. One arises from the fact that no statistically valid epidemiological data are available. This is due to the virtually unsurmountable difficulties imposed by the long-term nature of epidemiological work that would need to be conducted under controlled conditions on a very large population, and over several generations in case of genetic effects. Even epidemiology studies on comparable populations living exposed to different natural radiation levels have proved difficult and inconclusive. It also would be imperative to establish a unique cause and effect relationship, even if correlations between morbidity and radiation exposure were apparent, to rule out other causes such as economic conditions, nutritional factors, or improved diagnostic facilities due to better medical services. Another problem arises from the fact that, although it is easy to observe deaths and major malignacies such as may be observed at dose levels above 50-100 rad, it is more difficult to observe minor incidences of illness in a much larger population, such as might result from low-level exposures, and to distinguish them unambiguously from similar illnesses that could be caused by a host of other causes.

<u>The Safety of Nuclear Power Plants</u>:

Much thought has been given to ensuring operational safety and reliability of reactor systems under all conditions. These methods include the provision of engineering safety systems. Public concern about reactor operation has concentrated on the reliability and effectiveness of these systems (particularly on the Emergency Core Cooling System), and more generally on the possibility of an accident leading to the release of radioactive material to the environment.

Several studies have been made in which the probabilities of reactor accidents of various degrees of severity were estimated, using information on the failure rates of engineering components of the reactor and other scientific information, and in which the environmental impacts of the radioactive releases associated with these hypothetical accidents were calculated. WASH-1400* estimates the probability of a meltdown in a pressurized water reactor as one in 20,000 per reactor per year, and that most meltdowns would not breach the main containment above the reactor. According to the study, the individual risk of early fatality from nuclear power plant accidents in 2×10^{-10} per year averaged over approximately 15 million people located within 40 km of the first 100 LWR's and 2×10^{-11} per year if averaged over the entire United States population of 200 million. The worst accident, which WASH-1400 estimates might happen once per 10,000,000 years if there were 100 reactors, might cause 3300 early fatalities, about 10 times that number of early illnesses, additional genetic effects and long-term cancers, and perhaps $\$14\times10^9$ in property damage. The worst accident considered might cause as many as 57,000 latent cancer deaths in the exposed population over 30 years. A large accidental release could cause large numbers of early fatalities and large numbers of latent cancers, but the probability of such a large release is very low. The Rasmussen report estimated roughly a factor of uncertainty of 5 in its estimates (WASH-1400 Appendix 6, pp. 13-42 to 13-48).

---

* Also known as Rasmussen Report.

The Reactor Safety Study (WASH-1400) has been extensively reviewed and criticized; a recent Risk Assessment Review Group concluded that: (see Lewis et al., 1978)

> "We find that WASH-1400 was a conscientious and honest effort to apply the methods of fault-tree/event-tree analysis to an extremely complex system, a nuclear reactor, in order to determine the overall probability and consequences of an accident. We have reviewed the methodology, the data base, the statistical procedures, and the results. We have found a number of sources of both conservatism and non-conservatism in the probability calculations in WASH-1400, which are very difficult to balance. Among the former are inability to quantify human adaptability during the course of an accident, and a pervasive regulatory influence in the choice of uncertain parameters, while among the latter are nagging issues about completeness, and an inadequate treatment of common cause failure. We are unable to define whether the overall probability of a core melt given in WASH-1400 is high or low, but we are certain that the error bands are understated. We cannot say by how much. Reasons for this include an inadequate data base, a poor statistical treatment, an inconsistent propagation of uncertainties throughout the calculation, etc. Also, both the dispersion model for radioactive material and the biological effects model should be improved and up-dated before they are applied in the regulatory and licensing process."

Based on the findings of Lewis et al., the U.S. Nuclear Regulatory Commission decided that the error bounds in WASH-1400 are understated and advised its staff not to rely on the data given in the report (NRC, January 18, 1979).

In a recent review by the National Academy of Sciences of the risks associated with nuclear power (NAS, 1979), it has been concluded that:

> "It is now generally agreed that there is a large range of uncertainty in the numerical results quantifying the risks of an accident with large consequences in a light-water nuclear-power reactor. The data base needed for assessing probabilities of various failures in reactor components is at present inadequate in a number of respects. In the view of some, the risk of a major accident is not quantifiable, and living with uncertainty surrounding a possible significant accident is itself a cost that technology imposes on society. This report finds that, with a minimum of debatable assumptions about accident scenarios, one can draw a few useful conclusions limiting the probabilities of accidents with extreme consequences. Several lines of research that are in progress or can soon be undertaken offer hope both for reducing the consequences of accidents (e.g., by prevention of overpressure rupture of the containment building), and for reducing the uncertainty in our estimates of the probabilities of serious accidents (e.g., reliability of core-cooling systems, likelihood of steam explosions)."

The Three Mile Island Accident: The Three Mile Island nuclear power station, Harrisburg, Pennsylvania has two pressurized water reactors: Unit 1 with a maximum rated power of 2535 MW(t) and Unit 2 with a rated power of 2772 MW(t). The first Unit went into commercial operation on 2 September 1974, while the second Unit went into operation on 30 December, 1978. On 28 March, 1979, the Three Mile Island Unit 2 facility was operated at about 97% power when the plant experienced a loss of normal feedwater supply that led to a turbine trip and later to a reactor trip. Subsequently, a series of events took place that resulted in significant damage to portions of the reactor core. It is believed that the sequence of events that led to core damage involved equipment malfunctions, design related problems and human errors that contributed to varying degrees to the consequences of the accident (NUREG-0600, 1979).

The principal radioactive materials released to the environment, in the period from 28 March through 7 April, 1979, appear to be xenon-133 (half-life 5.3 days), xenon-135 (half-life 9.2 hours), and traces of radioactive iodine, primarily iodine -131 (half-life 8 days). Some of the radioactive krypton isotopes such as krypton-87 (half-life 76 min), krypton-85 m (half-life 4.5 hours) and krypton-88 (half-life 2.8 hours) may also have been released (NUREG-0558, 1979). Radionuclides in particulate form such as strontium-90, uranium isotopes, and plutonium would either have been retained in the fuel or if released from the fuel would remain in the coolant water. These nuclides have not been detected either in the environment (above pre-existing levels of natural background) in the vicinity of Three Mile Island or in the reactor containment atmosphere or gas decay tanks.

The collective dose to the total population (about 2 million)within a 50 mile radius of the plant has been estimated to be 3300 person-rem (for the period from 28 March through 7 April, 1979, during which releases occurred that resulted in exposure to the offsite population). The average dose to an individual in this population is 1.5 mrem. The projected number of excess fatal cancers due to the accident that could occur over the remaining lifetime of the population within 50 miles is approximately 1. Had the accident not occurred, the number of fatal cancers that would be normally expected in a population of this size over its remaining lifetime is estimated to be 325,000 (NUREG-0558, 1979). The projected total number of excess health effects, including all cases of cancer and genetic ill health to all future generations, is approximately 2. It should be noted that these estimations of the health consequences of the Three Mile Island accident are of preliminary nature, awaiting the results of extensive investigations under way in the U.S.A. However, the estimations are most likely to be of the same magnitude (NUREG-0558, 1979).

It is misleading to evaluate the consequences of the Three Mile Island accident in terms of expected health effects only. The accident has had more far reaching effects. First, it has certainly increased the sensitivity of the public and many policy-makers about the risks of nuclear power. In the U.S.A. a special Task Force has identified 23 specific requirements in 12 areas related to reactor operation whose implementation is judged to provide substantial additional protection which is required for the public health and safety (NUREG-0578, 1979). Both in U.S.A. and in other countries, resistance to nuclear power programmes has increased. Several countries have delayed such programmes and/or halted the construction of nuclear power plants (e.g. the Philippines) to re-assess their safety requirements. Second, the Three Mile Island accident has several economic impacts. Not only the costs of rehabilitating the damaged unit will be staggering, but the accident has also affected both local and international reactor markets. As a consequence of the accident the costs of nuclear power plants might increase to allow for additional safety measures. Last but not least, the Three Mile Island accident has shown that however sophisticated and advanced a technology may be, there is no fool-proof technology and no safeguard against human error.

Nuclear reactor safety is, therefore, a continuing issue. Even though the likelihood of a major accident at a carefully designed, maintained, and managed reactor is very small, its precise value is uncertain and nonzero. The actual degree of risks under various conditions of design, maintenance and management continues to be the subject of intensive analysis.

Radioactive Waste Management:

Radioactive wastes are generated in practically all areas of the nuclear industry and accumulate as liquids, solids or gases with a wide range of radiation levels. The total global amounts of nuclear wastes have grown steadily. The bulk of the wastes occurs at the front-end of the nuclear fuel cycle which includes mining and milling, while the more radioactive wastes occur at the back-end of the cycle which includes reactor operation and fuel reprocessing (in case of re-cycling). The former type of wastes (especially inactive mill tailings) present a potential for exposure to radiation; the doses to the general population near inactive mill sites can be greatly reduced by adequate design of the tailings pond and by the use of a clay cover. However, the radioactivity in the piles will continue for very long periods of time (the key ancestor of the radon in the tailings is a thorium isotope with an 80,000-year half-life) and sound techniques for accomplishing the long-term stability and isolation of tailings have not yet been demonstrated.

Wastes at the back-end of the nuclear fuel cycle are generally considered as low, intermediate, high-level and wastes contaminated with transuranic elements. Low-level and intermediate wastes are normally disposed of by shallow land burial or by dumping in the deep ocean in specially designed containers. High-level and transuranic wastes must be conditioned and subsequently disposed of, normally in a solidified form, at a suitable repository. Such material has sufficiently persistent biological hazards and require special long-term isolation. Several options have been proposed but none has yet been established as fully satisfactory.

In a recent review of the problem of disposal of radioactive wastes, the NAS (1979) noted that:

> "the long-term risk from wastes that have been separated and buried is determined by the product of two factors: the likelihood that radioactivity may escape from the repository into the environment, e.g., by leaching into groundwater, and the radiation dose that would be delivered to humans if such escape were to occur. The latter parameter has not received as much attention as it deserves. The consensus of serious studies to date is that, even if high-level wastes buried at sites of the type now envisioned were to be leached into groundwater, the time-integrated exposures would not exceed the present-generation doses from normal operation of the fuel cycle. Nevertheless, it is clearly prudent to try to minimize risks to future generations by selecting sites for nuclear-waste-disposal facilities that will minimize the likelihood of escape, especially during the first few hundred years required for the bulk of the fission products to decay. Considerable attention is being given to the development of scientifically based criteria for site

selection, but there remain gaps in our knowledge. Geological characterization of suitable media for disposal presents a serious issue. Some scientists are concerned about the stability of glass as a container and about the stability of salt under the conditions that would obtain in a repository for high-level nuclear wastes."

Nuclear Proliferation and Safeguards:

Much of the concern that is felt about the environmental impact of nuclear power relates not only to the effects arising from normal operations but also to those that might be created by illicit activities directed towards nuclear installations or materials. It is, therefore, of utmost importance that during the entire nuclear fuel cycle, including transport of nuclear material, strict vigilance and care must be ensured, both on national and international levels, so that nuclear material does not fall into unauthorized hands which may use it for uncontrolled activities leading to damaging effects to the population and the environment.

It is appreciated that the establishment and implementation of physical protection systems against diversion of nuclear material at the national and sub-national levels are the primary responsibility of Governments, but the effectiveness of such systems has been questioned (SIPRI, 1979). At the international level, the most important of the institutional arrangements to reduce the danger of proliferation of nuclear weapons is the Non-Proliferation Treaty (NPT). In ratifying this Treaty, nuclear weapon states agree not to transfer nuclear weapons or manufacturing technology to non-nuclear weapons states, and non-nuclear weapons states equally undertake not to manufacture or acquire these weapons, and to accept a system of controls and safeguards designed and operated by the International Atomic Energy Agency (IAEA) to prevent diversion to non-peaceful use of fissile material produced and used in their civil nuclear programmes. Ratification of the Treaty indicates acceptance of its principles by the states concerned. By March 1979, 107 nations had ratified the Treaty, and a further eight have signed but not yet ratified it (Rotblat, 1979).

The main limitations and weaknesses of the present safeguards arrangements which give cause for environmental concern can be summarized as follows: the failure of many states to become parties to the NPT; the inability of safeguards to prevent the transfer of nuclear technology from nuclear power production to the acquisition of nuclear weapons competence; the fact that many nuclear facilities are covered by no safeguards; the existence of a number of loopholes in safeguards agreements regarding re-transfer of materials to a third state; the absence, in practice, of safeguards for source materials; the practical problems of maintaining effective checks on nuclear inventories; the ease with which states can withdraw from the NPT and from most non-NPT safeguards agreements; deficiencies in accounting and warning procedures; and the absence of reliable sanctions to deter diversion of safeguarded material.

The measures taken so far, both technological and institutional, to remove the proliferation threat have not proved adequate or acceptable (Rotblat, 1979), but this does not mean that they should be abandoned. Although technical fixes will never be foolproof any additional security measure helps, and the search for more proliferation-resistant variants forthe nuclear fuel cycle is therefore warranted. The same applies to the institutional controls; those would be much more effective if more teeth were given to them, and if national and multinational arrangements were backed by international agreements. One of the important anti-proliferation attempts on an international scale is the International Nuclear Fuel Cycle Evaluation (INFCE), which was set up in 1977, to tackle the issues from the point of view of the proliferation danger. INFCE may come up with acceptable, more proliferation-resistant fuel cycles. Another attempt is the setting up of regional or multinational fuel centres as an advantageous way to satisfy, safely and economically, the needs of many countries, to initiate or expand their nuclear energy programmes, while at the same time facilitating physical protection and the application of safeguards. However, there are several environmental and political problems associated with the establishment of such centres*.

CONCLUSION

There is no doubt that the future development of nuclear energy depends to a large extent on finding acceptable solutions to the main issues summarized in this paper, and one cannot find a better conclusion than that reached by the Royal Commission on Environmental Pollution (1976):

> "We conclude that development of fission power on the scale we have described earlier** carries implications and potential risks for society which are too serious to be disregarded on the grounds that they are necessarily speculative and of a kind that we have not hitherto expected to address in decisions on technological development. Decisions should not be taken simply on the basis of technological or economic advantage and the assumed necessity of securing steadily increasing energy supplies. The social and ethical issues involved are real and important, and should be widely appreciated and discussed."

REFERENCES

Bross, I.D. and Natarjan, N.: Leukemia from Low-Level Radiation; identification of susceptible Children. New England J. Medicine, 287, 107 (1972).

Davis, J.E.O. et al.: Canadian Attitudes to Nuclear Power. Proct. Inter. Conf. on Nuclear Power and its Fuel cycle, paper IAEA-CN-36/580, IAEA, Vienna (1977).

Døderlein, J.M.: Nuclear Power as a Public Issue; Protection of the Public Interest. Proc. Inter. Conf. Nuclear Power and its Fuel Cycle, paper IAEA-CN-36/451, IAEA, Vienna (1977).

Edsall, J.T.: Further comments on Hazards of Nuclear Power and the Choice of Alternatives. Environ. Conservation, 2, 205 (1975).

---

* Several formal organizations, each comprising a number of countries, have indeed been set up in the past, but they were all based on commercial interest, and entailed collaboration in developing specialized nuclear facilities, which would have been too expensive, or beyond the technological capabilities, of the individual partners. Most of these organizations involve European countries, for example, Eurodif, Urenco and Eurochemic (Rotblat, 1979).

** The scale mentioned in the report is about 3000 GW by the year 2000. However, the same conclusion applied to other scales indicated in the present paper (author's note).

Edsall, J.T.: Hazards of Nuclear Fission Power and the choice of alternatives. Environ. Conservation, 1, 21 (1974).

El-Hinnawi, E.E: Review of the Environmental Impact of Nuclear Energy. Nuclear Power and Its Fuel cycle, Vol. 7, paper IAEA-CN-36/361. IAEA, Vienna (1977).

Evans, H.J. et al.: Radiation-induced Chromosome Aberrations in Nuclear Dock-year Workers. Nature, 277 531 (1979).

Kneale G. et al.: IAEA Symposium Late Biological Effects of Ionising Radiation. IAEA, Vienna (1978).

Kneale, G.W. et al.: Radiation Exposure of Hanford Workers Drying from Cancer and Other Causes. Health Physics, 36, 87 (1979).

Lewis, H.W. et al.: Risk Assessment Reviews Group. Report to U.S. Nuclear Reg. Comm., NUREG/CR-400, Washington, D.C., (1978).

Liverman, J.L. and Thorne, R.D.: Public Acceptance of Nuclear Power Generation in the U.S.A. Proc. Inter. Conf. on Nuclear Power and its Fuel cycle, paper IAEA-CN-36/507. IAEA, Vienna (1977).

Mancuso, T.F. et al.: Radiation Exposure of Hanford Workers Dying from Cancer and other sources. Health Physics, 33, 369 (1977).

Marks, S. et al.: Cancer Mortality in Hanford Workers. IAEA Symposium on the Latent Biological Effects of Ionizing Radiation IAEA SM-224 (1978).

Mole, R.H.: Lancet, 1156 (1978).

Najarian, T. and Colton, T.: Mortality from Leukemia and Cancer in Shipyard Nuclear Workers. The Lancet, 1018 (1978).

NAS: Risks Associated with Nuclear Power. Summary and Synthesis Chapter. National Academy of Sciences, Washington, D.C. (1979).

Nelkin, D. and Fallows, S.: The Evolution of the Nuclear Debate. Annual Review of Energy Vol. 3, 275. Annual Reviews Inc. Palo Alto, California (1978).

NRC: Policy Statement on Reactor Safety Study; U.S. Nucl. Reg. Commission, Washington, D.C. 19 January (1979).

NUREG-0600: Investigation into the March 28, 1979 Three Mile Island Accident. U.S. Nuclear Regulatory Commission, Washington, D.C. (1979).

NUREG-0578: TMI-2 Lessons Learned Task Force Status Report and Short-Term Recommendations. U.S. Nuclear Regulatory Commission, Washington, D.C. (1979).

NUREG-0558: Population Dose and Health Impact of the Accident at the Three Mile Island Nuclear Station. U.S. Nuclear Regulatory Commission, Washington, D.C. (1979).

Otway, H.J. and Pahner, P.D.: Risk Assessment, Futures, 2, 122 (1976).

Reissland, J.A. and Dolphin, G.W.: National Radiological Protection Board, Harwell, May 18, (1978).

Rotblat, J.: Nuclear Energy and Nuclear Weapon proliferation. In: SIPRI, Nuclear Energy and Nuclear Weapon Proliferation. Taylor and Francis, London (1979).

Rothman, K.J.: J. Amer. Med. Assoc. 238, 1023 (1977).

Royal Commission on Environmental Pollution: Nuclear Power and The Environment, Her Majesty's Stationery Office, London (1976).

SIPRI: Nuclear Energy and Nuclear Weapon Proliferation. Taylor and Francis, London (1979).

Slovic, P. et al.: Images of Disaster; Perception and Acceptance of Risks from Nuclear Power. In: G. Goodman: Impacts and Risks of Energy Strategies; Their analysis and role in management. Academic Press, New York (1979) b.

Slovic, P. et al.: Rating the Risks. Environment, 21, 14 (1979) a.

Smith, P.G. et al.: Brit. Med. Journal, 2, 482 (1973).

Surrey, J. and Huggett, C.: Opposition to Nuclear Power. Energy Policy, 286 (1976).

WASH-1400: Reactor Safety Study (the Rasmussen Report); U.S. Atomic Energy Comm., Washington, D.C. (1975).